"Dr. Bonime-Blanc provides a practical guid
mental, social, governance and technologica
In a world of turbulence and chaos, she arms executives with a positive and actionable road map to resilience and value protection and creation."

Angel Alloza, CEO, Corporate Excellence – Centre for Reputation Leadership, Spain

"Dr. Bonime-Blanc has always been a step ahead of emerging risk, shining a light on the key integrity and compliance problems and offering straightforward solutions to our global community. She proves that effective, ethical leadership is grounded in a firm awareness of what can go wrong, combined with the moral courage to step up and fight. This book provides historical context and a call to action filled with cautious optimism."

Amii Barnard-Bahn, Executive Coach,
Fellow – Harvard Institute of Coaching,
Former Chief Compliance & Ethics Officer,
McKesson US Pharma, USA

"A book as impressive in its breadth of content as its optimism, Bonime-Blanc offers a compelling analysis of the most pressing issues of the day, pairing them with concrete strategies that a wide variety of organizations are capable of adopting. For those in the field, a book not to be missed."

Ian Bremmer, President, Eurasia Group, USA

"Bonime-Blanc's *Gloom to Bloom* is an impressive tour de force, a timely and useful guide to our turbulent times and practical ways for organizations and leaders to build resilience and transform risk into value. Ambitious in scope, intellectually rigorous and practical in application, her book plows new ground in ethics, compliance and risk management, taking us on a journey from today's disruptive megatrends to our resistance and inability to adapt to them – and a reconciliation of these opposing forces at a higher level of meaning, purpose and action. An impressive achievement!"

Earnie Broughton, Senior Advisor, Ethics & Compliance Initiative, USA

"Dr. Bonime-Blanc takes us on a virtual roller coaster ride, first pinning us down in the tight turns of the '10 megatrends of our turbulent times,' then lifting us up the steep slopes with a leadership integrity blueprint, then dropping us into an inversion exploring the dark corners of environmental, societal, governance and technological realities and finally concluding with a master class of optimism on how to achieve organizational resilience and value. In her inimitable and practical style, Dr. Bonime-Blanc proves that it really is possible to turn risk into value."

Jacqueline E. Brevard, Chief Ethics & Compliance Officer,
Merck & Co., Inc. (Retired), USA

"*Gloom to Boom* is an enlightening work of art with predictive insights where the author brilliantly weaves the fabric of modern corporate governance's often-structured framework together with the entrepreneurial-resilient can-do spirit of successful, transformative corporate leaders. Confronting a fast-changing world full of possibilities and opportunities, *Gloom to Boom* expands the new horizon for corporate thought-leaders to confidently embrace the challenges ahead and thrive. The book is almost prophetic, visionary and compelling and should be a constant companion in the hands of transformative corporate leaders."

Paul W. Chan, President, Malaysian Alliance Of Corporate Directors, Iirc Ambassador of the International Integrated Reporting Council (UK), Executive Member of the Global Network of Director Institutes (GNDI), Malaysia

"Bonime-Blanc has written a revelatory book that is both practical and hopeful. She provides numerous tools for transforming global ESGT risk into sustainable value by building organizational resilience and a strong ethics culture. *Gloom to Boom* is an invaluable resource: it should be read slowly and returned to often. We are truly fortunate that Bonime-Blanc's *ikigai* is to share her strategy for transforming risk to value."

Donna Costa, Former President, General Counsel and Chief Compliance Officer of Mitsubishi Chemical Holdings America, USA and Japan

"*Gloom to Bloom* is an extraordinary and comprehensive treatise regarding the chaos and disruption we are currently experiencing. Dr. Bonime-Blanc offers extremely practical solutions – indeed, a clarion call to action – to transform risk into resilience and action, while achieving value. This monumental undertaking has numerous, very current, and practical case study examples, making this book a real resource for practitioners. Each chapter draws upon significant sources and resources, adding to the value of this book as a truly useful reflection on how we can survive and thrive in today's world."

Keith T. Darcy, President, Darcy Partners, Former Executive Director Ethics & Compliance Officer Association (ECOA), Member of the Faculty of the Wharton School, USA

"In *Gloom to Boom*, Dr. Bonime-Blanc, one of the most formidable minds on ethical leadership and corporate governance, lays out a powerful and practical case for how leaders can flip the script on risk, adverse market forces and disruption, while charting a course for long-term prosperity and ESG equilibrium. All too often books about risk and governance are cliffhangers. Heavy on the doom and gloom and light on practicality. Dr. Bonime-Blanc corrects this trend and pours her years of board, legal, ethical and governance experiences in this rich and essential work."

Dante Alighieri Disparte, Head of Policy and Communications, Libra Association, Geneva, Switzerland

"Thank you, Andrea, for expending all your energy to write this book – it is a testimony to who you are as a person. While I find many business books to be a lot of air around a sometimes interesting single premise, *Gloom to Boom* is rich in content, wide-ranging, in touch with reality, grounded in facts, positive, on message, and tangible and actionable as a bonus! But most of all, I hear you speak your passion, commitment and conviction when I read the words – a deeply personal legacy which I'm sure will be a milestone contribution to a better world. You should be very proud of it. And I'm sure I'll pull from your tour de force and refer to it in my speaking and teaching in the years ahead. Thank you so much and congratulations!"

John Metselaar, P&G Veteran (Retired), Professor of Management Practice "Leading and Living Innovation" at the Solvay Business School in Brussels, Belgium, and Leader of the Global Innovation and Digital Transformation Institute at the Conference Board

"In my experience, there are two kinds of people who write about corporate risk: those who have actually operated many different kinds of businesses all over the world, and then a bunch of other people. All of Dr. Bonime-Blanc's years of global experience in enterprise c-suites are on full display in *Gloom to Boom*'s novel approach vector. Lots of people might *talk* about the merits of turning risk into opportunity for stakeholders, but to actually *show* business leaders how and why to integrate integrity and resilience into the root fabric of their organizations – isn't for beginners. *Gloom to Boom* is a masterfully researched elixir to a turbulent zeitgeist."

Adam J. Epstein, Founder, Third Creek Advisors, USA

"Bonime-Blanc's new work should be required reading for every board member and governance professional seeking a timely, holistic understanding of the modern risk landscape. Her years of work on this topic shine through here. We live in a brave new world where technology, geopolitics, climate change and social/demographic shifts have upended traditional notions of industry, competition and risk. Corporate leaders seeking to understand how these forces impact ERM and, if understood, can provide a new strategic advantage, would be well served to read this book."

Erin Essenmacher, Chief Programming Officer, National Association of Corporate Directors, Board Director, USA

"In this impeccably researched book, Bonime-Blanc weaves together narratives of social and economic mega-trends confronting leaders today (the 'gloom') with practical case studies to equip executives with ideas for creating a better future. She also offers great advice for what not to do. Passionately demonstrating that no institution today – government, corporate, or civil society – is immune from challenges to leadership, this work calls for a new paradigm to lead with integrity."

Azish Filabi, Executive Director, Ethical Systems, Adjunct Professor, NYU Stern Business School, USA

"In any organization or body politic, things are the way they are for a reason. When we see breaches of integrity, parochial thinking and worse, it's usually not because no one has thought about a legal or more efficient, ethical, socially conscious, inclusive or otherwise better alternative. The reason is that leadership wants or allows it to be that way. This is a book for leaders who want to do it the right and successful way for an ever-increasing base of global stakeholders."

Kenneth Frankel, President, Canadian Council for the Americas, Former Legal Advisor, Organization of American States, Adjunct Professor, Ghent University (Belgium) and Western University (Canada), Canada, USA and Colombia

"Transforming risk into sustainable value has been Andrea's 'lionhearted' motto since her childhood - *Gloom to Boom* is her legacy. In her fifth book, she gets to the quintessential nature of corporate values and behaviors – and corporate *mis-values* and misbehaviors. Hers is a call to arms (in her own words) to leaders. They should understand – and act. If leaders love their companies, they should also love their stakeholders. Deep, full of insight, comprehensive and essential to the everyday organizational life, this book is great – you should have it at your desktop, always at hand."

José A. Herce, Senior Adviser at Afi, Author and Lecturer, Spain

"This is your number one manual for successful leadership. It dives deep into risk management, connects the dots on the megatrends of our times and provides concrete tools for organizational resilience. If the value of your business is non-negotiable, this is your book!"

Annette Heuser, Executive Director, Beisheim Stiftung, Germany

"In this turbulent world where we find risks of all kinds all around us, Bonime-Blanc goes to the positive – turning risk into reward. While the subject matter is weighty, it is written with a personal voice, direct and at times playful, which makes it a joy to read. ESG and corporate citizenship executives will find this book a useful tool to develop strategies to guide their companies to navigate the world. This book is thought-provoking and covers many critical issues that need to be on our radar as board directors and executives."

Jeff Hoffman, President, Jeff Hoffman & Associates, Global CSR, Philanthropy & Civic Engagement, Founding Leader, Corporate Citizenship & Philanthropy Institute, the Conference Board, USA

"This is the perfect moment for Andrea's book. Demands from stakeholders have never been so high – this is a guide to meeting those demands. It is *booming* with ways to turn the challenges into opportunities with a road map for leaders to follow. Every section has a summary I wanted to print out and stick on my wall."

Tina Kirby, Head of Innovation at Beazley (a Syndicate at Lloyd's), UK

"*Gloom to Boom* is a practical guide and handbook for executives and an interesting read for academics. Dr. Bonime-Blanc guides the reader through the risks and challenges of our increasingly complex world – combining an impressive 'tour d'horizon' of today's and tomorrow's risks with 'hands on' recommendations for how to survive and transform these risks for the greater good. The author not only delivers a sound description of the subject matter and best practices, with her experience as an executive, she has written a book with a personal touch that adds value to what might otherwise be an abstract subject."

Matthias Kleinhempel, Professor of Business Policy,
Center for Governance & Transparency,
IAE Business School, Argentina

"Aristotelian in its scope, Dr. Bonime-Blanc's book *Gloom to Boom* is an encyclopedic journey through the metaphysics of risk. The key questions pop into relief with ontological brevity: ESGT—environmental changes from humans' interaction with their planet, social changes from climate changes of the cultural kind, governance changes from evolving notions of business roles and responsibilities in society, and technological changes that are transforming basically everything. With Dr. Bonime-Blanc's passion nurtured by her personal journey through Scylla and Charybdis, readers can appreciate patterns of risks and opportunities that predictably lead to reputation damage or value creation, incarceration or promotion, insolvency or resilience. Mark her words: luck is not a viable strategy. Practical tools reveal the essence of Dr. Bonime-Blanc's philosophy: it's enlightened leadership, stupid! One can find today many books on risk management, and others on the themes of ESGT. But for a global view of ESGT and risk with an eye toward strategy that leaders can deploy to navigate expertly through myriad risks to find opportunities, one need search no further than *Gloom to Boom*."

Nir Kossovsky, MD, CEO, Steel City Re, Author,
Reputation Stock Price and You, *USA*

"We are living in one of the most turbulent times in modern history. The current incarnation of capitalism appears to have finally reached an inflection point. In the decades to come, I suspect there will be two distinct camps: those who thank goodness they read this prescient primer for preparing organizations to survive and thrive in these increasingly turbulent economic times – and those who regret not having done so. In this wonderful guide, Bonime-Blanc does far more than simply sound a clarion call warning of the dangers ahead, she offers constructive, practical, pragmatic guidance to those who are charged with ensuring the long-term success of their organizations – and advice on how best to fulfill their duties as fiduciaries of the future."

JT Kostman, PhD, CEO, Protectedby.ai, USA

"As time runs out to preserve our humanity, *Gloom to Boom* offers leaders a powerful recipe and ample food for thought to create resilient and sustainable governance. Throughout, Andrea reminds us of the quintessential strength and beauty of ethics in action."

Emmanuel Lulin, Senior Vice-President and Chief Ethics Officer of L'Oréal, 2018 UN Global Compact SDG Pioneer for Advancing Business Ethics, France

"What a *tour de force* full of timely and practical lessons from companies that have succeeded – or failed – to turn risk into opportunity. This is truly a 21st-century guide for global corporate directors and chief executives seeking to navigate a world full of digital disruption, enterprise risk, and unprecedented complexity. *Gloom to Boom* offers a hopeful and timely reminder that strong ethics and values continue to be at the heart of organizational resilience and sustainability. Riveting reading not just for corporate governance experts but also for investors and shareholders all over the world."

Michael T. Marquardt, Serial Entrepreneur, Global Corporate Director, USA, Asia and Europe

"In *Gloom to Boom* Andrea Bonime-Blanc has managed the seemingly impossible: how to take difficult, complex concepts, explain them in ways that every-one can understand and chart an actionable path for businesses to follow and thrive. Andrea does this in a natural, systematic way, a way that intuitively makes sense and compels the reader to follow-on. Andrea's chapter on Leadership reflects her direct, no-nonsense, 'get it done' style, and provides any board member with an irreplaceable diagnostic resource as they navigate these turbulent times. Finally, her treatment of 'Environment,' 'Society,' 'Governance' and 'Technology' is masterful. Andrea brings all this together in the final section of her book, aptly named 'Boom' where she demonstrates how business leaders, using the concepts presented, can achieve resilience and deliver sustainable value. I couldn't put this book down! I consider this book required reading for any investor, board member, executives across industries, government, nonprofits and academia."

Chris Moschovitis, CEO, TMG-Emedia, Author Cybersecurity Program Development for Business, *USA and Greece*

"Andrea Bonime-Blanc's *must-read primer* provides a solid conceptual framework with strongly documented case studies to help business leaders detect, administer and mitigate nonfinancial risks and, then, to turn them into strategic growth opportunities. One of its most novel contributions is the incorporation of the 'T' into the ESG equation in light of the increasing challenges posed by the current technological revolution. It is a must-read for investors who should embark in a risk due diligence process to decide whether or not to invest in a company.

The examples described in the book – including those of Equifax, Volkswagen and Wells Fargo, among many others – showcase the negative financial and reputational consequences of not doing so. Also, the description of how Microsoft, L'Oréal and Salesforce have addressed their respective ESGT challenges demonstrate the positive impact that ethical leadership, transparency and setting the right tone from the top have on corporate culture."

Karen Poniachik, Former Minister of Mining for Chile, Board Director, Director, Columbia University Global Center in Santiago, Chile

"Drawing on personal experience and professional observation, Dr. Bonime-Blanc has produced a thought-provoking call to action that deserves to find a place in the briefcase or on the bedside table of every C-Suite executive. At a time when political, economic and social norms are being stressed to the breaking point, she offers calm analysis and reasoned solutions. This book is both prescriptive and groundbreaking. I highly recommend it."

Jay Rosenzweig, CEO, Rosenzweig & Company, Executive Recruitment & Tech Entrepreneur, Canada

"As we near the third decade of the 21st century, it is clear that companies are struggling to respond to profound social, environmental, and technological disruption: Established structures and mindsets will no longer do the job. Bonime-Blanc has written a new operating manual for business in our era. Bonime-Blanc fuses the latest thinking from sustainability, ethics and compliance, and leadership and culture without ever succumbing to the jargon that pervades all these fields. She illustrates her new conceptual models with lively up-to-date case studies and practical guidance. This is essential reading for anyone who wishes to lead organizations with integrity – without sacrificing commercial opportunity."

Alison Taylor, Managing Director, Business for Social Responsibility, Professor at Fordham Law School, USA

Gloom to Boom

Leaders – whether in business, government or the nonprofit sector – take risks but often without fully understanding risk at a strategic level. Expanding upon the well-known "ESG" risks, this book explains the key nonfinancial (environmental, social, governance and technological or ESGT) risks. For many leaders (including board members), taking risk without knowledge or preparation can lead to organizational crisis, scandal and value destruction. For those who are prepared, resilience follows and so does the ability to transform ESGT risk into opportunity and value for stakeholders.

In this book, global governance, risk, ethics and cyber strategist, author and board member, Andrea Bonime-Blanc shows practitioners at all levels how to effectively identify and manage their top ESGT risks to avoid crises and transform risk into sustainable long-term resilience and value.

Gloom to Boom is a book for everyone – from the highest levels of leadership in an organization (the board, CEO and C-Suite), to other senior leaders (the chief risk officer, CFO, general counsel, head of CSR and sustainability, CISO, CHRO), and midlevel leaders, students and folks simply interested in current affairs and the role and impact of strategic risk and opportunity on their lives.

Andrea Bonime-Blanc, PhD/JD, is a global strategic governance, risk, ethics and cyber advisor to business, government and nonprofits. She is a former global corporate executive, Founder and CEO of GEC Risk Advisory, Ethics Advisor to the Financial Oversight & Management Board for Puerto Rico and an Advisory Board Member to several green finance and tech firms. She is author of several books, a global keynote speaker and adjunct faculty at New York University.

Gloom to Boom

How Leaders Transform Risk into Resilience and Value

Andrea Bonime-Blanc

LONDON AND NEW YORK

First published 2020
by Routledge
2 Park Square, Milton Park, Abingdon, Oxon OX14 4RN

and by Routledge
52 Vanderbilt Avenue, New York, NY 10017

Routledge is an imprint of the Taylor & Francis Group, an informa business

© 2020 Andrea Bonime-Blanc

The right of Andrea Bonime-Blanc to be identified as author of this work has been asserted by her in accordance with sections 77 and 78 of the Copyright, Designs and Patents Act 1988.

All rights reserved. No part of this book may be reprinted or reproduced or utilised in any form or by any electronic, mechanical, or other means, now known or hereafter invented, including photocopying and recording, or in any information storage or retrieval system, without permission in writing from the publishers.

Trademark notice: Product or corporate names may be trademarks or registered trademarks, and are used only for identification and explanation without intent to infringe.

British Library Cataloguing-in-Publication Data
A catalogue record for this book is available from the British Library

Library of Congress Cataloging-in-Publication Data
A catalog record has been requested for this book

ISBN: 978-1-78353-815-7 (hbk)
ISBN: 978-1-78353-733-4 (pbk)
ISBN: 978-0-429-28778-7 (ebk)

Typeset in Bembo
by Deanta Global Publishing Services, Chennai, India
Printed and bound by CPI Group (UK) Ltd, Croydon CR0 4YY

Cover art: "Night Fight" by Peter Blanc. 1957. Reprinted with permission from the collection of Roger A. Blanc and Andrea Bonime-Blanc.

For my father

At Hokkaido Lake, Japan, circa 1947

Contents

Foreword

It was an honor for the two of us to be asked by Andrea Bonime-Blanc to co-author the foreword for her new book, *Gloom to Boom*, and it has been a pleasure to work together with the goal of sharing our joint perspective on "how leaders transform risk into resilience and value" and why Andrea's book matters.

While we come from different backgrounds and have vastly different professional experiences, we now find ourselves working side by side with a joint focus firmly on strengthening the organization's ethics and risk culture with a view toward identifying emerging risk issues early enough to take forward-looking action.

When leadership combines emotional intelligence, integrity, and ethics with a more holistic approach to risk management and applies this mind-set to the development of more resilient and sustainable business models, the outcome should be better performance for the whole organization and greater value creation not only for the business itself but also for its stakeholders and the community at large.

The stronger the ethics and risk culture, the easier it becomes to respond effectively to unexpected risk events, to ensure that new recruits and seasoned professionals feel that the institution has their back so that they can deploy their talent in frontier areas to the best of their ability, to allow the institution to make better risk-informed decisions in the implementation of its mission, and ultimately to allocate scarce resources where they have the most meaningful impact.

From our joint perspective and that of the institution we serve, corporate values are no longer seen as an aspirational goal but as a necessary tool to bring about an increase in value creation and achieve better resilience in an ever-more challenging operating environment. Indeed, the key point made by Andrea in her book is that the main reason why companies go from boom to gloom is a leadership lacking in ethical values, risk awareness, and agility in responding to major risk events.

Now, why is Andrea's book important? All of us in management positions in the fields of ethics and risk face a similar conundrum: how do you increase risk awareness across the universe of stakeholders and ensure that colleagues at all levels continue to strive to "do the right thing"? Furthermore, how do you ensure that enough resources get allocated to support our work when the organization hasn't faced any debilitating risk event(s) for some time and we start believing in our magic touch or our good fortune? The simple answer is that we need our colleagues to be aware of what has happened elsewhere lest we be accused of playing the role of Cassandra, the mythical doomsayer. We need to be able to share real-life stories as warnings and subjects of study. So, what could be better than Andrea's book to serve such a purpose?

In *Gloom to Boom*, Andrea reflects on 30 years of multi-sector governance, risk, and ethics experience and gives us the opportunity to refresh our understanding and renew our perspective on a broad collection of actual environmental, social, governance and technology risk events. In other words, this is a rich resource for leaders at all levels and across all industries who are confronted with business problems, big and small – as such problems must be carefully have from a joint risk and ethics perspective in order to make the best possible decisions.

While any risk event calls first for an immediate response of a technical and tactical nature, Andrea underscores the need for leaders to take a step back, reflect more broadly on how such an event could change the trajectory of the enterprise, and ultimately test the validity of the current business strategy going forward.

As we both read through the manuscript, this is how we imagined our colleagues would use the book: to be kept on one's desk as a reference guide to help develop a holistic risk strategy and to be delved into when facing risk events similar to those documented and analyzed by Andrea.

As Andrea covers so much material, readers will have the opportunity to treat the book as a library of risk events, to choose specific case studies, to reflect on lessons learned and to use what are compelling stories to run desktop emergency response exercises. It will thus help leaders be better prepared and hopefully make better decisions if and when a crisis hits – as it probably will at some unexpected point in the future.

Ultimately, these materials should help readers not only improve their ability to understand the specifics of a risk event and manage its aftermath effectively but also to identify the opportunities which a risk event might unlock in terms of value creation. As its title indicates, Andrea's book is full of examples of successful transformations which came about as a result of human rigor, open mindedness and ingenuity in the face of a risk event which could have been catastrophic. It is therefore also a story of hope!

To conclude, we would sum up the value proposition of Andrea's book by saying that it will allow leaders in the public, private and nonprofit sectors to develop a holistic approach to leadership based on a clear strategic and ethics-based risk governance. We trust that it will serve as a key component of the management toolkit for long-term value creation.

Jorge Dajani (Chief Ethics Officer)
Amédée Prouvost
(Director, Operational Risk)
The World Bank

Preface

A government prosecutor's perspective on a culture of compliance

There are many reasons organizations need to pay attention to the ethics and compliance function that is the procedural mechanism that an entity puts in place to assure that its employees, agents and business partners are adhering to the internal corporate values and external rules that govern its conduct. Ethics and compliance can serve, as Andrea so cogently notes, multiple purposes that align with a successful company. Being a more profitable company, being a more sought-after company for recruiting desired applicants and being better at the retention of valued employees are just some of the important reasons for having a healthy corporate culture.

But there is another reason, one that I have spent a good part of my career focusing on. Good compliance is good government. And thus in evaluating corporate criminal liability, the company's corporate culture is a critical factor. A company's ethics and compliance function is an important factor in assessing that culture. There are many companies that can be criminally charged, but the key issue for the government is whether a company should be charged, not just can it be charged. There are many alternatives to prosecuting a company, and a company's ethics and compliance culture come into play in assessing those alternatives.

At one extreme a company can be indicted, along with its culpable employees. At the other extreme, the company can receive a declination from prosecution, and only the criminal individuals can face indictment. In between, the government can decide to offer the company a Deferred Prosecution Agreement, a Non-Prosecution Agreement, or a declination but with the disgorgement of profits. In all of the above circumstances the government can insist on a corporate monitor to oversee the company. By US federal policy, compliance is a component of all those decisions and can be the decisive factor. In my experience, for instance, a company that has an excellent compliance program reflective of the corporate culture of the organization will rarely if ever have a monitor put in place. There would be no need for a monitor where the government has faith in the company and its ethics and compliance function to perform the requisite oversight.

Although the role of compliance in assessing these issues has long been a factor in the written Department of Justice policies, in the last few years the government has been increasing its attention on the compliance factor.

Further, it has provided more and more guidance to companies on how the role of compliance will be assessed. No longer is compliance an after-thought for a federal prosecutor or a mitigating factor that a company can think it can obtain credit for with a cursory presentation.

The US Department of Justice has gotten smarter about compliance. It has brought in expertise, first through hiring an experienced compliance officer to provide training and more recently bringing in outside experts to provide more widespread training within the Department. And the Department's focus increasingly looks to objective metrics – the use of data by the company – in assessing programs – that is, hard, objective evidence is required and scrutinized. The Department has also done more to give companies guidance on what the government looks at in assessing compliance programs. Under my leadership of the Fraud Section, the Department of Justice published a set of questions it routinely thinks about in assessing a compliance program. Those issues were published so companies were not kept in the dark about governmental expectations and criteria. That effort has continued more recently with additional guidance being made available to companies to increase transparency. In short, the government has gotten smarter about compliance and has sought to share that newfound knowledge. Thus, even under different leadership at the Department, the increased scrutinizing of compliance has remained constant.

This is just one reason that Andrea's book is so timely. It provides all business leaders with an important perspective and tools to understand how ethics and compliance promote good governance, model leadership and a healthy work environment. It is a welcome addition to the growing commitment of the public and private sector to the salutary effect of ethics and compliance in today's business world and beyond.

Andrew Weissmann
Distinguished Senior Fellow,
NYU Law School

Former Senior Assistant Special Counsel to Robert S. Mueller III,
Special Counsel's Office, US Department of Justice
Former Chief, Fraud Section, US Department of Justice
Former Director, Enron Task Force, US Department of Justice
Former General Counsel, Federal Bureau of Investigation,
US Department of Justice
Former Chief, Criminal Division,
Eastern District of New York, US Department of Justice

Acknowledgments

For a book that became so much longer than I originally intended it to be, the sweep of folks I am deeply grateful to has also become broader and deeper than expected. I apologize to anyone I have not properly acknowledged – hopefully I can make amends in future editions of this tome should I be so lucky!

First and foremost, I must thank my husband and son – my two "Rogers" – because they truly suffered the most (after me) with my many out-of-town disappearances for 1–2 weeks at a time during 2018 and my constant angst about finishing writing this book. Thank you, boys – I love you more than anything in this world!

Next are the wonderful folks who agreed to be reviewers of the first manuscript and endorsers of the book – they all appear in the front matter of this book through their endorsements. All I can say to each of you individually and collectively is that your early vote of confidence and helpful feedback meant everything to me!

I want to thank several organizations and their people whose support and encouragement of my professional work over the last few years fueled my book-writing fire. They include (in no particular order!): Erin Essenmacher, Stephen Walker, Judy Warner and Candace Rothermel of the National Association of Corporate Directors; Anita Allen of NACD New Jersey; Philipp Aeby, Alexandra Mihailescu Cichon, Jenny Nordby and Gina Walser of RepRisk (to whom I am doubly grateful for the permission to use and for the meticulous feedback for the RepRisk data deployed in this book); Larry Clinton of the Internet Security Alliance; Matthias Kleinhempel of IAE Business School in Buenos Aires, Argentina; Tina Kirby of Beazley; Dante A. Disparte of Risk Cooperative and Libra Association; Pat Harned, Moira McGinty, Tia Berry and Earnie Broughton at the Ethics and Compliance Association; Karen Poniachik of the Columbia Global Center in Santiago, Chile; Andrijana Bergant of IECE in Slovenia; Angel Alloza and Saida Garcia of Corporate Excellence – Centre for Reputation Leadership in Madrid; Stephane Martin and his team at Swiss-based Risk-In 2018 and 2019; several folks over the years at The Conference Board, especially Chuck Mitchell; Dr. Nir Kossovsky of Steel City Re; Richard Haass of the Council on Foreign Relations; Michael Abramowitz of Freedom House; Valmiki Mukherjee of the Cyber Future Foundation;

Laurent Bernat and Andras Molnar of the OECD; Dora Gomez of ACFE and HTCIA; Lisa Roberts of NASDAQ; Cindy Fornelli and Erin Dwyer previously of the Center for Audit Quality; Terry Slavin and Liam Dowd of Ethical Corporation; Ed Amoroso of Tag Cyber; Jack Fingerhut of Kaplan Education; Coco Brown of Athena Alliance; Vera Jelinek of NYU; Ben Boyd and Tonia Ries of Edelman; Dr. Maya Bundt of Swiss Re; Mark Spelman of the World Economic Forum; Susan Gray of S&P Global; Jeff Pratt of Microsoft; Bonnie Green of the World Food Programme; and Olajobi Makinwa of the UN Global Compact. Countless other individual friends and mentors deserve my gratitude for their encouragement, counsel, wisdom and thoughtfulness, including Jose Antonio Herce, Hans Decker, Clara Durodie, Alex Newton, Shari Siegel, Friso Van den Oord, Simon Franco, Reatha Clark King, Martin Coyne III, Dr. Kerry Sulkewicz, Jaime El Koury, Lisa Davis, Willem Punt, Rene Castro, Eduardo Brunet, Gabriel Cecchini, Michael Thomson, Craig Carroll, Leonard J. Ponzi, Prakash Sethi and Mark Brzezinski (and his dearly departed dad, Zbigniew, my first and greatest mentor).

I want to express my deepest gratitude to the three writers of this book's Foreword and Preface (because an almost 500-page book requires more than one introduction, of course!). Jorge Dajani and Amedee Prouvost of The World Bank kick off with a "Gloom to Boom" Foreword from the viewpoint of the worlds of ethics and risk management and their dedicated work at the World Bank. In his Preface, my long-time friend, Andrew Weissmann, provides the "Gloom to Boom" view of someone who has dedicated his entire career life to putting "the bad guys away", from individuals to corporations in furtherance of the deep values of ethics, compliance, the rule of law and the protection of democracy.

I couldn't not thank my "rat pack" of Spanish girlfriends who are always, and I mean always (via WhatsApp) there for me – Maribel Garcia, Margarita Garcia, Susana Villanueva, Geles Medialdea, Mayte Santos, Mariche Segovia, Esperanza Romero, Isa de Haro and Teresa Arce. Con cariño.

Thank you to the Routledge folks – especially Rebecca Marsh and Sophie Peoples who were nothing but encouraging and wonderful throughout this long and winding road!

A great big shout out to my amazing, now departed, father-in-law, the great artist, painter and sculptor, Peter Blanc, who was on this earth almost 104 years! One of his great paintings, "Night Fight", graces the cover of this book. Wow! What an honor and a privilege.

In closing, I must reach back to my own two parents who are no longer with us (though they will always be with me). My mother gave me love, focus and resilience and my dad gave me adventure, discovery and creativity. I have dedicated past books to my mom but never to my dad, who remains (as Churchill once said) "a riddle wrapped up in an enigma" for me to this day since I lost him when I was very young. The photo on the dedication page says it all about him.

Gratefully,
Andrea
New York City

Introduction

A time of dangerous opportunity, attractive risk and necessary invention

Mater artium necessitas. (Latin)
Maybe Plato

Necessity is the mother of invention.
Roger Ascham, *Toxophilus*, 1545

Art imitates Nature, and Necessity is the Mother of Invention
Sir Walter Scott, 1658

★★★★★★★★★★★

الحاجة ام الاختراع. (Arabic)
which literally means "necessity is the mother of invention"

Голь на выдумки хитра (Russian)
which literally means "poor people are crafty".

Potrzeba jest matką wynalazków (Polish)
which literally means "necessity is the mother of the inventions".[1]

1 Background

1.a "WEIJI"

A few years ago, I was visiting Seoul, South Korea, to give a keynote speech on reputation risk and opportunity at the Good Corporation Annual Conference. It was attended by about 700 participants, most of whom were men in suits, but there were a few women too. Among the organizers was Dr. Jong Nam Oh, an illustrious business and government leader, with whom I shared a lovely breakfast and my fascination with the notion of risk and opportunity, including the idea (which is more of an obsession, hence this book) of transforming risk to value.

It was then that Dr. Oh shared with me the Chinese word "Weiji" which loosely means both risk and opportunity or the idea that there is opportunity embedded in every risk – that there is "dangerous opportunity" or (my words) "attractive risk". I loved that there was such a word that captured the object of my obsession since we don't have any such word in the English language (at least that I know of). I ended up writing an article about my observations of the state of corporate integrity and ethics in South Korea shortly thereafter titled "Dangerous Opportunity" which was my best and yet insufficient attempt at exploring that wonderful concept.

So here we are a few years later, and my obsession with transforming risk into value has now materialized into this book, for better or for worse. It is a book containing many thousands of words instead of one but hopefully it will serve as a vehicle that puts the meat on the bone of this concept in a way that is illustrative, educational, practical and actionable. It is my attempt to transform theory into practice, or better yet to transform our thinking about challenge and adversity into constructive and sustainable action.

And, yes, I acknowledge that the title of this book *Gloom to Boom* is a little risqué and maybe a little funny as well – but without risk and without fun there isn't much opportunity and without opportunity there is little chance of positive value creation – whatever that value might be: social good, individual growth, corporate earnings or national and international peace and stability. So I'm going to go with the *Gloom to Boom* title! And, frankly, the subtitle "How Leaders Transform Risk into Resilience and Value" hopefully makes the point of the book clear anyway!

1.b A personal note on risk, resilience, opportunity and value

The theme of this book is the theme of my own life (and I'm sure that of many of my readers), and I think it bears a brief mention. Without going into gory detail, the first stage of my life (my first 20 years) was full of risk and opportunity – the risk and challenge of living in a dysfunctional home, losing a father to alcohol as a preteen, my mother's struggle to make ends meet when he left us penniless, being uprooted from the only place I knew and loved (Malaga, Spain).

But looking at all of those things that were challenges, there was also real opportunity – the opportunity of living in Spain as a child, which gave me all kinds of gifts for life (language, culture, wonderful friends), the gift of an amazingly resilient, supportive and loving mother who got us through the toughest of times and the gift (as much as I didn't consider it to be one at the time) of arriving in New York City at age 17 where opportunity (unbeknownst to me in my teenage depression) awaited. I have lived in New York City ever since, turning my early despondency into hard work and study, and have been blessed with an incredible family, friends, career and, in recent years, a new calling (see the next paragraph). And now I get to write this book. You might call me a glutton for punishment, but gratitude is all I feel in my heart.

Another personal note. Seven years ago, I left the corporate world rather abruptly after 18 years as a corporate executive in four diverse and exciting global companies (and previous to that almost a decade on Wall Street in the "greed was good" era). My sudden departure from the relative comfort, predictability and safety of the corporate world to start my own business presented an enormous risk to me professionally and personally – as the primary breadwinner of my extended family, how was I going to pay the bills? What was I going to do that was meaningful, let alone allow me to make ends meet? How the heck was I going to find clients, since the idea of doing sales made me cringe? What would my friends and family say – had I lost my mind?

Yet, as the saying goes: "necessity is the mother of invention" and that's what happened to me. Because I no longer had the comfort of the corporate mothership, I detected a life-changing opportunity – that of marching to my own drumbeat doing what I love to do and finding a way to make a living off of it. And almost seven years after launching my business I can say that (at least for now) my new calling – as an independent global strategic risk and opportunity adviser to business, government and NGOs, author, speaker and board member – has far exceeded my wildest dreams. I can truly say that I have reached my own personal "Ikigai". And to understand what that last sentence means you have to read this entire book (or skip to the last paragraph of the last chapter). She said, smiling.

1.c A risky world full of opportunities

We live in a risky world and I'm willing to take the risk that many of my readers are people who are looking to create positive change in this very turbulent, confusing and, dare I say, increasingly dangerous world. This is a hopeful book written during time of great chaos and turmoil. It is a book that promises a more positive future if we collectively will it that way.

This book is about a voyage – a turbulent but (hopefully) enthralling trip from Gloom to Boom – a virtual trip through the Scylla and Charybdis of our modern turbulent times. We will embark on a voyage that will take us around the world to various locations and situations in which companies, non-governmental organizations (NGOs), government agencies and other forms of organized human activity have, on the one hand, faced challenge and hardship in the form of a risk gone wrong, a crisis or scandal and, on the other, have learned from these challenges and created resilience, opportunity and even greater value for their stakeholders.

This book is about giving leaders – executives, board members, investors, advisors and students in business, nonprofits, academia and government officials – the tools, techniques and business case to understand their greatest challenges and risks, manage and mitigate them but, more importantly, take on these challenges and transform them into resilience and value for their organizations.

We are living in a time of possibly unprecedented global turmoil, turbulence and change in just about every sphere of life – technological, political, transnational, social, economic. Every organization needs both an anchor and a sail – the anchor to be grounded when and however necessary – and a sail to pursue the opportunities and even dreams of the collective entity. The leaders are the captains and sailors who calibrate how best to navigate the Scylla and Charybdis of this time of hyper-transparency, super-connectivity, fake news, profound geopolitical disruption and technological tectonic shifts.

This book aims at changing how leaders view risk and challenge so that they view them as opportunity and value. By using illustrative cases throughout the book, we look to inform on how this can be done – how risk and challenge can be integrated and embedded deeply into an organization's strategy and business plan. By showing companies and entities that have transformed their risk into opportunity and value, we hope to inspire organizations on the importance of integrating integrity, resilience, risk and strategy to create maximum value.

1.d Introducing the concept of ESGT

Companies have traditionally focused on financial and operational matters by developing a good understanding of how to run and measure the financial aspects of their business. The same cannot be said about how they understand or manage a host of nonfinancial issues and activities that are deeply intertwined with running a successful business (or another type of organization). And such nonfinancial issues ultimately have deeply impactful financial and reputational consequences.

These nonfinancial issues – often referred to as environmental, social and governance ("ESG") – are often relegated to second-class status in the hierarchy of running a business successfully. ESG matters are often siloed or worse, considered to be an unnecessary cost center (an albatross around the neck) to business. If they have a budget line item at all, often ESG matters are the first to be cut or exist in a disaggregated, uncoordinated manner, on budget life support without real depth or materiality to the business. The word "greenwashing" comes to mind.

Well – despite the not yet completely recognized importance of ESG, this book not only focuses on the deep and broad importance of ESG issues, risk, opportunities and value creation to all types of organization, it is about more than ESG. Because we are living in a time of unprecedented invention, innovation and technological change, I am offering a new, more complete moniker for our age – and it is "ESGT" – environment, social, governance and technology.

Technology pervades and suffuses everything we do and will do so even more, if that is possible, in the coming age dominated by artificial intelligence (AI), machine learning, deep learning, robotics, augmented and virtual reality, 5G, bioengineering, digitalization, cyber-insecurity, quantum computing, nanotechnology and god knows what else. And, frankly, technological change may be so profound and existential that we may be moving into the age of the "singularity" not so far from now where humans and machines are one. To not include technology in a classification of key ESG or ESGT issues is to be blinded by the light.

Thus, the topics we examine in this book are organized and bucketed into these four categories of "nonfinancial" or "ESGT" issues which nevertheless have a potentially lasting, pervasive, direct impact on the financial and reputational well-being of human ecosystems we call companies, NGOs, universities and government agencies and other organizational forms. And ESGT issues aren't issues that live in silos either – ESGT issues are interconnected and intertwined – something we repeatedly emphasize throughout this book both in content and visually through some of our figures and charts.

Table I.1 provides a sneak preview of what we mean by ESGT issues explored in great detail in Part II of this book, titled "Navigating the Scylla and Charybdis of environment, society, governance and technology ('ESGT')".

Table I.1 A sampling of ESGT issues – a sneak preview

Environmental	*Social*	*Governance*	*Technology*
• Climate change	• Human rights	• Corporate governance	• Cyber-security
• Sustainability	• Labor rights	• Leadership	• AI geopolitics
• Water	• Child labor	• Culture	• Data mining
• Air	• Human trafficking	• Business ethics	• Internet of things
• Earth	• Human slavery	• Geopolitics	• Artificial intelligence
• Carbon emissions	• Health and safety	• Corruption/bribery	• Machine learning
• Energy efficiency	• Workplace conditions	• Fraud	• Deep learning
• Natural fesources	• Workplace violence	• Money laundering	• Robotics
• Hazardous waste	• Product safety	• Anti-competition/anti-trust	• Automated robotic processing
• Recycled material use	• Fair trade	• Regulatory compliance	• Military robotics
• Clean technology	• Data privacy	• Conflicts of interest	• Surveillance
• Green buildings	• Discrimination	• Compensation disclosure	• Dark web
• Biodiversity	• Harassment		• Fake news
• Animal rights	• Bullying/mobbing		• Deep fakes – visual
• Pandemics	• Diversity and inclusion		• Deep fakes – audio
			• Biometrics
			• Wearables
			• Nanotechnology
			• Bioengineering
			• Crispr

Source: Author.

Let's turn now to what you will find in this book in a quick overview of its three parts and eight chapters.

2 The book – Overview

Gloom to Boom is organized as a journey from a place of turbulence (Part I – Gloom) to a place of constructive empowerment and resilience (Part III – Boom), traveling through a Scylla and Charybdis landscape (or seascape) of ESGT issues, risks, opportunities and value (Part II).

PART I – GLOOM: SURVIVING AND THRIVING IN TURBULENT TIMES

The purpose of Part I is to lay down the current factors that are contributing to our general sense of change and turbulence. The point is to assist those of us hell-bent on providing solutions and being optimistic about the long term with the context and perspective we need to take this on as a long-term project – to get past "gloom".

Chapter 1 – Gloom: the ten megatrends of our turbulent times

This chapter examines the ten megatrends of our turbulent times to paint the context and situational awareness that are defining our global moment. The purpose of this chapter is to lay the groundwork of the overall argument we make in this book: while we live in a time of turbulence that is unlikely to change dramatically for the foreseeable future, we also live in a time of unprecedented opportunity for value creation afforded especially by technology and science innovation.

Chapter 2 – Leadership: surviving and thriving with a culture of integrity

What is the indispensable be-all and end-all of organizational success? It is great leadership – at every level and in every context. This chapter has three main parts – leadership, culture and ethics. It starts by exploring leadership from the individual, psychological standpoint, looking at the leadership characteristics of toxic, sociopathic, hubristic leaders and those of great, high-integrity leaders. Then we look at the essential tools of leaders – culture and ethics – and how leaders, depending on who they are, affect and effect organizational culture – good, bad or indifferent. Finally, the role of organizational integrity and ethics is woven into this fundamental package, two things without which organizations are not fully equipped to navigate the shoals of Scylla and Charybdis. The chapter concludes with our typology of ESGT Leadership – from Irresponsible and Superficial Leaders to Responsible and Enlightened ones – something we revisit in the final two chapters of the book as a crucial component of what makes an organization resilient, sustainable and valuable to its stakeholders.

PART II – NAVIGATING THROUGH THE SCYLLA AND CHARYBDIS OF ENVIRONMENT, SOCIETY, GOVERNANCE AND TECHNOLOGY (ESGT)

Part II of *Gloom to Boom* takes on us on a journey around the world to look at a variety of risks and opportunities confronting all forms of human endeavor and community today. From companies and NGOs to government agencies and academia there is a world of ESGT issues that each organization needs to understand, navigate and address on behalf of its stakeholders.

Each of the four chapters in Part II begins with an introductory overview of definitions, sources and resources relating to the specific topic (environment, society, governance and technology) to give the reader a sense of the sweep of sub-issues, risks and opportunities that exist in each category. The second, meatier part of each of these chapters, analyzes a handful of key strategic issues, risks and opportunities with cases and examples to illustrate the importance of the topic.

Chapter 3 – Environment: transforming environmental risk into opportunity from global warming to sustainable resilience

This chapter examines a variety of environmental issues – from climate change and plastic waste to the creation of environmentally friendly products and services and the building of urban and island resilience – that demonstrate that with the proper focus, any form of entity – company, NGO, government agency, community – can take its current environmental issues and risks and transform them into opportunity and potential value to their most important stakeholders.

Chapter 4 – Society: transforming social risk into opportunity from slavery and trafficking to safety and diversity

This chapter examines a variety of social issues – human slavery, workplace health and safety and diversity and inclusion – applying a similar lens of looking at examples of such risk and crisis gone wrong and gone right, gathering lessons learned from what happened to not only avoid the damage that might occur to both stakeholders and principals but also to create shared value for such stakeholders.

Chapter 5 – Governance: transforming governance risk into opportunity from authoritarianism to stakeholder centricity

This chapter explores governance at every level – from global inter-governmental governance and national governance down to typical corporate or nonprofit board governance. The issues we examine range from geopolitical risk, authoritarian regimes and illiberal democracy to specific governance issues such as fraud, corruption, board governance. The examples explored include a review of waves of scandal and reform from the 1970s to 2020, industry-wide scandals (in the financial and auto sectors) and Silicon Valley start-up and more mature company governance issues.

Chapter 6 – Technology: transforming technology risk into opportunity from cyber-fear to trusted tech

This chapter captures our bold attempt to add a fourth category to the ESG discussion to bring about a concept of ESGT where technological risks and opportunities are considered in their very own category of impactful and serious ESGT issues. From the challenges and opportunities that all forms of new technology – from AI and machine learning to nanotechnology and IoT – present to society, this Chapter tries to illustrate how each of these (and many more) technologies affect, intersect and pervade our everyday lives for good and for bad, in some cases predictably but mostly unpredictably. The chapter especially focuses on how to transform cyber-fear into cyber governance and how the big tech players can achieve tech trust in an increasingly complex world.

PART III – BOOM: ACHIEVING SUSTAINABLE RESILIENCE AND VALUE

Part III of this book takes us to the other side of Scylla and Charybdis, to a world of opportunity and value creation for the greatest number – i.e., all stakeholders with a stake in an organization's ESGT issues. We do so in two concluding chapters that talk about the vicious cycle of the non-resilient organization, the virtuous cycle of the resilient entity and what it takes to become the highest achieving and most valuable types of organization for the greatest array of stakeholders – the Responsible and Transformational Organizations.

Chapter 7 – Metamorphosis: achieving organizational resilience

Chapter 7 lays out the foundations of a resilient organization. What does it mean to have organizational resilience? What are the key elements? What are the roles of leaders in achieving resilience? In this chapter we contrast the "Vicious Resilience Lifecycle" of non-resilient organizations to the "Virtuous Resilience Lifecycle" of resilient ones. We argue that regardless of the type of entity – government agency, start-up, multinational *Fortune 50* or university – there is a customizable way, even a formula, to get to organizational resilience. It consists of eight key elements that are interdependent and begin with foundational elements of "Lean-In Governance", a "Culture of Integrity", "Strategic ESGT" and culminate with "Crisis Readiness" and an "Innovation Ethos". All of these concepts are illustrated with actual examples.

Chapter 8 – Boom: transforming resilience into sustainable value

Our final chapter represents the epitome of everything we have talked about in the preceding seven chapters. There are three main parts to this chapter: First, we combine the typology of ESGT Leadership explored in Chapter 2 with our typology of organizational resilience from Chapter 7 to produce a Typology of Organizational Resilience and Sustainability with six main categories – from the most frail and unsustainable – the "Outlaw" and "Compromised" Organizations to the most robust and resilient – the "Responsible" and "Transformational" Organizations. And in the process we name names of actual companies, nonprofits and governments that fit each category. Second, we examine the concept of "Value" and "Values" with examples from a variety of top-notch resources (McKinsey, ECI, BCG, Catalyst, Ethical Systems and others), because at the end of the day it is value with values that should be the ultimate objective of any organizational strategy and its leadership for the benefit of their stakeholders. Finally, we offer a third and closing section to this chapter – future-proofing technology – to discuss a vitally important topic of our day: how do the highest levels of leadership – governance and C-Suites – tackle and ensure that tech fear is transformed into sustainable and valuable tech trust.

One last comment or "caveat emptor" to my readers: I am not an academic (though I have a PhD); I am and have always been a practitioner in my 30 plus years of work – first as a lawyer, then as a corporate executive and now as a strategic advisor, board member and author. But I have always had a passion for writing and capturing whatever I have learned in the real world on paper or virtually. So what you see in this book is a combination of accumulated lessons learned from a practical perspective throughout my career with a smattering of my own research and semi-scholarly work as well as a reliance on others' scholarly or practical work which I widely reference throughout the book. Thus, what you will find won't be original academic-style case studies and findings but my original and hopefully useful interpretation of what is going on in the global marketplace (broadly defined) with a call to arms to leaders everywhere to work on the better angels of our human nature by integrating ESGT, resilience and sustainability into their organizational strategy to the ultimate benefit of the broadest array of stakeholders.

With all that said, let us now suit up for our journey from *Gloom to Boom*!

Note

1 Wikipedia. https://en.wikipedia.org/wiki/Necessity_is_the_mother_of_invention. Accessed on November 30, 2018.

Part I

Gloom

Surviving and thriving in turbulent times

1 Gloom

The ten megatrends of our turbulent times

The risk

The only thing necessary for the triumph of evil is for good men to do nothing.

Edmund Burke, 18th-century Irish political philosopher[1]

The opportunity

Per aspera ad astra.

Through hardship to the stars.

Latin phrase and motto of the State of Kansas[2]

(Also, as spoken by the character of Wernher von Braun in the Epic Theatre Ensemble production of the original James Wallert play *The Winning Side* produced off-Broadway in New York City in the Fall of 2018.)[3]

Chapter 1: Gloom – summary overview

1.1 Overview: the ten megatrends of our turbulent times

We are living in turbulent times – of that there is no doubt.

The real question is: will humanity properly recognize this and be able to get its act together to meet the oncoming dangers and risks and convert them into promise, opportunity, solutions and value?

While I have been writing this book (2017–2019), the world has experienced an unusual period (for more reasons than one): the first time in a decade when global economic growth and prosperity appear to be synchronized almost everywhere. GDP growth, lower unemployment (indeed almost full employment in some places), low interest rates and corporate and stock market bullishness seem to have been on full throttle not only in the industrialized world but also in the emerging markets.

And yet we seem to be blowing it. With the advent of Trumpism in America, Brexit in the UK, political populism and tribalism in some of the heretofore "safer", "more advanced" and "stable" geopolitical zones, what could have been unprecedented global prosperity and the lifting up of more people around the world from poverty, illiteracy, hunger and violence may be in jeopardy as I type these words in mid 2019. Add to these developments disturbing domestic and international trends and actual warfare involving authoritarian regimes like Russia and Saudi Arabia and the mix doesn't augur well for the near-term future, at least.

At the same time, we seem to be experiencing a series of larger global strategic changes – threats, risks, opportunities, game-changers – that cannot be easily explained and will no doubt impact human life as we know it.

Before we embark in the "Gloom to Boom" journey that is the core theme of this book, it is important for us to understand our current context – what are the megatrends of these turbulent times as we see them today? Where do we find ourselves at the beginning of this journey? What are the big contextual realities of our world – the situational awareness today that has and will have a direct or indirect effect on how leaders lead their organizations and how organizations of every shape and size (for-profit, nonprofit, governmental) conduct themselves and their activities? How are leaders prepared to navigate through these largely uncharted, choppy, even turbulent, waters?

I believe that now more than ever we need to have a sense of these large, tectonic moving pieces that are enveloping us – whether technological, geopolitical, environmental, socioeconomic, ethical or biological. Frankly, no organization and its leaders will be able to deal with the coming onslaught of change without being prepared and resilient.

The driving force behind this book is this: leaders of all sorts need to be ready to not only deal with these changes and mega-changes by building organizational resilience but they also need to be equipped with the practical tools, ideas and survival equipment to tackle change favorably and in such a way that protects and preserves stakeholders and existing value and, indeed, creates new value.

In this Chapter – titled "Gloom" because I believe we are starting from a point of relative concern that is not going away quickly – we detail the ten megatrends of our turbulent times. Table 1.1 provides a summary overview.

Table 1.1 The ten megatrends of our turbulent times

1.	The fourth industrial revolution: hurtling through space at the speed of light
2.	Global trust collapsing: recession or depression?
3.	A world of extremes: 21st century ethical leadership paradox
4.	Complex interconnected risk rising: beware "the purple techno-swan"
5.	Ecological apocalypse: a decade to oblivion or salvation?
6.	The new geopolitical abnormal: rumblings or tectonic shift?
7.	The meteoric rise of "ESG" and the virtual stakeholder: mirage or reality?
8.	Business as the new global social/moral conscience: fact or fake?
9.	From hyper-transparency to super-opacity: reputation risk on steroids
10.	Future fear: utopia, dystopia, life on Mars?

What follows below is the equivalent of "speed-dating" these ten megatrends which I believe are defining our present and midterm future. There may be others, but these are the ones that most affect leaders of all types of organizations focused on having a holistic strategy of not only accomplishing the core mission, vision, values and strategy of their organizations but also doing so by incorporating key environmental, social, governance and technological (ESGT) issues, risk and opportunities into that strategy in an integrated and cohesive way.

We should keep these ten megatrends in mind as our departure point when we engage in Part II in our navigation through the Scylla and Charybdis of the key ESGT issues of our time. Once we have survived that journey, Part III provides tools, tips and typologies on what it takes to not only survive but to thrive by building organizational resilience and engaging in sustainable stakeholder value creation.

1.2 Megatrend #1: the fourth industrial revolution – hurtling through space at the speed of light

1.2.a Core concept

In the 21st century, the pace and footprint of technological and scientific change has accelerated and grown exponentially, affecting and changing everything about everything. The impacts – known and unknown – of game-changing discoveries and technologies like AI, robotics, quantum computing, 5G communications, nanotechnology and bioengineering are affecting everything from the macrocosm of international governance and relations to the microcosm of individual daily behaviors, psychology and biology.

1.2.b The danger and the risk

There will be intended and unintended – potentially existential – consequences from the technological and scientific innovation that is taking place which humans may be unable to control, to the long-term detriment of life on earth. Think sophisticated military drone robotic biological weapons . . . in the wrong hands.

1.2.c The promise and the opportunity

There may be resolution through new technologies and scientific discoveries of some of the most intractable global health, hunger, climate, communications and sustainability challenges of our time. Think genetically engineered cures to dreaded diseases.

1.2.d Thought bubble

We have always been hurtling through space faster than we knew or could certainly physically feel. However, it seems that only lately this speed has become more perceptible and palpable to the mere human – at least psychologically.

A quick look at the World Economic Forum (WEF) summary in Table 1.2 reveals that until the 18th century we were a largely agriculturally based world with pockets and waves of innovation and progress scattered over the centuries and millennia. It wasn't until the invention of mechanical production equipment with the steam engine around 1784 that we could say that the means of production had been leveraged to such an extent that we had become an "industrial" society.

Table 1.2 World Economic Forum: Navigating to the next industrial revolution[4]

Revolution	*Year*	*What happened?*
#1	1784	Steam, water, mechanical production equipment
#2	1870	Division of labor, electricity, mass production
#3	1969	The computer, electronics and the internet
#4	?	The barriers between man and machine dissolve

Since that fateful year of 1784, humanity has averaged an industrial revolution per century – more or less. The 19th century witnessed the advent of the division of labor and capitalism as well as electricity and mass production. The 20th century gave rise in its last third to computers, electronics and the Internet.

And what Klaus Schwab, the founder and chairman of the WEF has denominated "The Fourth Industrial Revolution" – the time when barriers between man and machine become obfuscated – seems to be upon us much faster than the previous three revolutions. While there is a question mark in Table 1.2 as to when that fourth revolution will hit us, there is no doubt that those obfuscations and barriers are being torn down presently.

The world must be prepared to deal with the Fourth Industrial Revolution, as it is unstoppable and may not be fully controllable (unless something even bigger hits us in the meantime – like a catastrophic meteor). As Schwab has stated:

> We stand on the brink of a technological revolution that will fundamentally alter the way we live, work, and relate to one another. In its scale, scope, and complexity, the transformation will be unlike anything humankind has experienced before. We do not yet know just how it will unfold, but one thing is clear: the response to it must be integrated and comprehensive, involving all stakeholders of the global polity, from the public and private sectors to academia and civil society.[5]

We are living through the fastest period of change mankind has ever experienced. We live this daily even in the most microcosmic ways: walking down the street, riding our bikes, even driving our cars fixated on our screens, oblivious to the world around us, engaging in new physical and psychological behaviors and creating novel health and safety dangers galore.

And this is only the beginning. Wait till we have biologically embedded chips and apps.

1.2.e Leadership to-do

Leaders must realize that technological change and the behavioral and other impacts it has on life on earth are happening at a pace and expanse heretofore unheard of. It is everyone's responsibility – but especially that of leaders of every kind of organization – to take responsibility for understanding this moment to the full extent possible and urgently incorporating this concept into organizational strategy, tactics and culture – one entity at a time, one location at a time, one nation at a time and internationally all the time.

1.3 Megatrend #2: global trust collapsing – recession or depression?

1.3.a Core concept

The entire world and its major institutions – government, business, media and the nonprofit sector – are experiencing a continuing downward trend in stakeholder trust. Is this a trust "recession" from which we will recover soon or a longer-term collapse, closer to a trust "depression", from which we will have to meticulously and deliberately build our way back to "normalcy" (whatever that might be) or some other "new normal"?

1.3.b The danger and the risk

A continuing deterioration in trust in the major existing human institutions that we are used to – government, business, nonprofits, media – will undoubtedly lead to a shorter and more brutish life (to paraphrase Hobbes) where some of the worst aspects of human nature will dominate – deceit, fraud, corruption, violence, war – or to paraphrase another source – where the seven deadly sins of pride, greed, lust, envy, gluttony, wrath and sloth would reign over the contrary concept of trust.

1.3.c The promise and the opportunity

Building back trust will require a deliberate, conscious and constructive effort by leaders in each of our existing institutions or the building of novel alternative institutions. Either way, leaders of all kinds have an opportunity to lead, building back stakeholder trust by creating social value, competitive advantage and reputational opportunity. Right now, business and parts of the media seem to be leading the way out of the trust collapse, but the jury will be out for a while.

1.3.d Thought bubble

For those who have been following the results of the annual Edelman Trust Barometer, it is clear that a steady decline in individual and social trust in four key institutions has been taking place over recent years up to and including trust in what might be the most trusted types of institution – NGOs.

Edelman focuses on the trust mandates of each of these sectors and differentiates them as follows:

- The trust mandate of Government includes:
  - Driving economic prosperity.
  - Investigate corruption.
  - Protect the poor.
- The trust mandate of Business includes:
  - Protecting privacy.
  - Driving economic prosperity.
  - Provide jobs and training.
- The trust mandate of NGOs includes:
  - Protecting the poor.
  - Calling out abuses of power.
  - Creating a sense of community.
- Finally, the trust mandate of the Media is to:
  - Guard information quality.
  - Educate, inform and entertain.
  - Protect privacy.

I would add that each of these sectors has one additional mandate that is part of the expectation of every one of their most important stakeholders: to carry out their mandate in an ethical, high integrity and stakeholder-centric way.

On the one hand, Figure 1.1 shows the overall negative trends in trust gathered by the 2018 Edelman Trust Barometer.

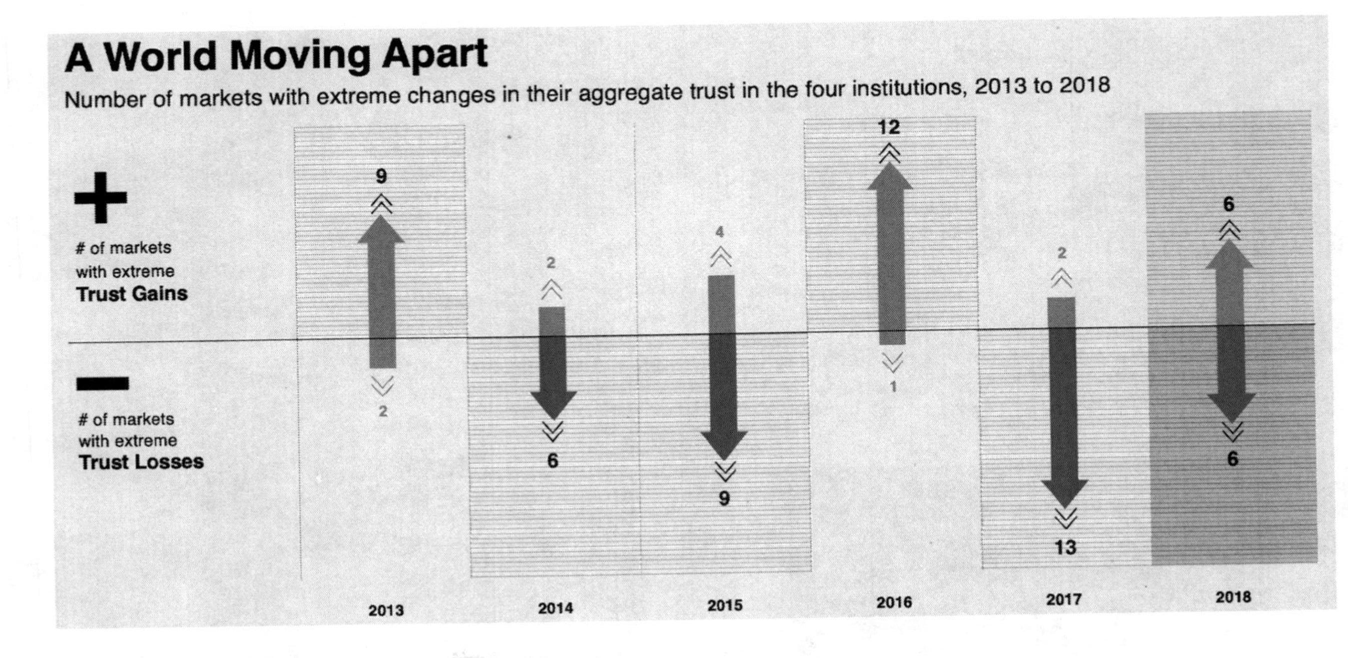

Figure 1.1 2018 Edelman Trust Barometer: a world moving apart.

Source: 2018 Edelman Trust Barometer.

On the other hand, the 2019 Edelman Trust Barometer reported a glimmer of possible better news, showing for the first time in years a slight increase in global trust (see Figures 1.2 and 1.3). Might this point to a trust recession instead of a depression? It is too early to say, but we can certainly not only hope so, but leaders of all kinds should keep working hard on whatever they were doing right in 2018.[6]

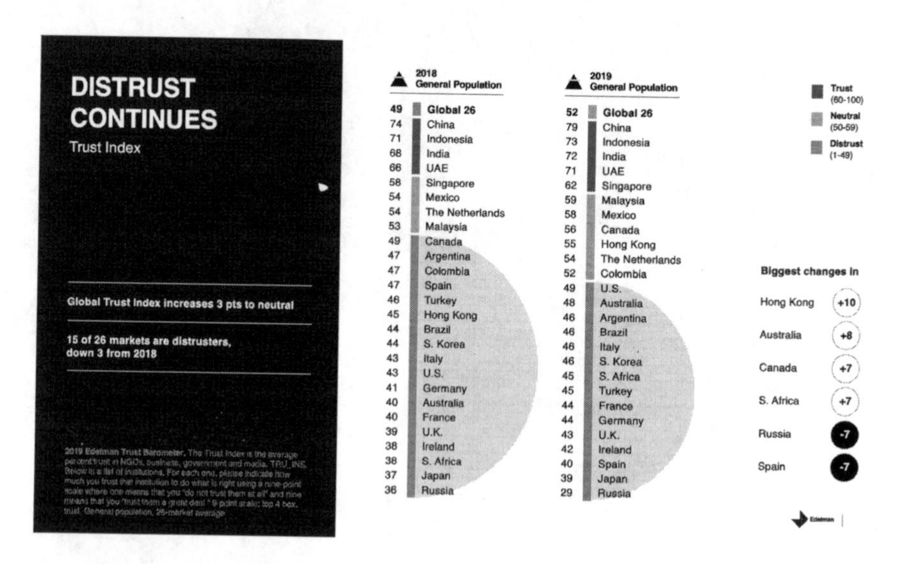

Figure 1.2 2019 Edelman Trust Barometer: distrust continues.

Source: 2019 Edelman Trust Barometer.

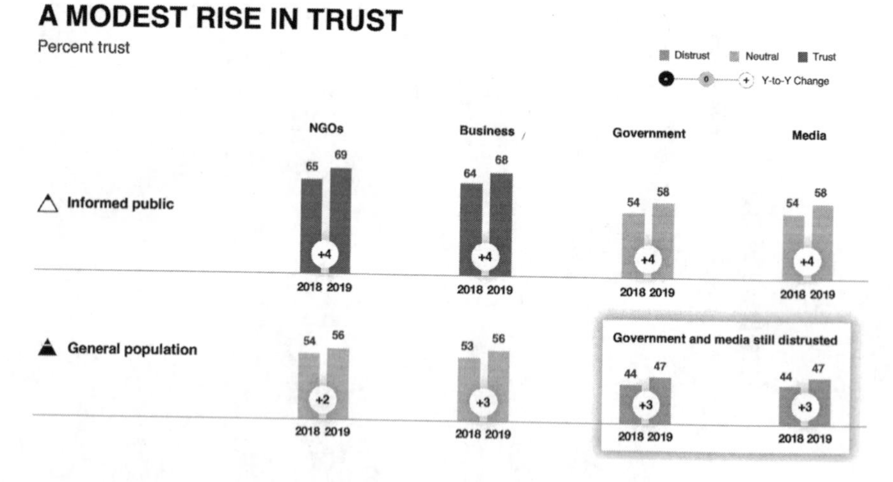

Figure 1.3 2019 Edelman Trust Barometer: a modest rise in trust.

Source: 2019 Edelman Trust Barometer.

What does all this mean in the longer-term context? Are we experiencing a long-term erosion of trust in our core institutions globally? It is probably not a wholesale collapse, but certainly pockets of erosion are somewhat synchronized given the reality of instantaneous global communications, which will only get even more instantaneous with the advent of 5G, quantum computing and other innovations.

1.3.e Leadership to-do

The message for organizational leaders: be aware of the fact that trust is a powerful and often impalpable intangible and that you may not be fully aware of its actual positive or negative tangible consequences until late in the game. Undertake an analysis of what your trust issues are by getting to know who your most important stakeholders are (above and beyond the most obvious one) and learn what their expectations are of you. Then live up to those expectations and own up to any breach of trust quickly, completely and transparently.

1.4 Megatrend #3: A world of extremes – the 21st century ethical leadership paradox

1.4.a Core concept

Because we are living in a world of extremes – extreme opportunity for good and extreme opportunity for bad – we are also living through an ethical leadership paradox. The ethical leadership paradox occurs because a good portion of leaders seek and hold leadership for unethical, narcissistic or selfish reasons (greed and/or power) and through technology they can commandeer even greater tools and assets to lead for the wrong purposes. And yet we also see the rise of ethical leaders in all walks of life (from business to government) driven by broader social responsibility, and they too can commandeer such tools and assets for the greater good. What type of leadership will prevail?

1.4.b The danger and the risk

The same discoveries that can make the world a better place threaten to make it a much darker place as well – exponentially, existentially and quickly. We are under the threat of out-of-control bad acts by bad actors (unethical, criminal or despotic leaders) that can become major, even existential, threats to the health and safety of life on earth. Unchecked, and unaddressed, unethical and nefarious forces will seize the lead in various pockets (or more extensively) around the world. Witness the 2020 "Social Credit" system in China where all of their 1 billion plus citizens will be continuously tracked and rated digitally socially, financially and reputationally. And this is certainly not the worst example of what leaders can do with technology.

1.4.c The promise and the opportunity

Responsible and ethical leaders – of which there are many in every sector – must seize technology and science for the good of humanity, creating responsible and ethical decision-making at all levels – from the inception of new technologies like AI to their implementation. The fact that technology and science can exponentially improve life on earth needs to be seized more systematically by responsible and conscientious multi-sector leaders above and over the darker impulses of some portions of mankind. But that doesn't mean that such leaders should be blind to the alternative uses of technology and science – indeed, we need to count on more ethical leaders to be uber-prepared in understanding, managing and countering the darker leadership arts and practices.

1.4.d Thought bubble

The 21st-century ethical leadership paradox is that in the midst of all of the turbulence we are living in we need responsible and accountable leadership – at every level of business, society and politics and more than ever before – because of the depth and extreme complexity of the technological, geopolitical, environmental, humanitarian and existential threats we are facing. Never has there been a time in human history where so many existential risks and opportunities have presented themselves simultaneously and rapidly. And never has there been a time when we need the right kind of leadership to help us navigate through this Scylla and Charybdis.

1.4.d.i The darker and the better angels of our nature

The 21st century ethical leadership paradox is dramatically illustrated by the following examples of both our darker and our better angels:

- The rise of authoritarian political leaders like Trump, Putin, Erdogan and Mohammad Bin Salman.
- The work of responsible political leaders like Merkel, Ardern, Mahathir Bin Mohamad and Guterres.
- The rise of illiberal democracies and populistic movements in several European (Hungary), Latin American (Brazil) and Asian (Philippines) countries.

- The rise of new democracy in unexpected places (Malaysia and Ethiopia and in 2019 potentially in long-standing authoritarian regimes like in Algeria and Sudan) and pro-democratic social movements within democratically compromised political systems (Russia, Hong Kong).
- The continuation of a terrible proxy war in Yemen that is becoming one of the greatest man-made famine-inducing conflagrations ever.
- The exponential impact of the humanitarian work that entities like the World Food Programme do every day around the world to help the most vulnerable.
- A variety of corporate leadership styles – the good, the ugly and the downright bad:
  - The "good" (ethical and responsible): Larry Fink of BlackRock and Marc Benioff of Salesforce.com.
  - The "ugly" (unethical or irresponsible): Les Moonves of CBS and Travis Kalanick of Uber.
  - The "bad" (potentially criminal): Elizabeth Holmes of the now defunct Theranos and Harvey Weinstein of the now defunct The Weinstein Company.

Since I tend to side with the better angel's side of the equation, I would also like to offer two additional positive data points from learned sources about the work impulsed by ethical and responsible leaders.

First, the following is data about a short history of the living conditions from the "World as 100 People" which provides a useful synthesis of some extraordinarily positive changes in the world over the past 200 years:

- **Extreme Poverty** has declined from 94% living in extreme poverty 200 years ago to 10% living in extreme poverty today. Of course, we have many more people populating the earth today, so relatively speaking, we have many more people living in extreme poverty on earth today, but the arc of history is leading us in a crystal-clear direction of progress on this issue.
- **Basic Education** has improved from only 17% of the human population having one 200 years ago to 86% having such education today.
- **Literacy** has gone from 12% of the population being able to read 200 years ago, to 85% being able to read today.
- **Democracy** has made incredible strides over the past 200 years – from 1% of the population living under some form of free elections and rule of law to 56% living in democracy today.
- **Vaccination** rates have made the most strides – from 0% 200 years ago (when no vaccines existed) to 86% of the human population being vaccinated today.

- **Child Mortality** – while 200 years ago 43% of all children born would not live past the age of 5 years old, today that figure has dropped to 4%.[7]

I would also like to cite Steven Pinker's extraordinary work – most especially his magnum opus, *The Better Angels of Our Nature: Why Violence Has Declined* – in which in agonizing detail Pinker makes the powerful case that violence and warfare have made a dramatic decline over the centuries because of the rise of education, rationalism, enlightenment and empathy (the "better angels of our nature"), as well as his most recent book, *Enlightenment Now: The Case for Reason, Science, Humanism and Progress*, in which he goes several steps further to make the case that the future is bright and that the present is brighter than we think it is.[8]

1.4.e Leadership to-do

At a time of such dramatic change where incredible inventions and initiatives are underway to address some of the most intractable global challenges, leaders should lead with integrity and think beyond the box on how their organization can be part of the solution. All of us have a role to play, both within our organizations and externally as part of communities, schools, workplaces, cities, countries and the globe. Likewise, the better angels (ethical and responsible leaders) should always be cognizant and prepared to deal with and check the lesser or darker angels (unethical, irresponsible and even criminal leaders) of our nature.

1.5 Megatrend #4: complex interconnected risk rising – beware the "purple techno-swan"

1.5.a Core concept

In a world of increasing risks, both natural and man-made risks are interconnecting and becoming more interdependent than ever before and creating a more complex environment requiring greater firepower to resolve. In addition, beware of the Black Swan of our times – the "Purple Techno-Swan", a risk so new, so different and so unpredictable, propelled by our age of exponential technological change.

1.5.b The danger and the risk

Intended and unintended potentially existential consequences of technological and scientific innovation that humans are unable to control for the betterment of humanity and the globe are ushering in the rise of new and intractable risk and existential risk. This includes risks we don't know already exist and risks we may never be able to predict or mitigate until it's too late. Think about what quantum computing might yield in terms of power and impact if in the wrong hands.

1.5.c The promise and the opportunity

There are multitudinous opportunities for responsible leaders to seize the moment presented by the vast array of new technologies of some of the most intractable global health, hunger, climate, communications and sustainability challenges of our time and mitigate the potentially existential risks that are associated therewith. By having greater technological firepower (including deploying AI, machine learning and eventually 5G and quantum computing), risk management might become much better at arming risk managers (and others in leadership positions) with the tools and knowledge to mitigate risk. Think progress on weather prediction techniques using satellites, other sensors and machine learning.

1.5.d Thought bubble

I love a graphic that philosopher Nick Bostrom of the Cambridge Centre for Existential Risk, put together and which I have recreated in Table 1.3, which shows in very simple terms the various layers of risk from the individual to the cosmic.

Table 1.3 Nick Bostrom's Risk Scope to Severity Table

SCOPE ↑					
	(COSMIC)				
	TRANS-GENERATIONAL	LOSS OF ONE BEETLE SPECIES	DRASTIC LOSS OF BIODIVERSITY	X	*EXISTENTIAL RISK*
	GLOBAL	GLOBAL WARMING BY 0.01 K	DESTRUCTION OF OZONE LAYER	AGING	*GLOBAL CATASTROPHIC RISK*
	LOCAL	CONGESTION FROM ONE EXTRA VEHICLE	RECESSION IN ONE COUNTRY	GENOCIDE	
	PERSONAL	LOSS OF HAIR	ONE'S CAR IS STOLEN	FATAL CAR CRASH	
		IMPERCEPTIBLE	*ENDURABLE*	*TERMINAL*	*(HELLISH)*
					SEVERITY →

Source: Nick Bostrom, Cambridge Centre for Existential Risk.

The World Economic Forum also provides a graphic that speaks louder than words to the point of this megatrend – the rise of man-made global interconnected complex risk – see Figure 1.4 for their powerful graphic visualization of this concept.[9]

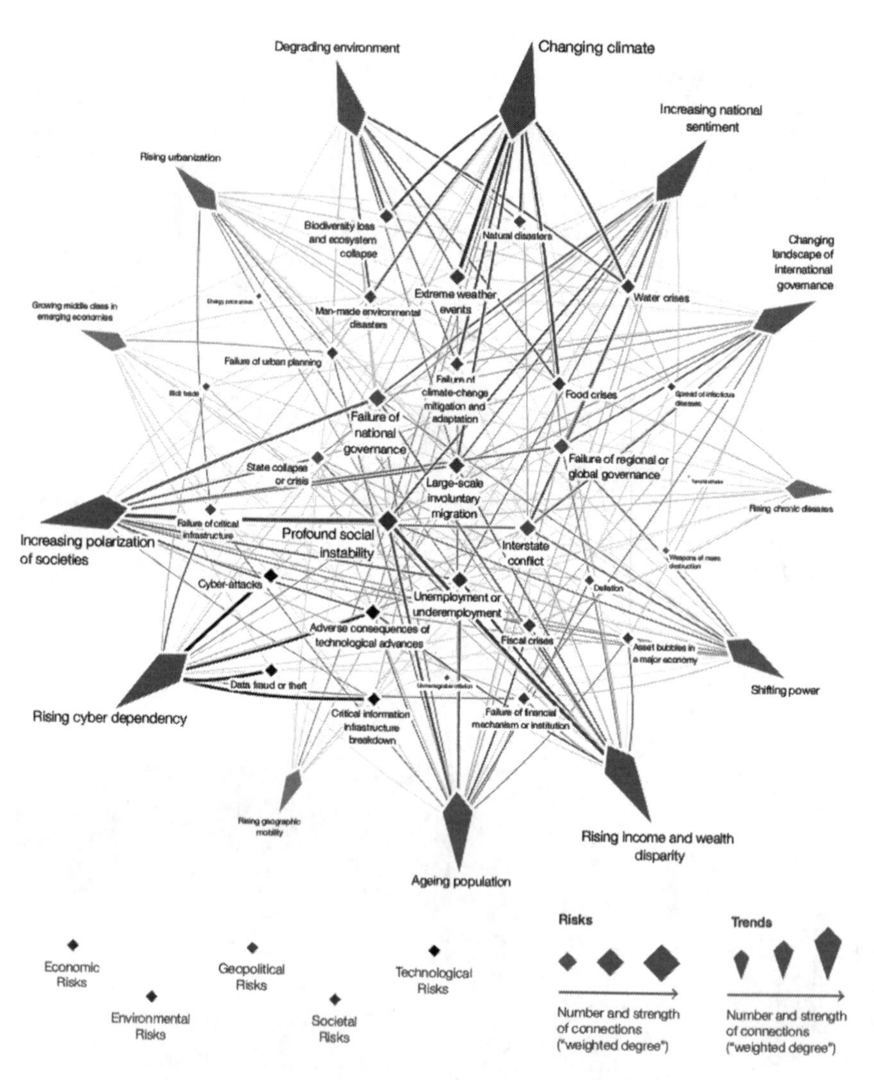

Figure 1.4 The World Economic Forum: the global risks interconnection map 2019.

Source: World Economic Forum 2019.

1.5.d.i The black swan transforms into the "purple techno-swan"

First coined by Professor Nassim Nicholas Taleb in his groundbreaking work, *The Black Swan: The Impact of the Highly Improbable*, the *Financial Times* summarizes the concept as follows:

> An event or occurrence that deviates beyond what is normally expected of a situation and that would be extremely difficult to predict.[10]

We are living through a time of multiple black swans or what I would perhaps dare to call the "Purple Techno-Swan" risk. First, there are no purple swans (unless someone uses Crispr to gene edit a regular swan from nature). Second, who knows what else a future swan might look like or be equipped with – maybe there will be purple-hued swans with tech implants that allow them to be used as intelligent cruising drones? And who knows what risks and opportunities such a purple techno-swan will present?

Another way to look at this trend is to consider the recently coined term of "disruptive risks" that references risks with severe impact occasioned by our fast-changing world. The National Association for Corporate Directors (NACD) produced a Blue-Ribbon Commission report in 2018 in which they suggested the "Taxonomy of Disruptive Risks" illustrated in Table 1.4, borrowed from Marsh & McLennan Companies.[11]

Table 1.4 Disruptive risk taxonomy

Government agenda	*Societal volatility*	*Technology advances*	*Hazards & accidents*
• Fiscal instability of foreign economies. • Protectionist programs. • Radical regulatory change. • Corruption and illegal discrimination. • Geopolitical conflict.	• Major demand shifts. • Popular and employee discontent. • Acts of terrorism. • Workforce unobtainability.	• New technology-driven business models. • Technology implementation problems. • Cyber-attacks. • False information.	• Natural disaster incidents. • Gradual natural environment depletion. • Public health crises. • Man-made disasters.

Source: NACD and Marsh & McLennan.

Finally, and relating more directly to everything we cover in this book, Figure 1.5 depicts in a purposefully oversimplified manner how the ESGT issues, risks and opportunities we discuss are also deeply intertwined, interdependent, interconnected and overlapping in various and sundry ways.

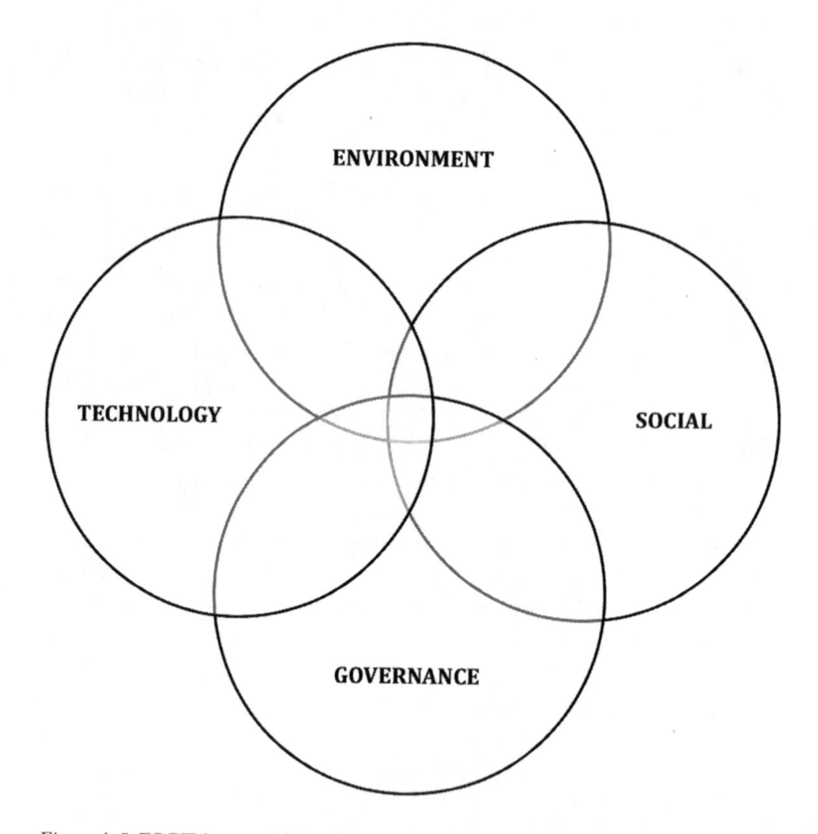

Figure 1.5 ESGT issues, risks and opportunities are interconnected.

Source: GEC Risk Advisory.

1.5.e Leadership to-do

We are most definitely in an age of complex, interconnected, man-made and other natural risks where ESGT risks and others overlap and are interesting in complex and novel ways. We may be reaching the age of the "purple techno-swan" risk – totally unpredictable to us today. Risk managers and leaders must be better equipped for risk management than ever before to do complex risk identification, coordination, mitigation, prediction and translation into strategic opportunity.

1.6 Megatrend #5: ecological apocalypse – a decade to oblivion or salvation?

1.6.a Core concept

Climate change is here and now and the time-fuse for addressing it properly has been lit – will humankind get its act together in time (over the next decade) or are we and future generations doomed to live in a more hostile, poor, nasty, short and brutish world (to paraphrase Hobbes once again)?

1.6.b The danger and the risk

Ecological damage, destruction and related existential risks are multiplying around the world because of climate change and other unleashed forces which may eliminate habitats, destroy species, damage coastal areas and create deserts and other uninhabitable areas where there was once life.

1.6.c The promise and the opportunity

The world knows what we are up against, and most leaders know this as well, despite some notable exceptions. There is a collective global vehicle for action – the Paris Climate Agreement – and there are also myriad local, urban, communal, social, business and governmental initiatives worldwide that represent thousands, even millions "points of light", that can come together to mitigate the climate onslaught that is coming. We can do this, together.

1.6.d Thought bubble

In Chapter 3 – "Environment: transforming environmental risk into opportunity from global warming to sustainable resilience" – we tackle in great detail this topic, including the Paris Climate Agreement and recent developments in the US with the publication in late 2018 of the US Climate Report, which painted a very dire picture for the US and the world. Instead of repeating something I will discuss in Chapter 3, I would like to share something a little different here that illustrates my point better than my own words can.

Table 1.5 titled "Bug apocalypse?" contains excerpts from a report on what is potentially happening in the world of insects in the rain forest. Please read this somewhat scientific description of the climate-induced disruption of our biosphere and what that means to the decline and end of life of some species – with consequences to all of us as part of a delicate interdependent ecosystem.

Table 1.5 Bug apocalypse? Excerpts from "Climate-driven declines in arthropod abundance restructure a rainforest food web"[12]

Arthropods, invertebrates including insects that have external skeletons, are declining at an alarming rate. While the tropics harbor the majority of arthropod species, little is known about trends in their abundance. We compared arthropod biomass in Puerto Rico's Luquillo rainforest with data taken during the 1970s and found that biomass had fallen 10 to 60 times. Our analyses revealed synchronous declines in the lizards, frogs, and birds that eat arthropods. Over the past 30 years, forest temperatures have risen 2.0°C, and our study indicates that climate warming is the driving force behind the collapse of the forest's food web. If supported by further research, the impact of climate change on tropical ecosystems may be much greater than currently anticipated.

(*Continued*)

Table 1.5 Continued

From pole to pole, climate warming is disrupting the biosphere at an accelerating pace. Despite generally lower rates of warming in tropical habitats, a growing body of theory and data suggests that tropical ectotherms may be particularly vulnerable to climate change. As Janzen pointed out, tropical species that evolved in comparatively aseasonal environments should have narrower thermal niches, reduced acclimation to temperature fluctuations, and exist at or near their thermal optima. Consequently, even small increments in temperature can precipitate sharp decreases in fitness and abundance. These predictions have been verified in a variety of tropical reptiles, amphibians, and invertebrates.
Given their abundance, diversity, and central roles as herbivores, pollinators, predators, and prey, the response of arthropods to climate change is of particular concern. Deutsch et al. predicted that, for insects living at mid-to-high latitudes, rates of increase should grow as climate warms, while in the tropics insects should decline by as much as 20%. Reduction in population growth, combined with elevated metabolic rates, could potentially lower abundances and raise arthropod extinction rates. If these predictions are realized, climate warming may have an even more profound impact on the functioning and biodiversity of the Earth's tropical forests than currently anticipated.

Source: National Academy of Sciences of the United States of America.

But it's not all doom and gloom when it comes to our environment, as we will see further in Chapter 3. The synopsis of the work that the Task Force on Climate-Related Financial Disclosures is doing to bring reporting and accountability to the highest echelons of leadership of businesses and other relevant entities speaks volumes and demonstrates how progress is and can be made (see Table 1.6).

Table 1.6 TCFD sample alignment assessment[13]

Governance	*Strategy*	*Risk management*	*Metrics & targets*
Describe the Board's oversight of climate-related risks and opportunities.	Describe the climate-related risks and opportunities identified over the short, medium and long-term.	Describe the organization's processes for identifying and assessing climate-related risks.	Disclose the metrics used by the organization to assess climate-related risks and opportunities in line with its strategy and risk management process.
NOT ADOPTED	ADOPTED	ADOPTED	ADOPTED
Describe management's role in assessing and managing climate-related risks and opportunities.	Describe the impact of climate-related risks and opportunities on the organization's businesses, strategies, and financial planning.	Describe the organization's processes for managing climate-related risks.	Disclose Scope 1, Scope 2 and, if appropriate, Scope 3 greenhouse gas emissions and the related risks.
ADOPTED	ADOPTED	ADOPTED	ADOPTED

(*Continued*)

Table 1.6 Continued

Governance	*Strategy*	*Risk management*	*Metrics & targets*
	Describe the resilience of the organization's strategy, taking into consideration different climate-related scenarios, including a 2-degree centigrade or lower scenario.	Describe how processes for identifying, assessing and managing climate-related risks are integrated into the organization's overall risk management.	Describe the targets used by the organization to manage climate-related risks and opportunities and performance against targets.
	ADOPTED	ADOPTED	ADOPTED

Source: S&P Global and the Financial Stability Board Task Force on Climate-Related Financial Disclosures.

1.6.e Leadership to-do

Climate change and everything that goes with it – from the lives and well-being of insects to that of humans – is everyone's responsibility. We can all do our part – from the individual who sorts her garbage properly for urban waste management and the Fortune 100 company seeking to achieve a neutral carbon footprint all the way up to (and especially) government leaders like the president of the US. We have ten years to address the ecological disaster that is coming, and even then, nothing will be easy. Everyone has an affirmative obligation to step up to this plate – nothing is guaranteed.

1.7 Megatrend #6: the new geopolitical abnormal – rumblings or tectonic shift?

1.7.a Core concept

Does the global regression in both domestic and international political governance to more forms of populism, tribalism, illiberal democracy and authoritarianism represent a fixable erosion of post-WWII norms of international governance and democracy or a more permanent reflection of larger change leading to a less stable international order and a rise in authoritarianism?

1.7.b The danger and the risk

The US does not recover its global political leadership and past global "moral" authority post-Trump. Instead, a new global disorder emerges where multiple national and cross-national power centers compete, and some of the key interests of humanity are set aside (human rights and other freedoms, climate, humanitarian initiatives, elimination of hunger, poverty, etc.). Authoritarian, even ruthless, leaders like Putin, Mohammad Bin Salman and Xi dominate their geographies (and beyond) and the world becomes generally more dangerous and less safe for larger swaths of the human population.

1.7.c The promise and the opportunity

Whether the US reemerges from its current Trumpian anti-internationalist, anti-immigrant, pro-authoritarian trends or another coalition of nations (e.g., led by the EU) are able to represent global "moral" political leadership, opportunities exist to move the global agenda forward. These opportunities include implementing the Paris Climate Agreement as well as moving forward on the overall Sustainable Development Goals (SDG) agenda with the help and collaboration of the United Nations and its many agencies, and other powerful global institutions like the IMF, World Bank and others, as well as the business and NGO communities globally, regionally and locally.

1.7.d Thought bubble

We discuss geopolitical governance issues in greater detail in Chapter 5. However, here I would like to underscore some of the key trends we have seen over the past few decades amounting to what now seems to be a series of tectonic geopolitical changes:

- The rise of democracy around the world from the late 1970s until recently.
- The end of the cold war in the late 1980s.
- The rise of populism in the 2000s.
- An increase in perceived or real macro and micro inequality.
- The rise of illiberal democracies.
- A steady decline in democracy worldwide in the past few years.
- The rise of global white nationalism and continuing metamorphosis of technology-induced global terror.
- Brexit drama.
- Trump drama.
- The rise of Kim Jun Un in North Korea.
- Major challenges to the EU from within.
- Enormous population shifts and migrations because of war (Syria), famine, loss of climate habitat.
- Rise of popular or national anti-immigrant parties and policies.
- China rises as an economic and technological superpower – "Belt & Road Initiative".
- The rise of overall cyber and other tech-insecurity.
- Cyber and social engineering interference by Russia (and others) in Western democratic elections.
- Cyber warfare targeting infrastructure – Russia v. Ukraine power plants.
- Other technological advances for military purposes (AI, robotics).
- Asymmetric technology-enabled power rises providing small non-state players with disproportionate power.

Among the tactical shifts that may have a longer-term impact is the rise of a different kind of geopolitical risk – that of the rise of populism in the democratic world, a populism that has suffused some of the leading long-standing democracies like the US and the UK. The real question here is: Is Trump the symptom or the cause, is Brexit the symptom or the cause, is the degradation of democracy the symptom or the cause? Are we living through a temporary shift or one that is longer-term, leading to a wholly different world from what we have grown accustomed to in the post-WW2 era? Perhaps democracies have become too complacent and lazy about how important it is to continually nurture civics, participation, grassroots knowledge and education?

Table 1.7 provides a bird's-eye view from the Council on Foreign Relations of the state of international governance at the end of 2018.

Table 1.7 Council on Foreign Relations: Global Governance Monitor[14]

Category	*Description*
Armed conflict	Preventing armed conflict, keeping peace, and rebuilding war-torn states remain the most difficult challenges for policymakers and government officials throughout the world.
Crime	Over the past two decades, the global impact of transnational crime has risen to unprecedented levels as criminal groups use new technologies and diversify their activities.
Nuclear proliferation	The current nuclear nonproliferation regime must be reinforced to effectively address today's proliferation threats and pave the way for a world without nuclear weapons.
Global finance	Regulating market volatility and economic risk has become fraught with difficulty following the 2008 financial crisis that plunged developed economies into recession.
Oceans	Nations around the world need to embrace multilateral governance to protect the world's oceans, which play a critical role in global climate, provide an avenue for commerce, and sustain life on earth.
Climate change	Avoiding the consequences of climate change will require large cuts in global greenhouse emissions and significant efforts to mitigate and adapt to changing weather patterns.
Public health	Despite medical advances and improvements in water and sanitation, nutrition, housing, and education, poor health still plagues hundreds of millions around the world.
Terrorism	The unprecedented reach and threat of terrorist networks constitute a new danger to states and requires innovative counterterrorism efforts.
Human rights	In shaping a human rights policy for the twenty-first century, states must carefully craft tactics consistent with their interests and values to protect victims of abuse.
The Internet	Collaborative Internet governance structures are emerging, but they are being outpaced by policy challenges arising from the Internet's rapid expansion and development.

Source: The Council on Foreign Relations.

While there is much turbulence and change taking place and a lot more may be in store, one reality is clear: the stability of the Cold War period (mutual assured destruction – a binary world where everyone knew where they fit), is giving way to a multipolar world where the name of the game will be jockeying for position, asymmetric power and potentially more conflicts. Table 1.8 showing the Global Peace Index 2018 seems to confirm these sentiments.

Table 1.8 Global Peace Index (GPI) 2018: trends in peace[15]

- The average level of global peacefulness deteriorated 2.38% since 2008 – over that period 85 countries deteriorated, 75 improved.
- The average country peacefulness level deteriorated for 8 of the past 10 years.
- Gap between least and most peaceful countries continues to grow.
- Of the three GPI domains, two deteriorated (Ongoing Conflict and Safety & Security) and one improved (Militarization).
- In Europe, the world's most peaceful region, 61% of nations have deteriorated since 2008.
- The indicator with the largest deterioration – Terrorism Impact – shows 62% of countries with increases in terrorist activity and 35% experiencing large deterioration.
- 2014 marked a 25-year high in battle deaths.
- Refugees made up almost 1% of the global population in 2017 for the first time in history at a rate 12 times higher than in 1951.

Source: The Global Peace Index 2018.

1.7.e *Leadership to-do*

The time of orderly, predictable international relations is over for the time being. The worldwide growth of democracy has also atrophied and actually experienced a reversal. The impact of technology and capitalism is affecting stakeholders differently. Business and NGO leaders need to be prepared for greater geopolitical risk and some of it from unexpected places and sources. Geopolitical risk needs to be part of everyone's risk management. Crisis management and business continuity need to be in place as well. Country leaders must rethink what leadership means and what alliances mean and for what. Leadership tone from the top has never been more important, as we have a new strain of leaders – Trump, Putin, Xi, Erdogan, Bin Salman – who are not shy about throwing their authoritarian weight around and are setting a different and more dangerous cultural tone globally. Geopolitical risk is real and different from what it has predominantly been in recent decades. Buckle up and prepare for a wild ride.

1.8 Megatrend #7: the meteoric rise of "ESG" and the virtual stakeholder – mirage or reality?

1.8.a Core concept

The rise of all things corporate responsibility (CSR, ethics, compliance, sustainability and ESG) over the past two to three decades reflects the rise in power of previously ignored or excluded stakeholders – beyond the primary stakeholders – and their primary interests. What that means for business is that, in addition to shareholders, customers and employees matter, a lot. What that means for nonprofits and NGOs is that, in addition to beneficiaries, donors and suppliers are super-important. What that means for governments is that, in addition to citizens and residents, others such as transients and immigrants matter, a lot. Is the rise of the concept of ESG and its expanded base of "stakeholders" an evanescent mirage or a long-term reality?

1.8.b The danger and the risk

Leaders and organizations that do not amplify their lens to incorporate the views, expectations and consequences of ignoring (or even damaging) their full spectrum of key stakeholders on ESG (or what I am coining as ESGT) issues, risks and opportunities run the risk of losing to competitors, engaging in misadventures, increasing reputation risk, liabilities and losses or even losing their license to operate.

1.8.c The promise and the opportunity

The opportunity to create greater resilience and new value by truly understanding your most important stakeholder and the next group of key stakeholders on their full spectrum of ESGT issues relevant to an entity cannot be underscored enough. It is the difference between resilience and weakness, successful and unsuccessful strategy, value creation and value destruction.

1.8.d Thought bubble

Leaders of any type of organization should understand who their most important stakeholders are and undertake a proper and rigorous analysis of such stakeholders' issues, risks and expectations. The only way an organization can fully understand its reputation risk and opportunity is to understand who these stakeholders are and what they want.

Figure 1.6 provides an overview of the ever-expanding universe of potential organizational stakeholders.

Figure 1.6 The universe of potential stakeholders.

Source: GEC Risk Advisory.

Figure 1.7 illustrates the meteoric rise of ESG regulations globally since 1997 as assembled by S&P Global.[16]

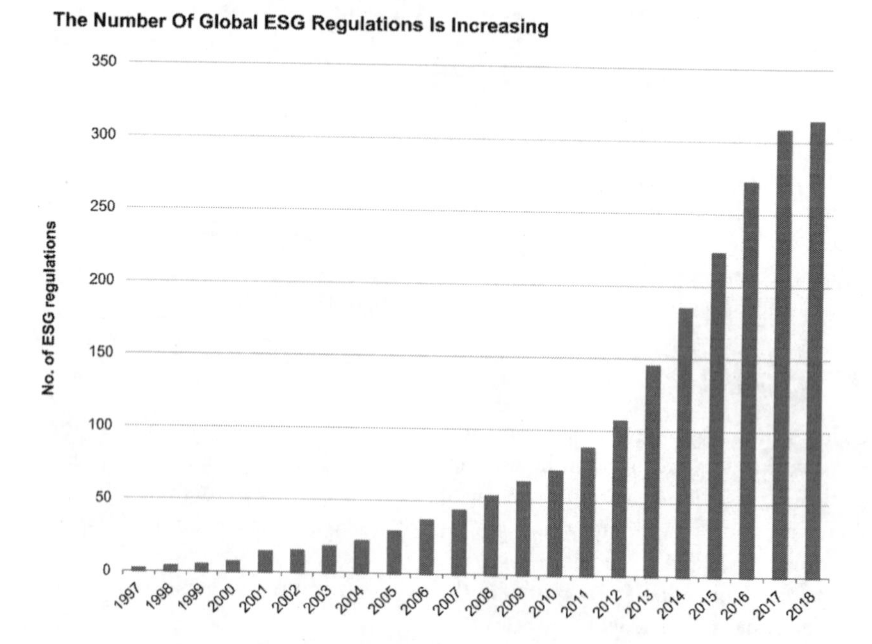

Figure 1.7 Number of ESG regulations globally 1997–2019.

Source: S&P Global.

1.8.e Leadership to-do

We are quickly approaching the time when every organization – including businesses that have always been exclusively focused on and beholden to their shareholders as the only stakeholders that matter – have to broaden their lens to incorporate the more complex landscape of key stakeholders that have a stake in their organization in one way or another and their ESGT portfolio of issues, risks and opportunities. And this applies to all types of organization – business, nonprofit, governmental and academic – as we repeatedly stress throughout this book.

1.9 Megatrend #8: business as the new global social/moral conscience – fake or fact?

1.9.a Core concept

With deteriorating political leadership in both democratic and nondemocratic regimes around the world, and the decline in trust in all other key fundamental institutions, suddenly business finds itself in the possible role of global moral steward. Is this really happening and, if so, is business (given its mixed track record at best) really equipped to take on the mantel of global conscience? Or is business simply taking advantage of the leadership vacuum for narrower purposes?

1.9.b The danger and the risk

For business, not taking leadership on important ESGT, financial and even political issues of our day is tantamount to abdicating responsibility for the overall well-being of their key stakeholders – whether they be shareholders and owners, employees, contractors, customers or even the regulator. This may translate into losing stakeholder trust, reputation risk and financial losses. On the other hand, other stakeholders may not trust business as the moral arbiter of global issues and for very good reason given some of the corporate track records of exploitation, looking the other way and singular focus on financial gain and reward.

1.9.c The promise and the opportunity

Businesses that align themselves with the interests and expectations of their most important stakeholders when it comes to key ESGT, financial, economic and even political issues of our day are setting themselves up for a greater chance at success, value creation, reputation opportunity and long-term sustainability. In a world where political leaders and governments appear to be even more narrowly focused on sheer political gamesmanship for the sake of narrow winning and not for the sake of broader governing, the world may find itself in need of a different global moral force.

1.9.d Thought bubble

There have been a number of important cases of corporate moral leadership in recent times, especially since Trump became US president. Several examples of corporate moral leadership come to mind:

- Merck & Co. CEO Ken Frazier leading a corporate CEO walk-out on Trump's presidential Corporate Council in August 2017 because of what appeared to be his pro-white nationalist comments.
- Salesforce CEO Mark Benioff leading an effort to pass corporate taxes in the city of San Francisco to help address the mounting homelessness crisis in that city.
- Patagonia CEO Rose Marcario deciding to pass along the $10 million corporate tax break her company got from the Trump 2018 corporate tax reforms in the US to fund climate change and environmental groups and causes.

Table 1.9 offers a few more headlines along these lines.

Table 1.9 Corporate "moral" leadership headlines in the news

Choosing plan b:
Danone rethinks the idea of the firm
A tradition of pursuing lofty social goals is going further
The Economist – August 9, 2018

Saudi Arabia:
Khashoggi case prompts Branson to suspend $1bn Saudi talks
Global executives split on participation in projects after journalist's disappearance
The FinancialTimes – October 11, 2018

When business executives become reluctant statesmen:
Without a strong stance from Washington to give them cover, top executives have been on their own dealing with Saudi Arabia, a country that has shown itself capable of holding a grudge
New York Times – October 16, 2018

Sources: *The Economist* 2018, *The Financial Times* 2018, and *New York Times* 2018.

Much of the corporate moral leadership on these topics can also be seen in the rise of ESG from an investor and impact investor standpoint. Europe has been way ahead of the rest of the world, including the US, on this topic for a couple of decades. With the advent of the United Nations SDG movement, the convergence with ESG and other corporate responsibility and ethics and compliance movements, we are witnessing a convergence of these factors into global business.

Recently, no one has verbalized this convergence more clearly than Larry Fink, the founder, CEO and chairman of Blackrock, the largest asset manager in the world. Table 1.10 provides excerpts of his letter to corporate CEOs.

Table 1.10 Excerpts from Larry Fink's 2018 letter to corporate CEOs: "a sense of purpose"

Dear CEO,

As BlackRock approaches its 30th anniversary this year, I have had the opportunity to reflect on the most pressing issues facing investors today and how BlackRock must adapt to serve our clients more effectively.

We also see many governments failing to prepare for the future, on issues ranging from retirement and infrastructure to automation and worker retraining. As a result, society increasingly is turning to the private sector and asking that companies respond to broader societal challenges. Indeed, the public expectations of your company have never been greater. Society is demanding that companies, both public and private, serve a social purpose. To prosper over time, every company must not only deliver financial performance, but also show how it makes a positive contribution to society. Companies must benefit all of their stakeholders, including shareholders, employees, customers, and the communities in which they operate.

Without a sense of purpose, no company, either public or private, can achieve its full potential. It will ultimately lose the license to operate from key stakeholders. It will succumb to short-term pressures to distribute earnings, and, in the process, sacrifice investments in employee development, innovation, and capital expenditures that are necessary for long-term growth. It will remain exposed to activist campaigns that articulate a clearer goal, even if that goal serves only the shortest and narrowest of objectives. And ultimately, that company will provide subpar returns to the investors who depend on it to finance their retirement, home purchases, or higher education.

Your company's strategy must articulate a path to achieve financial performance. To sustain that performance, however, you must also understand the societal impact of your business as well as the ways that broad, structural trends – from slow wage growth to rising automation to climate change – affect your potential for growth.

The board's engagement in developing your long-term strategy is essential because an engaged board and a long-term approach are valuable indicators of a company's ability to create long-term value for shareholders. Just as we seek deeper conversation between companies and shareholders, we also ask that directors assume deeper involvement with a firm's long-term strategy. Boards meet only periodically, but their responsibility is continuous.

We also will continue to emphasize the importance of a diverse board. Boards with a diverse mix of genders, ethnicities, career experiences, and ways of thinking have, as a result, a more diverse and aware mindset. They are less likely to succumb to groupthink or miss new threats to a company's business model. And they are better able to identify opportunities that promote long-term growth.

Companies must ask themselves: What role do we play in the community? How are we managing our impact on the environment? Are we working to create a diverse workforce? Are we adapting to technological change? Are we providing the retraining and opportunities that our employees and our business will need to adjust to an increasingly automated world? Are we using behavioral finance and other tools to prepare workers for retirement, so that they invest in a way that that will help them achieve their goals?

Source: Blackrock.

In mid-2019, US based CEO association The Business Roundtable took a dramatic symbolic step declaring that shareholder weren't the only stakeholders businesses needed to pay attention to. Lastly, the 2019 Edelman Trust Barometer produced a very interesting finding which reflects and illustrates some of the megatrends we discuss in this Chapter – that business is increasingly being asked to step up to the plate of championing ESGT issues. See Figure 1.8.[17]

LOOKING FOR LEADERSHIP FROM CEOS

Percent who say that CEOs should take the lead on change rather than waiting for government to impose it

76% +11pts

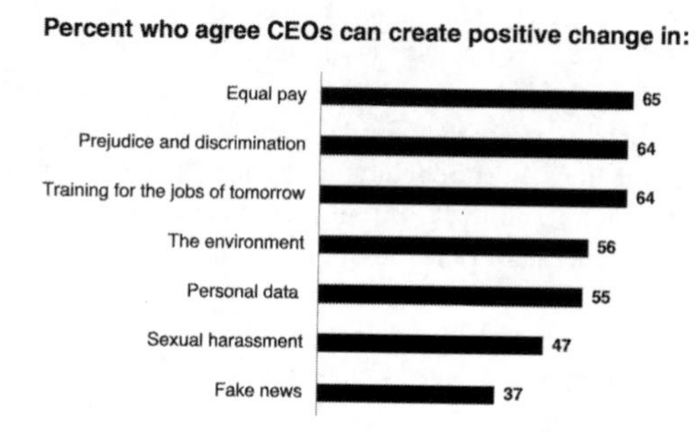

Figure 1.8 2019 Edelman trust barometer – looking for leadership from CEOs.

Source: 2019 Edelman Trust Barometer.

1.9.e Leadership to-do

The new generation of leaders is supposedly more conscientious than their precursors, or are they? If one looks at Silicon Valley, the track record there is decidedly mixed to downright negative when it comes to ESGT issues. However, at a time of such immense change and political leadership turmoil, other leaders – namely in the business community – have an opportunity for broader leadership by opening up to the gamut of issues that are key to their stakeholders. That requires a systematic approach to ESGT issues, risk, opportunity and crisis management, something we do in depth throughout this book and especially in the final part in Chapters 7 and 8.

1.10 Megatrend #9: hyper-transparency becomes super-opacity – reputation risk on steroids

1.10.a Core concept

The promise of hyper-transparency brought by the explosion of the Internet, telecommunications and social media has given way to an unexpected and unintended series of consequences – the creation and viralization of super-opacity where a myriad fake news and deep fake capabilities enabled by technological advances are creating a trust and reliability deficit that also translates into reputation risk on steroids.

1.10.b The danger and the risk

Social media and the Internet gave us the ability to spread the word super-fast, the democratization of information, but as with everything else in life, in the wrong hands, positive developments can also be twisted into negative tools. Denizens of the dark – whether nation state intelligence agencies, international crime rings or the 400 pound guy sitting in his mother's basement – have been able to deploy fake news, fake audio, fake video and all manner of other deep fakes – to trick, cheat, endanger and damage their targets, ranging from famous people to the average Joe on the street. This will continue and get worse before it gets better.

1.10.c The promise and the opportunity

Even though the mountain that needs to be climbed is very tall and very steep, there are opportunities to deal with the dangers brought about by the rise of hyper-opacity and downright fakeness. As we review in great detail in Chapter 6 on Technology, the tech, business and governmental worlds are starting to deploy countermeasures to the rise of hyper-opacity and all things fake with the creation, for example, in the media world, of a variety of technology-enabled countermeasures to identify and counteract fake things of all kinds. And, yes, there is money to be made in this as well.

1.10.d Thought bubble

"Is it real or is it Memorex?" For those of us who were around in the 1980s, that was a popular advertisement from an audio tape manufacturer, Memorex. They claimed their taped audio quality was so good that you couldn't tell the difference between the real/live version of the music and the tape recorded one.

We now live in a tech world where the light can be shone on almost anyone, anywhere, under just about any circumstances, especially because of the rise of the smartphone, which in so many ways has not only become an appendage to our bodies but also what I like to call my external brain, sadly, in many ways. Through these magnificent machines and associated technologies, we are also witnessing the rise of the fakes – fake news, fake visuals, fake audio and the supreme social media manipulation by the good, the bad and the ugly.

Over the past few decades we have moved from a long-standing historical concept of "reputation" to one of "reputation risk" as Figure 1.9 shows. Why? Because of the advent of the Internet and social media at the dawn of the 21st century. Because every organization's typical risks can be amplified and accelerated via social media, reputation risk has now become a thing in our world. I wrote a book about this a few years ago in which I defined reputation risk (and, by implication opportunity) as follows:

> Reputation risk is an amplifier risk that layers on or attaches to other risks – especially ESG risks – adding negative or positive implications to the materiality, duration or expansion of the other risks on the affected organization, person, product or service.[18]

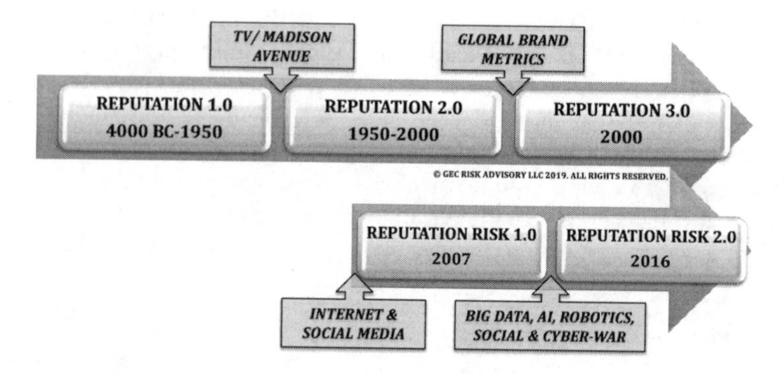

Figure 1.9 Reputation and reputation risk: a brief history.

Source: GEC Risk Advisory.

To manage the risks associated with the sometimes crippling mix of fake news, deep fakes and social media, organizations need to understand what reputation risk means and how it is managed. Because of the rise of the multi-stakeholder reality together with the rise in social media and technological changes that pair with it, this is a reality that is here to stay, not a mirage or a passing thing.

Indeed, this is why organizational leaders everywhere should pay heed to the fact that reputation risk is a layer that can amplify and worsen your other unattended risk as stakeholder expectations are obliterated, as I described in a piece for Ethical Boardroom for which I created the Reputation Iceberg picture reproduced in Figure 1.10.[19] Much of this is also described in greater detail in Chapter 7 and 8 of this book.

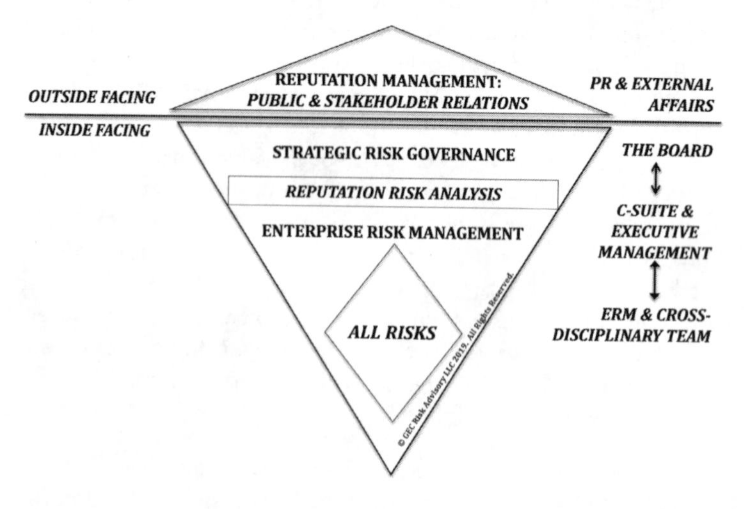

Figure 1.10 The reputation iceberg.

Source: GEC Risk Advisory.

1.10.e *Leadership to-do*

Leaders in every type of organization need to understand the power and the danger of the age of opacity where their own statements, images and audio can be manipulated for nefarious purposes. Each organization and their leaders – management and the board – must be savvy and prepared to deal with social media, media and the arts of the darker parts of our virtual and physical worlds to combat fake news, audio, video, products and services.

1.11 Megatrend #10: future fear – utopia, dystopia, life on Mars?

1.11.a *Core concept*

With the rise of paradigm-shifting technologies – AI, robotics, Crispr, 5G, quantum computing, synthetic bioengineering – what will happen to us humans – to how we live, how we work, how we interact with each other and our environment? Will our future be utopian – where everyone works a little, has some form of universal basic income (UBI) and is otherwise free to pursue their passions with enough time for leisure, entertainment, family and friends? Or will it be more dystopian, with large swaths of otherwise able-bodied people unable to find work because jobs have been taken over by "machines" and no real alternative roles are available? Or will be living on Mars with the descendants of genius futurist inventor Elon Musk?

1.11.b *The danger and the risk*

A dystopian future where marked labor unemployment has occurred in both the industrial and emerging markets with large swaths of unemployed mostly young people having little to do every day and the elderly, disabled, young and frail not being properly tended to, where there is spotty health care, welfare safety nets and thoughtful and responsible governmental decision-making. Such a world leads to greater strife, civil unrest, inequality and war.

1.11.c *The promise and the opportunity*

While we cannot expect a utopian future, we can certainly see a world where collaborative groups of virtual and actual cross-disciplinary experts and others work together to solve some of the greatest problems confronting this world – hunger, violence, inequality, poverty. Where the world has generally united to get behind the Sustainable Development Goals (SDGs) and reupped them when they came due in 2030. Where new jobs – heretofore unseen, unpredicted and unknown – evolve and are created as needed because all those new technologies need new kinds of human roles to manage, train and monitor them.

1.11.d Thought bubble

Older people are losing jobs younger, and younger people can't get them at all or leave or lose them quickly. Younger generations face the gig economy and/or joblessness, too much leisure time and the great divide of inequality – and that's just in the developed world.

What faces younger generations in less wealthy, poorer corners of the earth is even more daunting and dangerous. As Ian Bremmer brilliantly laid out in his book, *Us versus Them*, the divide is broadening and the dangers are being exacerbated by the vast technological and other changes taking place worldwide. If the effect of technologically driven joblessness (where AI and robotics take over many of the lower end or repetitious jobs that still exist today) becomes challenging in the industrialized world, with their relatively controlled populations, think about what will happen to joblessness and potentially greater inequality and poverty that may take place in less advanced economies with large and growing young populations (India comes to mind).

As we close this chapter and enter the future of technology every single day, it is worth mentioning some of the trends and opportunities that may also await us literally today and tomorrow. Data scientist and author JT Kostman provides a synthesis of the key technological trends, challenges and opportunities we are facing today. They are summarized in Table 1.11.

Table 1.11 JT Kostman's: "*five forces forging the future*"[20]

1. Artificial intelligence The economics of commerce can, essentially, be reduced to just two elemental factors: (1) better meeting consumers' ever-evolving wants, needs, desires, expectations, or demands; and (2) meeting them with greater efficiency. AI is poised to confer advantages to competing corporations like nothing we've ever seen before.
2. Blockchain Using a blockchain, the provenance of any product or thing can be established and maintained, indelibly, and in perpetuity. Food, equipment, medical records – anything. traditional approaches to bookkeeping, auditing, record keeping, and contract management will soon be obsolesced, as will supply chain management, project management, and the exchange of digital information and services. Companies that fail to get ahead of the blockchain curve will soon be at a distinct disadvantage in competing with their technologically enabled competitors.
3. Cybersecurity To underscore just how vulnerable our most precious data assets are, in August at Def Con 26, one of the two major annual events for cybersecurity professionals, an 11-year-old boy hacked into an exact replica of Florida's state election website and changed simulated voting results – in under 10 minutes. Board members are increasingly compelled to ask how safe their data – the most valuable and irreplaceable asset of any modern enterprise – really is.

(*Continued*)

Table 1.11 Continued

4. Hyper-connectivity The advent of 5G and beyond – the collective intelligence and abilities of the human network within organizations has far less to do with the intellect and efforts of any individual – and far more to do with the extent and strength of the connections between people who possess diverse perspectives, ideas, abilities, knowledge, and skills. Graph theoretic models demonstrate unequivocally that the breadth and density of connections directly correspond to organizational problem solving, decision making, and innovation abilities. In the corporate arena, this equates to bottom-line tenets, including increased productivity, probability, quality, service, sales, and employee engagement.
5. Symbio-tech systems "Moravec's paradox" – Moravec observed that problems which are trivially easy for computers tend to be exceedingly difficult for people – and vice versa: the sorts of problems people can solve without any effort prove utterly confounding to machines. What makes Moravec's paradox so interesting is that those same ai-enabled robots and systems excel at precisely the sorts of tasks that humans tend to dislike and avoid – while machines are incapable of doing precisely those things that make us most human. Moravec's paradox also provides us with a solution to one of the most presently pressing concerns the uninformed public has about technology: will AI and robots put people out of work?

Source: NACD Directorship Magazine.

The summary in Table 1.12 from the World Economic Forum's 2019 Global Risks Report serves as an erudite reminder of some of the key risks that are right around the corner for which we will need to apply our future shock armor and resilience and try to mitigate risk and amplify opportunity.

Table 1.12 World Economic Forum Global Risks Report 2019: Future Shocks[21]

Weather wars • Use of weather manipulation tools stokes geopolitical tensions
Open secrets • Quantum Computing renders current cryptography obsolete
City limits • Widening gulf between urban and rural areas reaches a tipping point
Against the grain • Food supply disruption emerges as a tool as geo-economic tensions intensify
Digital panopticon • Advanced and pervasive biometric surveillance allows new forms of social control
Tapped out • Major cities struggle to cope in the face of the ever-present risk of water running out
Contested space • Low earth orbit becomes a venue for geopolitical conflict
Emotional disruption • AI that can recognize and respond to emotions creates new possibilities for harm
No rights left • In a world of diverging values, human rights are openly breached without consequence
Monetary populism • Escalating protectionist impulses call into question independence of central banks

Source: World Economic Forum.

1.11.e Leadership to-do

Leaders of every shape and form have the responsibility today to understand the short-term and longer-term future implications of their organization's strategy, mission, vision, values and tactics. Even more than that, leaders have an obligation – a moral and ethical duty – to assess the impact of future trends on key stakeholders, especially workforces and customers – and their expectations and to plan accordingly. They should try to fulfill expectations or change them with the changing times in a manner that is as responsible, ethical, inclusive and constructive as is possible under our rapidly changing circumstances.

1.12 Endnote: a "gloom to boom"/"risk to value" leadership blueprint

Now that we have the birds-eye view of the ten megatrends of our turbulent times, each of which has yielded a "Leadership to-do", before we jump into our next chapter (Chapter 2 – Leadership – Surviving and Thriving with a Culture of Integrity"), I offer Table 1.13 as a summary blueprint for leaders everywhere – from the biggest to the smallest, from the most junior to the most senior, women and men, from every ethnic and racial and geographic corner of the earth and from every walk of life – business, government, academia, nonprofits – the following "Gloom to Boom" or "Risk to Value Leadership Blueprint" to start you thinking about the challenges, decisions, opportunities and actions that face you today.

Table 1.13 A "Gloom to Boom" or "Risk to Value" Leadership Blueprint

Megatrend	*Leadership to-do*
1. The fourth industrial revolution: hurtling through space at the speed of light	Leaders must realize that technological change and the behavioral and other impacts it has on life on earth are happening at a pace and expanse heretofore unheard of. It is everyone's responsibility – but especially that of leaders of every kind of organization – to take responsibility for understanding this moment to the full extent possible and urgently incorporating this concept into organizational strategy, tactics and culture.
2. Global trust collapsing: recession or depression?	The message for organizational leaders: be aware of the fact that trust is a powerful and often impalpable intangible and that you may not be fully aware of its actual positive or negative tangible consequences until late in the game. Undertake an analysis of what your trust issues are by getting to know who your most important stakeholders are (above and beyond the most obvious one) and learning their expectations of you. Then live up to those expectations and own up to any breach of trust quickly, completely and transparently.

(*Continued*)

Table 1.13 Continued

Megatrend	*Leadership to-do*
3. A world of extremes: the 21st century ethical leadership paradox	At a time of such dramatic change where incredible inventions and initiatives are underway to address some of the most intractable global challenges, leaders should lead with integrity and think beyond the box on how their organization can be part of the solution. All of us have a role to play both within our organizations and externally as part of communities, schools, workplaces, cities, countries and the globe. Likewise, the better angels (ethical and responsible leaders) should always be cognizant and prepared to deal with and check the lesser or darker angels (unethical, irresponsible and even criminal leaders) of our nature.
4. Complex interconnected risk rising: beware the "purple techno-swan"	We are most definitely in an age of complex, interconnected, man-made and other natural risks where ESGT risks and others overlap and are interesting in complex and novel ways. We may be reaching the age of the "purple techno-swan" risk – totally unpredictable to us today. Risk managers and leaders must be better equipped for risk management than ever before to do complex risk identification, coordination, mitigation, prediction and translation into strategic opportunity.
5. Ecological apocalypse: a decade to oblivion or salvation?	Climate change and everything that goes with it – from the lives and well-being of insects to that of humans–is everyone's responsibility. We can all do our part from the individual who sorts her garbage properly for urban waste management and the Fortune 100 company seeking to achieve a neutral carbon footprint all the way up to (and especially) government leaders like the president of the US. We have ten years to address the ecological disaster that is coming and even then, nothing is guaranteed. Everyone has an affirmative obligation to step up to this plate – yesterday.
6. The new geopolitical abnormal: rumblings or tectonic shift?	The time of orderly, predictable international relations is over for the time being. The worldwide growth of democracy has also atrophied and actually experienced reversal. The impact of technology and capitalism is affecting stakeholders differently. Business and NGO leaders need to be prepared for greater geopolitical risk and some of it from unexpected places and sources. Geopolitical risk needs to be part of everyone's risk management. Crisis management and business continuity need to be in place as well. Country leaders must rethink what leadership means and what alliances mean and for what. Leadership tone from the top has never been more important as we have a new strain of leaders – Trump, Putin, Xi, Erdogan, Bin Salman – who are not shy about throwing their authoritarian weight around and are setting a different and more dangerous cultural tone globally. Geopolitical risk is real and different from what it has predominantly been in recent decades. Buckle up and prepare for a wild ride.

(*Continued*)

Table 1.13 Continued

Megatrend	*Leadership to-do*
7. The meteoric rise of "ESG" and the virtual stakeholder: mirage or reality?	We are quickly approaching the time when every organization – including businesses that have always been exclusively focused on and beholden to their shareholders as the only stakeholders that matter – have to broaden their lens to incorporate the more complex landscape of key stakeholders that have a stake in their organization in one way or another and their ESGT portfolio of issues, risks and opportunities. And this applies to all types of organizations – business, nonprofit, governmental and academic – as we repeatedly stress throughout this book.
8. Business as the new global social/moral conscience: fact or fake?	The new generation of leaders is supposedly more conscientious than their precursors, or are they? If one looks at Silicon Valley, the track record there is decidedly mixed to downright negative when it comes to ESGT issues. However, at a time of such immense change and political leadership turmoil, other leaders – namely in the business community – have an opportunity for broader leadership by opening up to the gamut of issues that are key and important to their stakeholders. That requires a systematic approach to ESGT issue, risk, opportunity and crisis management, something we do in depth throughout this book and especially in the final Part in Chapters 7 and 8.
9. Hyper-transparency becomes super-opacity: reputation risk on steroids?	Leaders in every type of organization need to understand the power and the danger of the age of opacity, where their own statements, images and audio can be manipulated for nefarious purposes. Each organization and their leaders – management and the board – must be savvy and prepared to deal with social media, media and the arts of the darker parts of our virtual and physical worlds to combat fake news, audio, video, products and services.
10. Future shock fear: utopia, dystopia, life on Mars?	Leaders of every shape and form have the responsibility today to understand as best they can the short-term and longer-term future implications of their organization's strategy, mission, vision, values and tactics. Even more than that, leaders have an obligation – a moral and ethical duty – to assess the impact of future trends on stakeholders and their expectations and to plan accordingly. They should try to fulfill expectations or change them with the changing times in a manner that is as responsible, ethical and constructive as is possible under such changing circumstances.

Notes

1 Edmund Burke quoted in Brainy Quotes. https://www.brainyquote.com/quotes/edmund_burke_377528. Accessed on September 18, 2018.
2 Wikipedia. https://en.wikipedia.org/wiki/Per_aspera_ad_astra. Accessed on October 14, 2018.
3 Epic Theatre Ensemble. James Wallert. "The Winning Side". Produced Off-Broadway, Fall 2018. http://www.epictheatreensemble.org/the-winning-side/. Accessed on October 14, 2018.
4 Klaus Schwab. "The Fourth Industrial Revolution: What It Means, How to Respond". January 14, 2016. World Economic Forum Agenda. https://www.weforum.org/agenda/2016/01/the-fourth-industrial-revolution-what-it-means-and-how-to-respond/1/2. Accessed on November 28, 2018.
5 Klaus Schwab. "The Fourth Industrial Revolution: What It Means, How to Respond". January 14, 2016. World Economic Forum Agenda. https://www.weforum.org/agenda/2016/01/the-fourth-industrial-revolution-what-it-means-and-how-to-respond/1/2. Accessed on November 28, 2018.
6 Edelman Trust Barometer 2018. https://cms.edelman.com/sites/default/files/2018-01/2018%20Edelman%20Trust%20Barometer%20Global%20Report.pdf. Accessed on November 28, 2018. Edelman Trust Barometer 2019. https://www.edelman.com/sites/g/files/aatuss191/files/2019-03/2019_Edelman_Trust_Barometer_Global_Report.pdf?utm_source=website&utm_medium=global_report&utm_campaign=downloads. Accessed on April 12, 2019.
7 Max Roser. "The Short History of Global Living Conditions and Why It Matters That We Know It". Our World in Data. https://ourworldindata.org/a-history-of-global-living-conditions-in-5-charts. Accessed on December 1, 2018.
8 Steven Pinker. *The Better Angels of Our Nature: Why Violence has Declined*. Viking 2011. *Enlightenment Now: The Case for Reason, Science, Humanism and Progress*. Penguin 2018. https://www.amazon.com/default/e/B000AQ3GGO/ref=dp_byline_cont_book_1?redirectedFromKindleDbs=true. Accessed July 4, 2018.
9 World Economic Forum. *Global Risks Report 2019*. http://www3.weforum.org/docs/WEF_Global_Risks_Report_2019.pdf. Accessed on April 12, 2019.
10 FT. FT Lexicon. "Black Swan". http://lexicon.ft.com/Term?term=black-swan. Accessed September 7, 2015.
11 NACD Blue Ribbon Commission. "Adaptive Governance: Board Oversight of Disruptive Risks". 2018.
12 Bradford C. Lister and Andres Garcia. "Climate-Driven Decline in Arthropod Abundance Restructure a Rainforest Food Web". *Proceedings of the National Academy of Sciences of the United States of America*. October 30, 2018. http://www.pnas.org/content/115/44/E10397.full. Accessed on December 1, 2018.
13 S&P Global Ratings. "The ESG Advantage: Exploring Links to Corporate Financial Performance". April 8, 2019.
14 Council on Foreign Relations. Global Governance Monitor. https://www.cfr.org/interactives/global-governance-monitor#!/global-governance-monitor. Accessed on August 20, 2018.
15 Global Peace Index 2018. http://visionofhumanity.org/app/uploads/2018/06/Global-Peace-Index-2018-2.pdf. Accessed on April 8, 2019.
16 S&P Global Ratings. "The ESG Advantage: Exploring Links to Corporate Financial Performance". April 8, 2019.
17 Edelman Trust Barometer 2019. https://www.edelman.com/sites/g/files/aatuss191/files/2019-03/2019_Edelman_Trust_Barometer_Global_Report.pdf?utm_source=website&utm_medium=global_report&utm_campaign=downloads. Accessed on April 12, 2019.
18 Andrea Bonime-Blanc. *The Reputation Risk Handbook*. Sheffield, UK: Greenleaf, 2014.

19 Andrea Bonime-Blanc. "Deploying Reputation Risk 2.0". Ethical Boardroom. January 2017. https://ethicalboardroom.com/deploying-reputational-risk-2-0/. Accessed on December 1, 2018.

20 JT Kostman. "Future Tech: Unprecedented Opportunities, Unrivaled Threats". *NACD Directorship Magazine*. September/October 2018.

21 World Economic Forum. *Global Risks Report 2019*. http://www3.weforum.org/docs/WEF_Global_Risks_Report_2019.pdf. Accessed on April 12, 2019.

2 Leadership

Surviving and thriving with a culture of integrity

The risk

"The fish rots from the head".
Ancient saying, unknown origin[1]

"Qui cum canibus concumbunt cum pulicibus surgent".
Unreliably attributed to Seneca, 1st-century Roman philosopher

Or, as later adapted to English:

"If you lie down with dogs, you wake up with fleas".
Reliably attributed to Benjamin Franklin,
18th-century US Founding Father[2]

"One who deceives will always find those who allow themselves to be deceived".
Niccolò Machiavelli, 15th–16th-century Italian diplomat[3]

The opportunity

"With great power comes great responsibility".
Voltaire, 18th-century French writer[4]

"When I do good, I feel good. When I do bad, I feel bad. That's my religion".
Abraham Lincoln, 19th-century US President[5]

Chapter 2: Leadership – summary overview

2.1 Chapter overview

In 2014, I wrote a piece for *Ethical Corporation* magazine called "The Biggest Risks Nobody Talks About" (until it's too late, I would often add when sharing the article) and subtitled "the failure of leaders and the business cultures they encourage can have devastating consequences".[6] It is one of my most downloaded articles – because (I believe) it hit a raw nerve. It actually goes back to that famous quote from Baron Acton, the 19th-century English historian, who famously said: "Power tends to corrupt and absolute power corrupts absolutely. Great men are almost always bad men".[7]

For years, I had pondered the phenomenon of major scandals and value destruction connected to leaders' behaviors without being able to properly synthesize why the phenomenon occurred. There was Enron, and WorldCom, and Adelphi, and Tyco, and Parmalat, etc. – you get it. And then I had an "Aha!" moment that helped me to encapsulate the following thought: if you look at the long list of scandals and crises that have hit organizations (and countries, for that matter), the most consequential and devastating risks are the risks of bad leadership (unethical, illegal, criminal) and the toxic culture that such leaders create, impart, emanate, encourage and by their own misbehaviors propagate – consciously or unconsciously, intentionally or negligently.

In other words, tone at the top is the be-all and end-all that determines in one way or another everything else about an organization. And yet, ask any company chief risk officer (until just a few years ago) if either leadership or culture risk were even on their radar (let alone risk register) and you would have gotten a blank stare. It took the likes of the second largest global financial crisis ever (of 2007/2008) to rattle some cages within the financial sector (where leadership and culture risk seem to be the greatest) to start focusing the mind on this set of tremendously important issues. And yet we continue to have a parade of horrible behaviors in the financial sector to this day – Wells Fargo and

Deutsche Bank for a while, and more recently, Danske Bank and Swedbank come to mind.

I wrote this in the aforementioned "Biggest Risks" article:[6]

> the failure of leaders, the risk that leaders have the wrong motivations and incentives, the risk that motivations other than the common good of the enterprise are driving CEOs, their hand-picked executive teams and their rubber-stamping boards, to allow excess, short-termism and, in the worst cases, illegal behaviors …
>
> But it also goes beyond leadership. It goes to the culture of the organization. And it may even go to a systemic culture in an entire sector. It is a culture that instead of encouraging a race to the top encourages a race to the bottom. This can best be seen in the financial sector over the past two decades.

In that same article, I defined "leadership risk" as follows:

> Leadership risk is … the risk of a top executive either engaging in direct criminal or civil legal violations or engaging in unethical or borderline practices that may come back to haunt not just him or her but the entire organization. This is a form of reputation risk.

It is worth thinking about leadership and culture risk in a wide variety of organizational contexts:

- **As the CEO of a business.** Bernie Ebbers of WorldCom, serving a life sentence for a massive fraud that destroyed the company is in prison, WorldCom goes bankrupt, millions of stakeholders (employees, pensioners, customers, shareholders) adversely affected.
- **As the head of a university.** Graham Spanier, president of Penn State University, serving time for ignoring and covering up a long-standing sexual abuse scandal in their leading football program. The reputation of a historically admired University is lost, but worst of all, sexually abused children and young people are forever damaged.
- **As the head of an international organization.** Sepp Blatter, long-standing president of FIFA, resigns in the midst of one of the largest and most intricate corruption scandals ever. Millions of stakeholders (local teams, players, fans, officials, others) adversely affected.
- **As the leader of a major charity.** William Aramony, CEO of the United Way of America for 22 years, convicted of fraud and embezzlement in 1995. The reputation of a leading US charity is sullied for a long time, and the long-term trust of many stakeholders is obliterated.
- **As high-level leaders of a global church.** While this is still an evolving story and may or may not reflect negatively on the current pope,

Pope Francis, the vastness and depth of the leadership (at the Cardinal, Bishop levels) and culture failure in the Catholic Church globally is now incontrovertible and speaks to a very long-standing, widespread, hubristic culture of secrecy, fear, and degradation given the now massive, proven track record of cover-ups and protection of sexual abuse and pedophilia by priests in many parts of the world.

- **As leaders of both autocratic and democratic governments.** Vladimir Putin of Russia and Donald J. Trump of the US come to mind. When elected or imposed leaders breach the trust of a majority or a large portion of their citizenry by acting despotically, deliberately, capriciously or otherwise to hurt the interests of particular stakeholder groups (internal or external to their country), contrary to national or international law and norms with potentially long-term and widespread negative and damaging reverberations, that can echo, in the case of a government leader's actions, for years to come.

In the "Biggest Risks" article, I also define "culture risk" as follows:

> Culture risk is the risk that the culture that suffuses an organization (reflects) the culture of a leader who is more narcissist than true leader, more self-enriching than steward of his/her company; someone who may (a) not be true to the principle of legal compliance, (b) not be living up to common ethical behavioral standards, or (c) be driven by a performance incentive structure that encourages overly risky or even illegal behaviors.

And it goes beyond leadership and culture to encompass ethics and integrity, as they are core parts of a strong culture. As a long-standing practitioner in the field of ethics, compliance and culture, I view ethics or "integrity" (a word I often prefer to use) as an integral part of leadership and culture and, what's more, of a successful environmental, social, governance and technology (ESGT) strategy (something that we will explore in detail in Chapter 7). That's why we will explore the essence of what an effective and high-performing ethics program looks like later in this chapter as well.

But let's go back for a moment to the title of this chapter. Oxford's English Dictionary provides these simple definitions of "survival" and "surviving" as well as "thriving":[8]

> Survival is the "state or fact of continuing to live or exist, typically in spite of an accident, ordeal, or difficult circumstances".
>
> Surviving is "Remaining alive, especially after the death of another or others".
>
> "Continuing to exist; remaining intact".
>
> Thriving is being "Prosperous and growing; flourishing".[9]

Surviving and thriving as an organization (and maybe even as a country) rest heavily on the leaders of the organization (or the country). It is for that reason that this chapter on leadership is placed at the beginning of this book – and at the beginning of our ESGT, resilience and sustainability journey.

Without at least competent, well-intended and capable (let alone highly effective and visionary) leadership, culture and ethics, organizations will have a hard time doing what they are supposed to do in the shorter term or surviving intact in the longer. When leaders are competent, smart, ethical, stakeholder-centric and mission-driven, the opportunity for a successful and sustainable strategy will also go up, as we will discuss in Chapter 8. Good leadership (and the culture and ethics that hopefully go with it) is the quintessential ingredient – indeed the sine qua non – of organizational success.

2.2 Leadership, culture and ethics: the essential survival kit

This chapter is all about leadership, culture and ethics and why a strong culture and ethical leadership allow for a more successful deployment of a holistic and beneficial ESGT strategy. After we do a deep dive into our four main categories of issues in Part II, Chapter 7 in Part III will pull together a practical blueprint on what steps leaders need to take to deploy a successful ESGT strategy to achieve the purpose of the organization, its goals and objectives and deliver benefits to stakeholders and in the process not only build organizational resilience but perhaps even a transformative organization (the topics of our last Chapter 8 – "Boom").

Leadership is about various kinds of leadership – all the way from the very top level of a board of directors to the front lines of an organization at the grassroots. However, the most important determinant of culture and ethics is tone at the top, and that tone is determined by the CEO (or other leader) and the board of directors, trustees or supervisors the CEO reports to.

I view ethics as the beating heart of a culture and of good leadership – or the reverse, unethical behaviors as the reflection of the soul (or absence thereof) of a darker leadership which then also suffuses and one might say infects the overall culture of an organization or entity (or even country when we're talking about the most powerful leader of a nation). As we all know, history is littered with the damage inflicted by such leaders. Figure 2.1 represents the interrelationship I believe exists between leadership, culture and ethics.

Some readers may react to this and say that this is overly simplistic or a limited view of the connection between leadership and culture in that there can be different leadership styles and outcomes even within an organization. I wouldn't disagree but would like to illustrate my point with a couple of true-to-life hypothetical examples that probably resonate with a few of my readers.

- **Example 1: Global Company X.** The CEO of Global Company X, headquartered in London, is known for her authoritarian, command-and-control form of leadership, while the head of the company's

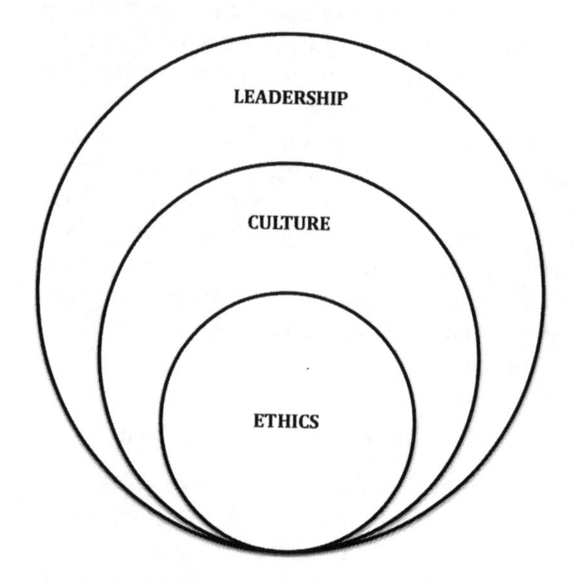

Figure 2.1 Leadership, culture and ethics.

Source: GEC Risk Advisory.

Australian operations couldn't be more empathetic and emotionally intelligent. Global Company X thus appears to have very different cultures or "micro-cultures" within the same organization – how does one reconcile that?

- **Example 2: Global NGO Y.** The Executive Director of Global NGO Y has a "live and let live" leadership style that provides for "degrees of freedom" to his trusted lieutenants throughout the organization. There are, however, a couple of lieutenants who don't reflect this ethos within their division or business unit but in fact reflect the opposite. In this case, while the overall organization CEO promotes a culture of trust and openness, a couple of specific leaders at the front lines are not fostering this culture locally, once again creating a different, competing and potentially less desirable local micro-culture.

What I would point out from these two examples is that in both cases there are culture issues that can be summarized as follows:

- With regard to Global Company X, it would be interesting to understand the employee culture survey and financial results for both the Australian business unit and overall Global Company X – would a comparison and deep dive show differences in both culture and financial results? Might there be a correlation between the two or are there differences?

- A similar question should be asked about the employee culture survey results for Global NGO Y – is there a difference between the headquarters and the local operations? Although financial metrics might not be the right form of measurement for the purpose of an NGO (although they could be), maybe other metrics measuring impact and/or expectations fulfillment (or lack thereof) of critically important stakeholders (like beneficiaries and partners) would shed light on the relative performance of the different micro-cultures within the NGO.
- Finally, and most importantly in both cases, if culture is not proactively considered in the mix of how the organization is measured and how the HQ and local leaders are held accountable, there may be some other negative consequences to the organization that no one is identifying and that in both cases for different reasons may result in money or value being left "on the table" or even being destroyed for both the organization and its stakeholders, as follows:
    - In the case of the Global Company X authoritarian CEO, is the culture that this CEO is promoting having negative effects on the financial and other results (demoralized employees, heavy staff turnover, loss of partner business) of the organization?
        - If so, where is the board of directors in exercising what should be proactive oversight of company leadership and culture? We discuss a variety of tools and techniques boards, as ultimate stewards of an organization, should deploy to understand their entity's culture in great detail later in this chapter.
    - In the case of the trusting Executive Director of Global NGO Y, are the degrees of trust he places in all leaders throughout the organization warranted and, especially in the case of the lieutenants who are doing the opposite locally, resulting in reputation risk and other potentially harmful results for the NGO?
        - Once again, where is the board of directors or trustees of this NGO? Asleep at the switch, uninformed, lulled into a sense of complacency because of their charismatic and highly emotionally intelligent Executive Director? Or are they paying attention to details, stories and metrics that will help them gauge the relative health of their organization on behalf of stakeholders?

This digression is actually central to the case we make in this chapter and throughout this book: no matter how great or poor a leader an organization has, the culture that leader promotes (actively, passively or negligently) has reverberations for the organization as a whole and for its stakeholders. It is one of the most critical responsibilities of a board (who ultimately are the leaders of leaders) to oversee management's performance not only in delivering the financial and other "hardcore" operational results but in understanding the culture that the leader is promoting or allowing.

As the earlier example of Global NGO Y hopefully illustrates, it isn't good enough to say that a leader has all the great qualities of a leader from an emotional intelligence and leadership qualities standpoint – if that leader is too nice, or too hands-off or too laissez-faire about culture, that leader may be unwittingly hurting the organization and its stakeholders – perhaps not as much as a more negative and destructive leader but still to an unacceptable degree.

Let me go a step further. I would argue (and try to do so repeatedly throughout this book) that when you have effective and positive leadership, a good culture usually follows and that good culture almost always includes a heavy dose of ethics and integrity. What would follow from this statement is that when you have ineffective or negative leadership, a more compromised, limiting or downright negative culture would follow, often bereft of integrity or ethics at its core. Think Jeff Skilling and Ken Lay of Enron, Bernie Ebbers at WorldCom, Richard Fuld of Lehman Brothers. Or think Bashar Al Assad of Syria, Vladimir Putin of Russia, Recep Erdogan of Turkey or Donald Trump of the US. Historically, think Hitler, Stalin, Mussolini and Franco.

Of course, there is a vast spectrum between the extremes of "saintly" leadership and governance on one end, and psychopathic leadership on the other, with many variations and permutations possible in between, relating not only to the actual personality and character of a leader but also to many other environmental and situational factors that affect a given organization, situation and time.

But I remain singularly convinced of this precept: positive culture follows from effective leadership, and most effective leaders also have integrity and ethics at their core. Think Sir Richard Branson of Virgin, Paul Polman at Unilever, Marc Benioff at Salesforce and Satya Nadella at Microsoft as corporate leaders trying to combine high integrity leadership with positive financial results. Historically, think Mahatma Gandhi, Nelson Mandela, Mother Teresa and Martin Luther King Jr. as social and political leaders who combined strong leadership qualities with powerful causes.

Consistent with the theme of "the fish rots from the head" introduced at the top of this chapter (for which I would offer my hopeful counterpart: "The fish can also rejuvenate from the head"), it is important to start at the very top of the food-chain (no pun intended). The world is experiencing, perhaps for the first time ever, what it is to have what well may be "dangerous leadership" at the very highest levers of world power – the dangers of an unstable, psychologically hampered and perhaps even criminal president of the US – with both severe and dangerous domestic and global implications.

Indeed, much is happening while I am writing this book and much might have happened by the time this book sees the light in late-2019. I will nevertheless refer readers to an important book published six months into President Trump's term in October 2017 titled *The Dangerous Case of Donald Trump: 27 Psychiatrists and Mental Health Experts Assess a President* as well as the link to the video op-ed by my good friend, renowned psychiatrist and leadership adviser Dr. Kerry Sulkowicz (which at the time of this writing had over 10 million views). Both of these resources explain the dangers of unstable and perhaps psychologically challenged people in positions of great power from the perspective of those who are experts in this area – psychologists and psychiatrists (of which I am neither).[10]

With that said, let's delve into a more detailed exploration of leadership, culture and ethics by beginning with a review of individual leadership characteristics gleaned from the fields of psychology, psychiatry, behavioral economics and business. We will then pivot to a consideration of what the leaders of leaders – the board of directors or similar oversight body – should do to hold the leaders they oversee accountable for not only financial and operational results but the ethics and culture of the organization. We can then turn to the heart of the matter – ethics and integrity – and discuss how a robust organizational ethical culture and program not only builds resilience for that entity but gives it a leading edge in the marketplace.

2.3 Leadership – from toxic to enlightened

> One immoral act begets another, brain research shows: The more dishonest acts we commit the easier they become to do.[11]

Sometimes it's about bad behavior spreading – i.e., bad behavior begets bad behavior not only within a company (the erstwhile slippery slope we often refer to) but within an entire industry as well (as we explore in Chapter 5 on governance where we examine the financial sector and the auto sector as two sectors that have had their fair share of industry-wide bad behaviors and "races" to the bottom).

But other times we are talking about a specific high-level leader – mostly the CEO or president of an organization (or leader of a country) who, with often vast amounts of power, is able to either do a lot of good or wreak a lot of havoc on behalf of (or against) his or her stakeholders. The following section takes an evolutionary look at types of leaders, from the most toxic (psychopathic and hubristic leaders) to the most desirable (those who can lead in a positive, ethical and effective way). Let's start with the bad and the ugly and work our way up to the good and the beautiful.

2.3.a Toxic leaders: is your boss a psychopath?

About 15 years ago, I accidentally stumbled across a short but powerful book – one that I have probably recommended more times than any other book because of its deep insights and practical application to the world of leaders and leadership, and because it resonated so much with my own experience with a wide variety of toxic to great leaders in my own career. The book is by psychologist Martha Stout and its title is *The Sociopath Next Door*. In it, the author profiles a number of unidentified but real patients she has treated in private practice over a 30-year period, ranging from a very young cruel sociopathic boy to the mistreated girlfriend and then desperate wife of a sociopath.

The purpose of her book is simple: to help those (purportedly 96%) of the human population who are not sociopathic identify and deal with the possible 4% of people who are sociopaths among us. If 4% of humankind are sociopaths, that means that for every 25 humans in a room, at least one is a sociopath. Sociopaths are not necessarily violent psychopaths a la Anthony Perkins's character in Hitchcock's famous thriller *Psycho* or the Christian Bale character in *American Psycho*. Indeed, some authors and studies have shown that it is generally beneficial for society to have some people with sociopathic tendencies as long as those tendencies are in check, not harmful to society's better interests and deployed in furtherance of a beneficial social cause. In *The Wisdom of Psychopaths: What Saints, Spies and Serial Killers Can Teach Us About Success*, author Kevin Dutton makes the case that certain professions – neurosurgeons, special forces, and others – provide benefits that not everyone can give. I for one agree, I would be the worst spy on earth, feeling fear and remorse at every turn even if I was doing the right thing for "god and country".

However, sociopaths most of the time are people who share certain traits that can be damaging and hurtful to others – lack of empathy, inability to love, singularly driven to achieve success, smarter than the average, willing to step on others. Sound like any of your bosses? A family member perhaps? Someone who has had power over you? A CEO? A politician, a president, a prime minister?

Table 2.1 contains a pop quiz I have used as an icebreaker from time to time in ethics trainings which can help one get to the heart of the matter rather quickly, though it is clearly unscientific.

Table 2.1 Is your boss a psychopath?[12]

1. Is he glib and superficially charming?
2. Does he have a grandiose sense of self-worth?
3. Is he a pathological liar?
4. Is he a con artist or master manipulator?
5. When he harms other people, does he feel a lack of remorse or guilt?
6. Does he have a shallow affect?
7. Is he callous and lacking in empathy?
8. Does he fail to accept responsibility for his own actions?

For each question, score two points for "yes," one point for "somewhat" or "maybe," and zero points for "no." Key to answer: "If your boss scores 1–4 – be frustrated; 5–7 – be cautious; 8–12 – be afraid; 13–16 – be very afraid."

Source: Fast Company.

But this is not a joking matter and is indeed something that has concerned and troubled me and many others over the years. There is now some research to back up the idea that at the highest levels of power – for example, CEOs – there is a greater preponderance of sociopaths than in the average population. How true and rigorous these findings are may be up for debate, but any scan of the news headlines will quickly point out how bad leadership behaviors – whether sociopathic or something else – may be interfering with good organizational and social stewardship.

Indeed, this is clearly something that those in charge of governance – boards of directors, boards of trustees, oversight committees, etc. – need to know about when dealing with organizational leadership dynamics but often don't as they don't see it as being within their scope of responsibility. Of course, it is also possible that in the less diverse, high-powered world of the boardroom there may be a greater preponderance of such behaviors as well.

In *The Psychopath in the C Suite: Redefining the SOB (Seductive Operational Bully)*, author Manfred F.R. Kets de Vries of INSEAD focuses on the SOB or "psychopath lite" and offers the following description (emphasis mine):

> SOBs can be found wherever power, status, or money is at stake. Outwardly normal, apparently successful and charming, their inner lack of empathy, shame, guilt, or remorse, has serious interpersonal repercussions, and can destroy organizations. **Their great adaptive qualities mean they often reach top executive positions, especially in organizations that appreciate impression management, corporate gamesmanship, risk taking, coolness under pressure, domination, competitiveness, and assertiveness.** The ease with which SOBs rise to the top raises the question whether the design of some organizations makes them a natural home for psychopathic individuals.

In "Corporate Psychopathy: Talking the Walk", authors P. Babiak, C.S. Neumann and R.D. Hare make the following findings (emphasis mine):

> In this study, we had a unique opportunity to examine psychopathy and its correlates in a sample of 203 corporate professionals selected by their companies to participate in management development programs … The prevalence of psychopathic traits … was higher than that found in community samples. The results … indicated that the underlying latent structure of psychopathy in our corporate sample was consistent with that model found in community and offender studies. **Psychopathy was positively associated with in-house ratings of charisma/presentation style (creativity, good strategic thinking and communication skills) but negatively associated with ratings of responsibility/performance (being a team player, management skills, and overall accomplishments)**.[13]

The "offender" sample the authors refer to here is imprisoned populations where the prevalence of psychopathy has been proven to be higher than in average society, not surprisingly.

Clearly, when we talk about bad behaviors in the C-Suite we are not always talking about sociopathic behavior but other kinds of negative or damaging behaviors like narcissistic, bullying and other abusive behaviors. Along these lines, there is an interesting study that connects the dots between CEOs who are narcissistic and their boards, as many CEOs have the power to load their boards up with their chosen "friends and family" as I like to refer to them – people, usually similar in background, diversity and station as themselves; in other words, they create a very homogenous, group of yes-men or yes-people. Indeed, this study showed the following:[14]

> The push for great corporate governance has failed to identify problems which occur when narcissistic CEOs load their board with images of themselves ... we explain why CEOs tend to favor new directors who are similar to them in narcissistic tendency or have prior experience with other similarly narcissistic CEOs. Because more powerful CEOs are more able to select such individuals onto their board, CEO power is predicted to be positively associated with the above characteristics of a new director.

Indeed, we can point to several scandals where this seems to be the case – the Harvey Weinstein case and the board of The Weinstein Company, the Steve Wynn case and the board of Wynn Resorts, the Les Moonves case and the board of CBS and, most recently, the Carlos Ghosn case and the boards of Nissan and Renault. There are several common threads among these cases, as follows:

- Very powerful, long-standing, controlling CEOs.
- In two cases – Weinstein and Wynn – founder CEOs.
- In most cases, boards made up of "friends and family", board members with mostly similar backgrounds and demographics.
- In most cases, CEOs accused of sexual harassment, bullying, retribution, imperious and despotic behaviors toward various parties, and, in more than one case, ongoing criminal investigations.
- Interestingly, most of these companies are in similar fields – media and entertainment.
- In all cases, the CEOs received extraordinary pay packages – indeed in one case (Ghosn) there is an ongoing criminal investigation into alleged fraud and embezzlement.
- Little to no evidence of performance accountability beyond financial metrics (indeed, in the Weinstein case, evidence that the board approved the CEO contract with coverage for sexual harassment claims).

In the midst of the #MeToo movement, the handling by the CBS board of sexual harassment allegations against their CEO, Les Moonves provides a textbook case of how not to do it. Indeed, reporting showed that for many months the board unwaveringly believed Moonves against a first set of serious sexual harassment and assault allegations before outside lawyers had concluded their investigation. It was only when a second set of very disturbing accusations came to light through a second major investigative report by Ronan Farrow (the Pulitzer prize-winning journalist who was part of the original reporting that uncovered the Harvey Weinstein story) that the board turned on Moonves, not because of allegations but because he had clearly lied to them:

> In the end, it was the evidence that Mr. Moonves had misled his board—even more than the allegations of abuse from multiple women—that doomed him.[15]

At the time of this writing, the Moonves case is still unfolding, with the CBS board having finally decided not to pay him a $120 million parting compensation package, despite the allegations and the damage inflicted by the whole sordid affair on the reputation of CBS and the interests of its stakeholders. Moonves is striking back through the courts to salvage that termination package.

Table 2.2 provides a few lessons learned from this situation.

Table 2.2 The CBS Les Moonves board – lessons learned about what to do when a CEO is accused of improprieties (or worse)

- Take allegations seriously at the beginning – when the first ones emerge.
- Don't wait to engage in a concerted and thorough investigation.
- Don't minimize, belittle or otherwise dismiss accusations or allegations about the CEO.
- Engage in a worse-case scenario exercise to be ready for the worst.
- Include crisis management training and scenario planning at the board level.
- Have competent inside and outside counselors as needed for legal, ethical, compliance, public relations and investor relations stakeholder management.

Source: Author.

Last but not least in this "bad and ugly" category of leaders is the only woman to achieve the vaunted title of Unicorn CEO in Silicon Valley – Elizabeth Holmes of Theranos. The following statement was attributed to none other than highly regarded ex-US Secretary of Defense and retired four-star general Mattis, who, upon retirement from the military, was lured onto the Theranos board by Holmes, as recounted by Wall Street Journal investigative reporter John Carreyrou in his runaway best-selling real-life thriller *Bad Blood: Secrets and Lies in a Silicon Valley Start-up*:

> "she has probably one of the most mature and well-honed sense of ethics, personal ethics, managerial ethics, business ethics, medical ethics that I've ever heard articulated," the retired general gushed.[16]

Needless to say, the rest is history – Holmes, after almost a decade of misrepresentations, fraud, deceit and what some observers might regard as sociopathic behavior toward anyone who would cross her ambition of becoming the first female

Steve Jobs, has been fined and is scheduled for federal criminal trial for the summer of 2020. And Theranos is no more, having been dissolved in late summer 2018.

2.3.b Hubris Syndrome: is your boss hubristic?

All of which brings us to another interesting behavioral take on powerful people. In his *Atlantic Monthly* magazine article "Power Causes Brain Damage", in part referencing the behavior of Wells Fargo ex-CEO Stumpf, who presided over the unraveling of a variety of ethics scandals under his watch, Jerry Useem states:

> Hubris syndrome … is a disorder of the possession of power, particularly power which has been associated with overwhelming success, held for a period of years and with minimal constraint on the leader.[17]

The author refers to the work of the Daedalus Trust that is focused on exploring "Hubris Syndrome", especially in political leaders. They define leaders as having this syndrome if they have three or more of the 14 clinical features of this syndrome, as summarized in Table 2.3. As you read this list, think about leaders – political, business, nonprofit, others – that you may know or know of and think about their profile against this list.

Table 2.3 14 clinical features of "Hubris Syndrome" from the Daedalus Trust[18]

1. The world is an arena in which to exercise power and seek glory.
2. A predisposition to take actions which seem likely to cast the individual in a good light – taken in part in order to enhance their image.
3. A disproportionate concern with image and presentation.
4. A messianic way of talking and a tendency to exaltation in speech and manner.
5. An identification with the nation or organization – to the extent that they regard the outlook and interests of the two as identical.
6. A tendency to speak of themselves in the third person or use the royal 'we'.
7. Excessive confidence in the individual's own judgement and contempt for the advice or criticism of others.
8. Exaggerated self-belief, bordering on a sense of omnipotence, in what they personally can achieve.
9. A belief that rather than being accountable to the mundane court of colleagues or public opinion, the real court to which they answer is much greater: history or god.
10. An unshakable belief that in that court they will be vindicated.
11. Loss of contact with reality; often associated with progressive isolation.
12. Restlessness, recklessness and impulsiveness.
13. A tendency to allow their 'broad vision', especially their conviction about the moral rectitude of a proposed course of action, to obviate the need to consider other aspects of it, such as its practicality, cost and the possibility of unwanted outcomes.
14. Incompetence in carrying out a policy, where things go wrong precisely because too much self-confidence has led the leader not to worry about the nuts and bolts of a policy.

Source: The Daedalus Trust.

The Daedalus Trust defines hubris, linking it back to the ancient Greeks, as follows:

> The concept comes from the ancient Greeks and their drama. Typically, the hero wins glory and acclamation by achieving great success against the odds. The experience then goes to their head: they begin to treat others, mere ordinary mortals, with contempt and disdain and develop such confidence in their own ability that they begin to think themselves capable of anything. This excessive self-confidence leads them into misinterpreting the reality around them and into making mistakes. Eventually they get their come-uppance and meet their nemesis, which destroys them.

The author then goes on to apply the Hubris Syndrome concept to modern-day leaders in a way that brings it alive:

> Hubris Syndrome is an "acquired personality change" *i.e.* brought on over a period of time. It is sparked by a specific trigger – exercising power. In other words, people who appear normal achieve positions of leadership, but once in power seem to alter their behavior … Hubris Syndrome seems to be driven from within the individual as a result of the act of exercising power, and not by outside factors … It only comes out when they are appointed to a job with high responsibility. When they have this power, great power, there comes a point when they can no longer accept any form of dissent and they take on a sort of messianic persona.

The author goes on to discuss the appearance of John Stumpf, the former Wells Fargo CEO who presided over the slow but shocking revelations about the bank when he was compelled to testify before Congress about the bank's culture of "eight is great" – a moniker that described the twisted incentive performance culture in which the more accounts (including fraudulent ones) the sales force could obtain from a single customer the better (and the bigger the bonus):

> But it was Stumpf's performance that stood out. Here was a man who had risen to the top of the world's most valuable bank, yet he seemed utterly unable to read a room. Although he apologized, he didn't appear chastened or remorseful. Nor did he seem defiant or smug or even insincere. He looked disoriented, like a jet-lagged space traveler just arrived from Planet Stumpf, where deference to him is a natural law and 5,000 a commendably small number. Even the most direct barbs—"You have got to be kidding me" (Sean Duffy of Wisconsin); "I can't believe some of what I'm hearing here" (Gregory Meeks of New York)—failed to shake him awake. What was going through Stumpf's head? New research suggests that the better question may be: What wasn't going through it?

Leadership begins and ends with the leaders that we have or that we choose or who are imposed on us. In the case of organizations or situations where leaders are chosen either via election (government roles) or appointment (boards of directors) or some other form of selection (acclamation for a political party leader), isn't it absolutely essential that we understand whom we are selecting? Wouldn't it have been better for the US to have seen Mr. Trump's tax returns and financial history before he was elected, or maybe even for a serious background check to have been part of the vetting process for the job of becoming a candidate for the most powerful office on earth?

As I am concluding the writing of this book, another case study has just popped up in the "hubris" department, and it involves what appears to be yet another in a long string of leadership failures, this time involving Carlos Ghosn, auto-sector chairman and CEO who oversaw for many years a global collaboration between Renault, Mitsubishi and Nissan and has found himself arrested in Japan for alleged financial malfeasance:

> Mr. Ghosn's trajectory resembles that of many high-flying business leaders, for whom hubris is a constant threat. His descent indicates that the flight path of the jet-setting, globetrotting corporate leader has peaked, as barriers to world trade go up and popular tolerance of their pay, perks and lifestyle declines.[19]

Moving on to other reasons why leaders might fail, subtitled "Transformational leaders are the exception, not the rule", the author in "Six Reasons CEOs Fail" enumerates six reasons why CEOs fail, summarized in Table 2.4. The research is based on data gathered from the *Harvard Business Review* (HBR) survey "Best Performing CEOs of 2017".

Table 2.4 Six reasons CEOs fail[20]

Tendency to grow stale in the saddle	When Jeff Immelt, regarded as one of the icons of American capitalism, became the CEO of General Electric in 2001, he inherited what was then the most valuable company in the US. Some 16 years later, the company's annual net earnings had shrunk 35 percent and the stock posted a negative total return. GE's weighting on the Dow even shrank under Immelt's watch. Many critics questioned why he was left in the top job for so long.
Response to stress and success	A string of successes or a good early start can fuel CEO narcissism and hubris. This has two possible consequences. One is more risky behavior, the other is complacency.
Top management team (TMT) problems	A TMT that works well is aligned on goals, shares information and makes joint decisions. Fragmentation arises when TMTs are misaligned, intergroup hostility grows and formal structures start to break down. TMTs can sway in the opposite direction, falling victim to groupthink, withholding crucial information from their CEOs, undermining their ability to make good decisions.

(*Continued*)

Table 2.4 Continued

Poor performance	This is usually characterized by an inability to respond to a changing economy, digitization, competition and evolving customer demands. From Blockbuster to Kodak, stories abound of corporate inertia. It is the CEOs' job to inspire their team with a vision and simultaneously read the landscape to capitalize on emerging trends.
Inadequate board vigilance	Boards add value when they're vigilant. According to a 2007 study, board vigilance made a difference to shareholder return on mergers and acquisitions …. Boards can shift the balance of power in a firm and temper CEO hubris. They can also coach their CEO by reading market change signals and tapping their invaluable networks and resources. CEOs should not ignore them, and boards should not be shy about guiding their CEO.
Scandal	Wrongdoing is now more easily exposed, and more companies are getting into trouble. Scandals can arise because of non-conformity to expectations (Volkswagen), financial irregularities … (Wells Fargo), social misconduct, ethical lapses or corruption. Former BP CEO Tony Hayward was felled by the Deepwater Horizon oil spill after being widely criticized for his handling of it. The famous line "I want my life back" did little to assuage those who had lost their livelihoods as a result of the spill.

Source: INSEAD.

This typology of reasons why CEOs fail revolves largely around the character and temperament of the CEO. Looking at these six categories, almost every one of them is closely associated with CEO personality issues or flaws that the board should be all over in terms of performance management oversight – a CEO's growing stale in the saddle, narcissism, hubris and complacency, lack of creativity/ability to inspire, inability to deal with or propensity for scandal.

And one of these categories actually blames the board for inadequate vigilance, which happens all too often, as many boards still view their role to be that of reactive caretaker rather than proactive conscience of the organization. Indeed, board vigilance is a thread that runs through these six categories, as the board's most important job is to make sure that leadership is not only competent to deliver the needed and desired results but able to do so in a manner that is respectful, ethical and holistic toward stakeholders' interests.

2.3.c Good to great leaders: achieving real leadership

> One of Nadella's first acts after becoming CEO, in February 2014, was to ask the company's top executives to read Marshall Rosenberg's *Nonviolent Communication*, a treatise on empathic collaboration. The gesture signaled that Nadella planned to run the company differently from his well-known predecessors, Bill Gates and Steve Ballmer, and address Microsoft's long-standing reputation as a hive of intense corporate infighting.[21]

There is no doubt that before Satya Nadella took over as CEO of Microsoft, whether deserved or not, the company had a reputation for being a suffer-no-fools, accept-no-mistakes, highly competitive place where the tone was pretty much set by exacting founder Bill Gates and hard-driving CEO Steve Ballmer.

Shockingly to the marketplace and happily for the company's stakeholders (including shareholders who have never seen a better run than the last few years under Nadella), a soft-spoken, emotionally intelligent and highly empathetic leader, Satya Nadella, was chosen new CEO in 2014. One of the first things CEO Nadella did when he took over leadership was to move the mission statement that Bill Gates had created to a new one as follows:

> From:
> "A PC on every desk and in every home, running Microsoft software".
>
> To:
> "To empower every person and every organization on the planet to achieve more".[22]

Beyond the new strategy, mission and vision that he has clearly set for Microsoft (which under his leadership has had spectacular financial results), Nadella has also demonstrated an uncanny and brilliant ability to set the tone from the top, change and renew the culture of the company – literally "walking the talk" and admitting his own failings or mistakes. This vignette illustrates this point beautifully:

> The CEO experienced his own difficult lesson in growth just eight months into his tenure. Invited to participate in a Q&A at the Grace Hopper Celebration of Women in Computing, a major annual event, he told the largely female audience that women in the tech industry should forgo asking for raises and instead trust that the system would reward them appropriately. The negative reaction was swift, with attendees quickly tweeting out their pushback.
>
> Nadella realized his mistake, and the next day issued an apology. "I answered that question completely wrong," he wrote in an email to Microsoft employees …
>
> But Nadella did more than deliver a *mea culpa*; he explored his own biases—and pushed his executive team to follow suit. "I became more committed to Satya, not less," says Microsoft chief people officer Kathleen Hogan, the former COO of worldwide sales, whom Nadella promoted into her current role soon after the kerfuffle. "He didn't blame anybody. He owned it. He came out to the entire company, and he said, 'We're going to learn, and we're going to get a lot smarter.'"
>
> It was a rare public falter for Nadella, but Microsoft got stronger. In the aftermath, one longtime rank-and-file Microsoftie told me, the company stepped up internal messaging that encouraged employees to respect

diversity and combat their unwitting biases. Nadella set an example for the rest of the company: We make mistakes, but we can learn to do better.[23]

Nadella even wrote a book about this whose title and subtitle summarize it all – *Hit Refresh: The Quest to Rediscover Microsoft's Soul and Imagine a Better Future for Everyone*. In the book, he not only recounts his own personal background and touchingly his family's struggles and triumphs raising a son with health challenges but also unfurls the next best thing to an ethical business manifesto for the digital age comprised of critical and central ideas like:

- Transforming the culture of a technology and engineering heavy company of "know-it-alls" to a company of "learn-it-alls".
- The idea of building coalitions and partnerships even with competitors under certain circumstances and not just competing for competition's sake.
- The central credo that Microsoft seems to be upholding better than some of its technology-sector competition (at least so far) – that there are critically important and "timeless values" in the digital age such as privacy, security and free speech (as we will explore at some length in the technology chapter later in this book).
- Developing a human ethical framework for new technologies like AI, which will upend all concepts we have known heretofore.
- The idea that these changes must be done in a manner that is consistent with ESG principles of corporate responsibility.

And all this goes back to the inner core of the kind of leader and the kind of culture Nadella wants to be and impart within his organization. He himself is quoted in the book as saying:

> Ideas excite me. Empathy grounds and centers me.

And none of Nadella's "culture" changes have been done to the detriment of the company or any of its stakeholders, including the almighty shareholder. Indeed, to the contrary. Table 2.5 and Figure 2.2 show the RepRisk reputation risk index of Microsoft over the past year as it compares to other tech companies (favorably) as well as the market performance of Microsoft's stock over the past five years, which has been enviable, to say the least.

Table 2.5 Tech sector RepRisk Index (RRI) and RepRisk Rating (RRR) as per RepRisk on April 12, 2019[24]

ESG Risk:	*Medium*	⟸	*High*		⟹	*Very high*
	Twitter	*Microsoft*	*Apple*	*Google*	*Facebook*	*Amazon*
Overall RepRisk rating	BB	B	CCC	CCC	CCC	CCC
Peak RepRisk index	48	57	64	65	66	68
Current RepRisk index	30	40	57	58	62	62

(*Continued*)

Table 2.5 Continued

ESG Risk:	*Medium*	⟸	*High*	⟹	*Very high*
	Twitter	*Microsoft*	*Apple*	*Google*	*Facebook*	*Amazon*

Explanation of RepRisk's risk metrics:
"The RepRisk Rating (RRR) is a proprietary risk metric that captures and quantifies a company's risk exposure related to ESG issues. It combines a company's own ESG risk exposure (Peak RRI) with the ESG risk exposure of the countries and the sectors in which the company has been exposed to risks. The RepRisk Rating ranges from AAA to D".
"The RepRisk Index (RRI): A quantitative measure (0 to 100) of a company's risk exposure to ESG and business conduct risks".

Key to each RRR:
AAA TO A – Low ESG risk exposure (best ratings)
BBB TO B – Medium ESG risk exposure
CCC TO C – High ESG risk exposure
D – Very high ESG risk exposure (worst rating)

RRI calibration:
The RRI ranges from zero (lowest) to 100 (highest). The higher the value, the higher the risk exposure:

- 0–24 generally denotes low risk exposure
- 25–49 represents medium risk exposure
- 50–59 denotes high risk exposure
- 60–74 denotes very high-risk exposure
- 75–100 denotes extremely high risk exposure

Source: RepRisk.

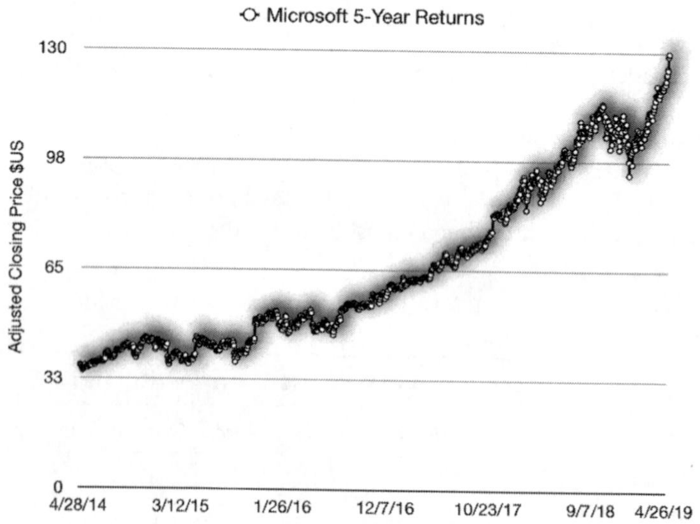

Figure 2.2 Microsoft 5-year stock performance.
Source: Steel City Re and Yahoo Finance.

Building on this truly transformative example of enlightened leadership and culture reset, let's turn to some useful data from around the world on what makes for a competent-to-great leader according to a *Harvard Business Review* study on the most important leadership competencies according to leaders worldwide.

Table 2.6 shows a summary of the results of this global survey, which very meaningfully lists "has high ethical and moral standards" as the most important and desirable leadership competency of all competencies as scored by leaders worldwide and by a significant margin to the next competency.

Table 2.6 Most important leadership competencies according to leaders from around the world surveyed by *HBR*[25]

Leadership competency	*Percentage of respondents*
1. Has high ethical and moral standards	67%
2. Provides goals and objectives with loose guidelines/ direction	59%
3. Clearly communicates expectations	56%
4. Has the flexibility to change opinions	52%
5. Is committed to my ongoing training	43%
6. Communicates often and openly	42%
7. Is open to new ideas and approaches	39%
8. Creates a feeling of succeeding and failing together	38%
9. Helps me grow into a next-generation leader	38%
10. Provides safety for trial and error	37%

Source: *Harvard Business Review*.

According to the *HBR* Best Performing CEOs, the top performing CEO for 2017 and 2018 was Pablo Isla of Inditex (the parent company of Zara and the largest retail company in the world).[26] They unequivocally stated that if this CEO's performance had been measured on financial results alone he would have come in 18th on the list – it was his performance on ESG metrics that propelled him to the top of the pile. Here's what they said:

> Measured on financial returns alone, Isla comes in 18th in our ranking; his company's performance on environmental, social, and governance (ESG) factors, which count for 20% of a leader's score, propelled him to the top spot. ESG-rating firms praise Inditex's transparency in managing, monitoring, and auditing its supply chain. The company encourages consumers to bring worn-out clothing to its stores for recycling (in Spain it runs an at-home-pickup recycling program), and the Join Life brand of Zara, its largest chain, is produced using recycled fibers and with careful attention to the consumption of water and other resources.

A final word of caution even for the best of the best leaders: even the most emotionally intelligent people can apparently lose their empathy as they grow in stature and power. A *Harvard Business Review* article documented this phenomenon and provided the following toolkit of questions that leaders should be asking themselves as they reach higher levels of responsibility and power:[27]

1. Do you have a support network of friends, family, colleagues who care about you without the title and can help you stay down to earth?
2. Do you have an executive coach, mentor or confidant?
3. What feedback have you gotten about not walking the talk?
4. Do you demand privileges?
5. Are you keeping the small, inconvenient promises that fall outside of the spotlight?
6. Do you invite others into the spotlight?
7. Do you isolate yourself in the decision-making process? Do the decisions you're making reflect what you truly value?
8. Do you admit your mistakes?
9. Are you the same person at work, at home and in the spotlight?
10. Do you tell yourself there are exceptions or different rules for people like you?

Finally, Table 2.7 provides a birds-eye view of the spectrum of leadership we have surveyed in this section - naming a few names from the "bad and the ugly" to the "good and the better".

Table 2.7 A leadership spectrum of "the bad, the ugly, the good and the better"

Hubristic and/or sociopathic	*Technocratic functionally effective*	*Exemplary/well rounded*
C. Ghosn (Nissan/Renault/Mitsubishi) E. Holmes (Theranos) T. Kalanick (Uber) L. Moonves (CBS) H. Weinstein (Weinstein Company) S. Wynn (Wynn Resorts) M. Bin Salman (Saudi Crown Prince) V. Putin (Russian President) D. Trump (US President)	The Majority of Leaders	M. Barra (General Motors) M. Benioff (Salesforce) R. Branson (Virgin) W. Buffett (Berkshire Hathaway) P. Isla (Inditex) S. Nadella (Microsoft) P. Polman (Unilever) Dalai Lama (Spiritual Leader) J. Ardern (New Zealand Prime Minister) A. Merkel (German Chancellor)
The "bad" and the "ugly" ⟸	⟹	the "good" and the "better"

Source: Author.

2.4 Culture: from toxicity to integrity

2.4.a What is culture? Why culture? Why now?

Culture experts consider culture to be an intrinsic and irreplaceable part of an organization's ability to succeed. Indeed, the leading business thinker Peter Drucker once coined the famous and oft repeated phrase "culture eats strategy for breakfast". Hence our treatment of culture as part of the leadership survival package.

Harvard Business Review published an exceptionally good issue on culture called "The Culture Factor" in January 2018. Several authors who penned a piece called "The Leader's Guide to Corporate Culture" remarked on the relationship between strategy and culture as follows:

> Strategy offers a formal logic for the company's goals and orients people around them. Culture expresses goals through values and beliefs and guides activity through shared assumptions and group norms.[28]

Culture wasn't talked about much until the aftermath of the massive financial crisis of 2007/2008. When all was said and done, the regulators and authorities began to get at what might be the root cause of these massive failures – the culture inside the banking sector – at least the US and UK/EU banking system, because this massive failure of trust wasn't happening in every banking sector everywhere. Amazingly, the Canadian banking sector – the closest neighbor to the US banking sector and one that by proximity or contagion could have easily been similar to the US banking system culture, didn't suffer contagion from the US financial crisis – they had next to no related or similar failures. Why? Because the Canadian banking sector had a very different history and regulatory structure, but even more importantly, had a very different culture toward stakeholders, mainly customers. Customers weren't there to be fleeced – they were there to be served and brought into a circle of mutual trust that would in turn be mutually beneficial. Canadian banks have a generally conservative culture that is customer-centric, where the type of systemic abuse and outright theft that took place massively in the US and parts of Europe leading up to the American and European financial crisis could not and did not happen.

The authors of the aforementioned article also provide these two useful definitions of culture:

> Culture is the tacit social order of an organization: It shapes attitudes and behaviors in wide-ranging and durable ways. Cultural norms define what is encouraged, discouraged, accepted, or rejected within a group. When properly aligned with personal values, drives, and needs, culture can unleash tremendous amounts of energy toward a shared purpose and foster an organization's capacity to thrive.
>
> Culture can also evolve flexibly and autonomously in response to changing opportunities and demands. Whereas strategy is typically determined by the C-suite, culture can fluidly blend the intentions of top leaders with the knowledge and experiences of frontline employees.

Table 2.8 contains useful parameters about culture for leaders to think about, before we turn to a very specific section in which we will lay out the ultimate responsible party for culture oversight: the role of the board.

Table 2.8 Culture change practices for leaders to consider[29]

Articulate the aspiration	[C]reating a new culture should begin with an analysis of the current one, using a framework that can be openly discussed throughout the organization. Leaders must understand what outcomes the culture produces and how it does or doesn't align with current and anticipated market and business conditions.
Select and develop leaders who align with the target culture	Leaders serve as important catalysts for change by encouraging it at all levels and creating a safe climate ... Candidates for recruitment should be evaluated on their alignment with the target. A single model that can assess both organizational culture and individual leadership styles is critical for this activity.
Use organizational conversations about culture to underscore the importance of change	To shift the shared norms, beliefs, and implicit understandings within an organization, colleagues can talk one another through the change. Our integrated culture framework can be used to discuss current and desired culture styles and also differences in how senior leaders operate. As employees start to recognize that their leaders are talking about new business outcomes—innovation instead of quarterly earnings, for example—they will begin to behave differently themselves, creating a positive feedback loop.
Reinforce the desired change through organizational design	When a company's structures, systems, and processes are aligned and support the aspirational culture and strategy, instigating new culture styles and behaviors will become far easier. For example, performance management can be used to encourage employees to embody aspirational cultural attributes. Training practices can reinforce the target culture as the organization grows and adds new people. The degree of centralization and the number of hierarchical levels in the organizational structure can be adjusted to reinforce behaviors inherent to the aspirational culture. Leading scholars such as Henry Mintzberg have shown how organizational structure and other design features can have a profound impact over time on how people think and behave within an organization.

Source: *Harvard Business Review*.

Some examples also gleaned from the research described in Table 2.8 that are based on the authors' eight different culture types with examples of eight organizations that exercise such culture as championed by their leaders are shown in Table 2.9.

Table 2.9 Culture and leaders – examples[30]

Learning: Tesla

I'm interested in things that change the world or that affect the future and wondrous new technology where you see it and you're like "Wow, how did that even happen?"

Elon Musk, cofounder and CEO

(*Continued*)

Table 2.9 Continued

Purpose: Whole Foods
Most of the greatest companies in the world also have great purposes .... Having a deeper, more transcendent purpose is highly energizing for all of the various interdependent stakeholders.
John Mackey, founder and CEO

Caring: Disney
It is incredibly important to be open and accessible and treat people fairly and look them in the eye and tell them what is on your mind.
Bob Iger, CEO

Order: SEC
Rule making is a key function of the commission. And when we are setting the rules for the securities markets, there are many rules we, the SEC, must follow.
Jay Clayton, chairman

Safety: Lloyd's Of London
To protect themselves, businesses should spend time understanding what specific threats they may be exposed to and speak to experts who can help.
Inga Beale, CEO

Authority: Huawei
We have a "wolf" spirit in our company. In the battle with lions, wolves have terrifying abilities. With a strong desire to win and no fear of losing, they stick to the goal firmly, making the lions exhausted in every possible way.
Ren Zhengfei, CEO

Results: GSK
I've tried to keep us focused on a very clear strategy of modernizing ourselves.
Sir Andrew Witty, former CEO

Enjoyment: Zappos
Have fun. The game is a lot more enjoyable when you're trying to do more than make money.
Tony Hsieh, CEO

Source: *Harvard Business Review.*

2.4.b Transforming culture risk into culture value – an urgent focus for boards

While developing and implementing a healthy organizational culture is the frontline responsibility of management in any type of organization – from the White House to the local fish and chip house – culture oversight is the domain and the responsibility of governance – a board of directors or trustees for most entities, or, for the US president, the US Congress. But time and again the scandals, crises and aberrations we have witnessed have another root cause – lack of oversight and an absence of leadership and culture risk governance, in other words, the breakdown of governance. In this section, we do a deeper dive into the roles and responsibilities of the board on culture and offer a toolkit for boards to use to get on with the job of proactive leadership and culture oversight.

The importance of the #MeToo movement that emerged in 2017 cannot be over-emphasized. It ushered in a unique time for leaders to step up to their responsibility for creating and owning a healthy workplace culture and for boards to acknowledge and embrace their responsibility: exercising proactive oversight of, and holding management accountable for, creating and maintaining a healthy workplace culture. And this applies across the board to all forms of organizations and institutions, from the corner mom-and-pop store and the largest global corporate behemoth to nonprofits, universities, international government agencies and the Vatican.

Culture isn't (and never was) a soft issue – it's a "hard" issue in that it is "complex" and requires focus and expertise. It's also a "hard" issue because it has tangible financial and reputational impacts which can spell the difference between sustainable value creation, mere survival, value destruction or even the unraveling of an organization.

Table 2.10 provides a summary and sneak preview of the three roles and seven tools of culture governance we suggest as part of this discussion.

Table 2.10 The three roles and seven tools of "culture governance"

Role	*Tools*
THE *CULTURALLY EQUIPPED* BOARD *The board is sensitive to and informed on issues of workplace culture*	1. The board is educated on the culture big picture by experts. 2. Every board has at least one independent director with broad and deep ESGT/culture expertise.
THE CULTURALLY TUNED-IN BOARD *The board thoroughly understands the culture of its organization*	3. The board insists on getting the right members of management to report on culture to the board periodically. 4. The board regularly reviews, is updated on, and asks questions based on a customized "culture" dashboard. 5. The board thoroughly understands the culture of the organization it oversees.
THE *CULTURALLY CONSCIENTIOUS* BOARD *The board is the guardian of a healthy culture and an instigator of culture change when necessary*	6. The board exercises proactive CEO ESGT/culture accountability oversight. 7. The board enforces CEO and management ESGT/culture accountability, including discipline and performance incentive packages.

Source: Author.

The October 2017 NACD Blue Ribbon Commission Report on "Culture as a Corporate Asset" was prescient in addressing the fact that boards and executive teams must immediately focus on understanding the culture of their workplaces as part of the value chain and strategy. In Table 2.11 is a summary of some of the key "culture red flags" that this study identified for boards to pay attention to.[31]

Table 2.11 NACD "culture as an asset" study: Culture red flags

- Focus on performance with little regard to how results were achieved.
- High performers are allowed to operate outside established policies. Behaviors that are not consistent with the company's stated values/code of conduct are rewarded.
- Frequent "near misses" of adherence to code of conduct, risk-appetite limits, etc. Or frequent requests for exceptions to these policies in order to meet performance targets.
- Excessive focus on consensus/collegiality leading to prevalence of "go along to get along" attitudes.
- Relationships outweigh skills and/or performance in determining promotions or other recognition to an inappropriate degree.
- Sharing bad news is discouraged (or, bearers of bad news are punished outright).

Source: NACD.

But boards and management must also understand how to get to the root of any workplace culture dysfunction that may exist. In this era, the excuse that only shareholders matter no longer holds. Boards and management are responsible to all of their stakeholders (shareholders, employees, customers and beyond) for ESG (or ESGT) results as well, which include proactively maintaining and nurturing a healthy workplace culture. It does not pay to turn a blind eye or to wait for the crisis to hit, as serious reputation risk in today's age of hyper-transparency can hit hard and fast. The rapid-fire downfall of not only Harvey Weinstein but of his entire company, including its reputationally damaged board and board members, is the cautionary tale of the day.

And on the positive side, there is plenty of evidence that while a toxic culture destroys value, a strong and resilient one fully championed and embodied by the very top of the organization (CEO and board) can and will add long-term sustainable value – both financial and reputational. Such value protects the organization from the crises that will inevitably come and adds bottom-line financial value as the famous J&J Tylenol case of the 1980s first demonstrated. We discuss this topic in great detail in our final chapter, Chapter 8.

2.4.b.i *Three governance culture roles and seven practical culture tools*

These are the three key roles of the board in supervising the culture of its organization:

1. **The culturally "equipped" board.** A board that is fully educated on culture and leadership risk and opportunity generally and willing to be educated on the nonfinancial ESG (or ESGT) topics, viewing them not as an alien, irrelevant or otherwise soft topic but as one that enhances and creates greater stakeholder value.
2. **The culturally "tuned-in" board.** A board that knows, and intimately understands the specific culture of the organization it oversees.
3. **The culturally "conscientious" board.** A board that acts as the guardian of a good culture and/or an instigator and overseer of culture change when necessary.

To be able to execute on these three key culture-related roles, the board must be equipped with a series of tools – grassroots, tactical, data-based ways to understand and perceive the existing organizational culture. These key tools are related to the three roles and are the following:

ROLE 1: The culturally equipped board: *the board is sensitive to and informed on issues of workplace culture*

- **TOOL #1: The board is educated on the culture big picture by experts.** The board is scheduled to have annual or periodic culture segments with inside and outside experts to educate the board on the latest and greatest culture tools and developments in their entity. The chief ethics and compliance officer (CECO), chief learning officer, chief human resources/talent officer are good resources to look to for inside expertise.
- **TOOL #2: Every board has at least one independent director with broad and deep ESGT/culture expertise.** This means not having only directors with financial, operational, CEO and CFO backgrounds but diverse, independent directors with broad and deep ESG (or ESGT) experience and expertise. They will know what questions to ask when it comes to culture issues.

ROLE 2: The culturally tuned-in board: *the board thoroughly understands the culture of its organization*

- **TOOL #3: The board insists on getting the right members of management to report to the board on culture periodically.** The CECO, chief learning officer, chief human resources/talent officer are good resources to report to the board from time to time and regularly. Indeed, the CECO should report to the board (or a committee thereof) on a quarterly basis with a dashboard that includes some of the key culture metrics described here. Moreover, the board/committee should have regular executive sessions with the CECO and maybe even develop a more informal phone call check-in between the CECO and the chair of the audit committee, for example.
- **TOOL #4: The board regularly reviews, is updated on and asks questions based on a customized "culture" dashboard.** Such a dashboard should be unique to each organization but should include many of the following qualitative and quantitative metrics:
  - *Ethics and Compliance (E&C) Metrics*:
    E&C risk assessments: key data, key topics.
    - Helpline/hotline trends and key issues.
    - Training and communications trends and topics.
    - Pulse surveys on ethics and compliance program.
    - Investigations: type, process and outcome.
    - Periodic internal and external evaluations of the effectiveness of the E&C program.

- *Employee/Culture Survey Metrics*:
    - Culture climate metrics geared at workplace issues, including supervisory relationships.
    - E&C benchmarking against peers.
- *Human Resources Data*:
    - Intake interviews.
    - Exit interviews.
    - Performance management results (with financial and non-financial or ESG metrics included).
    - 360 leadership assessments or the like.
- **TOOL #5: The board thoroughly understands the culture of the organization it oversees.** The board understands where its organization fits in the spectrum of workplace culture. An example of useful benchmarking may involve using the Ethics Research Center's Global Business Ethics Survey. Boards should see a culture survey, understand it, ask about the culture climate, the temperature and how it is reflected at different divisions, business units, etc. Do the company's culture surveys have consequences or are they merely window dressing? If the latter, why do them? If the former, what are the actual concrete consequences? Do "golden boys/girls" who are abusive get counseled, disciplined and/or terminated when infractions occur? Or are they ignored or merely slapped on the wrist for things that get others fired?

ROLE 3: The culturally conscientious board: *the board is the guardian of a healthy culture and an instigator of culture change when necessary*

- **TOOL #6: The board exercises proactive CEO ESGT/culture accountability oversight.** The board exercises proactive oversight over the CEO, requiring him or her to describe and prove the culture of the workplace as it exists. Boards should ask for culture surveys, risk assessments and evaluations from the experts. Require a gap analysis between what the CEO says and what the CEO does (what his/her description of culture is and that of the experts). Most of all, introduce culture accountability metrics and exercise holistic metric accountability by demanding the inclusion of both financial and nonfinancial/ESG (or total stakeholder) reporting metrics into the CEO, C-Suite and full company performance management system.
- **TOOL #7: The board enforces CEO and management ESGT/culture accountability, including discipline**. The board needs to exercise and enforce its own tone from the top when necessary – whether to intervene proactively when the CEO and/or executive team do not deliver on ESG/nonfinancial metrics (even if they do on financial metrics) or to proactively discipline and even terminate a CEO whose workplace culture behavior is in question, questionable or worse. It is the board's responsibility (in addition to the CEO's) to

get ahead of leadership and culture risk as the reputation risk associated with this type of risk can be strategic and even existential (e.g. The Weinstein Company). When the CEO and company attain new heights of robust and resilient workplace culture, the board should recognize and reward such achievements as well.

2.5 Ethics: the beating heart of good leadership and culture

2.5.a Organizational ethics: building resilience

Over the past 20 years (but stretching as far back as 30 and even 40 years ago, as Figures 2.3 and 2.4 illustrate), there has been a body of work best called "ethics and compliance" (E&C).[32] The "ethics" part of it refers largely to normative standards of behavior above and beyond what is required by law. The "compliance" part has been rooted fundamentally in required law and regulatory observance. The now fairly well evolved E&C field first emerged in the US rooted in two major defense industry scandals – a US defense industry foreign corruption and bribery scandal in the 1970s which gave way to the US Foreign Corrupt Practices Act of 1977 (FCPA) and another defense industry scandal involving directly defrauding the US government, which gave rise to the ethics-focused Defense Industry Initiative of 1985, which continues to exist to this day.

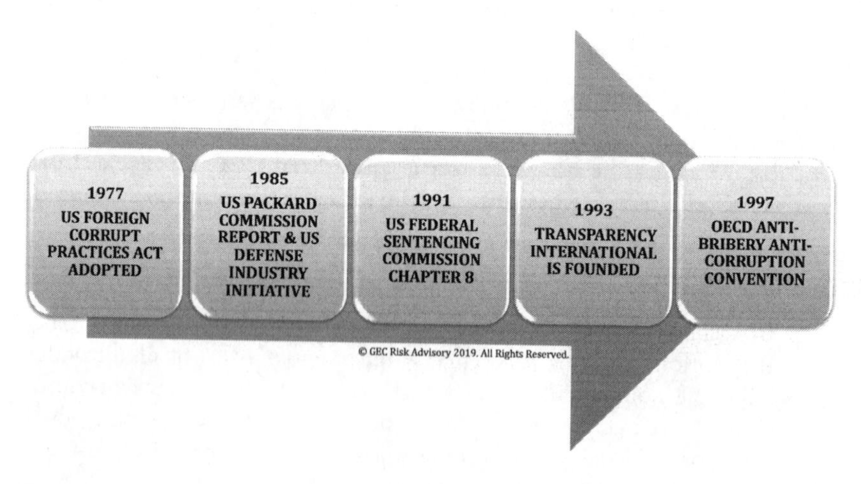

Figure 2.3 Evolution of ethics and compliance I: milestones 1977–1997.
Source: GEC Risk Advisory.

Those events led to the creation of a set of evolving E&C best practices formulated formally in the critically important US Sentencing Guidelines in 1991 (USSG). These guidelines are not law per se but provide guidance to the judiciary and the law enforcement community on how severely to sentence a corporate wrongdoer in the US. If that wrongdoer followed a semblance of the (initially) seven best E&C practices enshrined in the USSG, that company would get less severe punishment.

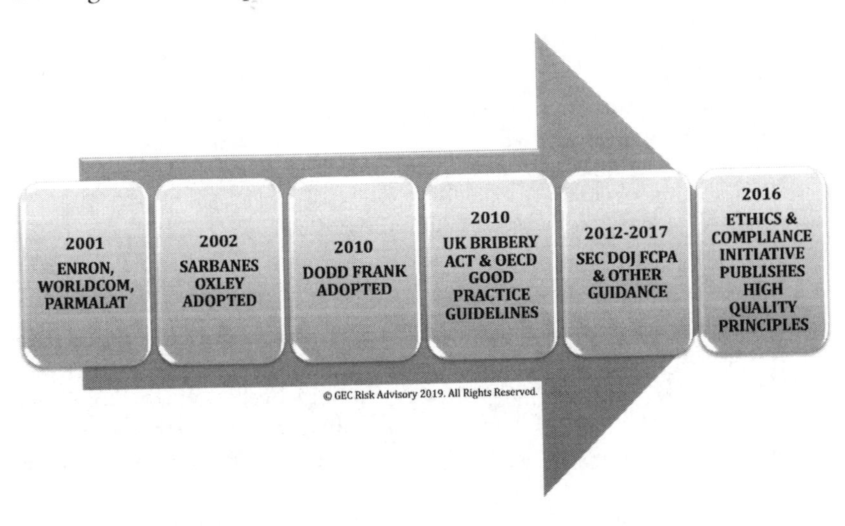

Figure 2.4 Evolution of ethics and compliance II: milestones 2001–2019.
Source: GEC Risk Advisory.

With the help of Transparency International (founded in 1993) and the adoption of the OECD Anti-Bribery and Anti-Corruption Convention of 1997 (now adopted by over 40 countries worldwide), the field of E&C continued to grow with both the private and the public sector recognizing that there were a set of E&C practices that could protect companies from severe liability and reputational damage if they were in place.

Figures 2.3 and 2.4 show a summary overview of the main milestones in the evolution of E&C programs over the past five decades. Likewise, the best practices that have emerged to date over this period of time are summarized in Table 2.12.

Table 2.12 Key elements of an effective global ethics & compliance program

1. Periodic E&C Risk Assessment	Knowing your risks allows you to address, mitigate and prevent them.
2. Code of Conduct and Policy Framework	Providing employees and third parties with guidelines informs them.
3. Resources and Budget	Having the right balance of resources allows risks and opportunities to be addressed.
4. Board and C-Suite Access	Access to and representation at the highest levels of organizational leadership informs them and legitimizes the program.
5. Education and Communication	Reaching your audience informs them and hands them tools for better ethical decision-making.

(*Continued*)

Table 2.12 Continued

6. Internal Controls Alignment	Proper internal controls alignment enables process and continuous policy improvement.
7. Safe to Speak-Up/ Listen-Up Culture	Employee speak-up and leadership listen-up creates a safe and empowering culture.
8. Organizational Justice and Consistent Discipline	A sense of internal justice creates a valuable culture of loyalty, safety and common purpose.
9. Auditing, Monitoring, Continuous Improvement	Lessons learned, root cause analysis and constant improvement creates stakeholder value.

Source: Author.

While these E&C practice standards have been mostly directed at the corporate sector, other sectors have adopted their general contours, including the US federal government (for example, in the US the FBI has one of the leading and best E&C organizational programs, first established when Robert S. Mueller III was their director). The nonprofit, university, research and healthcare sectors also all follow in one way or another the general precepts outlined in Table 2.12.

There are several nonprofit educational organizations at the cutting edge of work relevant to what I like to call ethical governance today. While we do not have space to go into great detail about their work and their areas of focus, I would like to share a snippet of the work they are doing, as it adds value to any organization's considerations of E&C governance.

2.5.a.i Ethics and Compliance Initiative

The Ethics and Compliance Initiative has broken new ground in the definition of what a High-Quality Ethics and Compliance Program (HQP) looks like.[33] Table 2.13 shows a very high-level view of the elements of the HQP.

Table 2.13 Ethics and Compliance Initiative HQP standards

Principle 1	Ethics and compliance is central to business strategy.
Principle 2	Ethics and compliance risks are identified, owned, managed and mitigated.
Principle 3	Leaders at all levels across the organization build and sustain a culture of integrity.
Principle 4	The organization encourages, protects and values the reporting of concerns and suspected wrongdoing.
Principle 5	The organization takes action and holds itself accountable when wrongdoing occurs.

Source: The Ethics and Compliance Initiative.

Table 2.14 provides a simple comparison of what the HQP E&C program would look like compared to the average basic E&C program which in and of itself is not a bad place to be for an organization, especially compared to organizations that have done nothing or next to nothing.

Table 2.14 Key differences between the "basic" E&C program and the ECI'S high-quality E&C program

E&C elements	*Basic program*	*High-quality program*
E&C is central to business strategy	No	Yes
Leadership & culture of integrity	No	Yes
Continuous improvement	No	Yes
Code of conduct & policy framework	Basic	Highly integrated
Chief E&C officer executive & board reporting	Basic/indirect	Robust/structured/periodic/dual
Delegation of approval authority	Basic/indirect	Structured
Risk assessments	Basic	Integrated w/ERM
Training and communications	Basic	Integrated w/risk
Monitoring, auditing, evaluations	Basic	Integrated w/risk
Helplines/hotlines/reporting systems	Basic	Lifecycle speak-up/listen-up
Discipline and enforcement	Consistent but unintegrated	Consistent/integrated "procedural justice"

Source: GEC Risk Advisory.

2.5.a.ii Ethical Systems

The work of Ethical Systems also bears mentioning for two major reasons: first, they focus on a critically important and cutting-edge area of ethics – behavioral ethics – and have a broad and diverse consortium of leading academics and researchers who have done groundbreaking work on behavioral ethics.[34] Second, they have developed a culture diagnostic tool that emerges from a vast and deep array of culture work that also dovetails nicely with that other huge component of governance – culture, which we explored earlier in this chapter.[35] Here is a sampling of some of their freely downloadable *Behavioral Ethics One Sheets*:[36]

- Bounded Ethicality – Identifying Common Pitfalls.
- Ethical Fading – Overlooking Ethical Considerations When Making Decisions.
- Nudging for Ethics – Applying Small Changes to Promote Ethical Outcomes.
- Speak Up Culture – Designing Organizational Cultures that Encourage Employee Voice.
- Ethics Pays – How an Investment in Ethics Translates to Profit, Productivity and Prestige.
- Goals Gone Wild – How Aggressive Goal Setting Can Lead to Unethical Behavior.
- Motivated Reasoning – The Use of Reasoning to Justify Self-Serving Decisions.

2.5.a.iii Institute of Business Ethics

The UK-based Institute of Business Ethics has long been a leader in providing important research and cutting-edge concepts to the field of organizational ethics as well including surveys, original research and the development of best practices. It also provides a wide variety of how-to tools for implementing different elements of an effective program like *The IBE 9 Step Model for Developing and Implementing a Code*.[37]

2.5.a.iv National Association of Corporate Directors

The work of the National Association of Corporate Directors (NACD), which is done from the perspective of boards of directors, also provides invaluable assistance to the highest levels of organizations (executives and boards) on their roles and responsibilities on issues of culture and ethics. Two notable resources they have produced in recent times include "Culture as a Corporate Asset" and "Director Essentials: Strengthening Compliance and Ethics Oversight".[38]

Finally, Figure 2.5, for the E&C professionals and enthusiasts among us, is a depiction of where I believe the compliance piece fits into the overall interrelationship between leadership, culture, ethics and compliance. Compliance is a fundamental and necessary (almost mechanical) part of good governance but in no way a sufficient one. Good and excellent governance and leadership can only come from an added and critical layer of ethical behaviors and standards that are part of a healthy culture that leaders (both operational and oversight) implement each and every day.

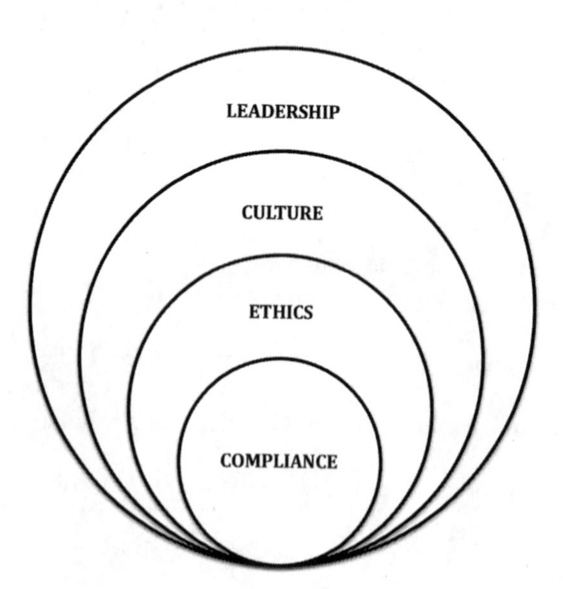

Figure 2.5 Leadership, culture, ethics and compliance.
Source: GEC Risk Advisory.

2.5.b From ethical culture gloom and doom to ethical culture reform

The past several years have witnessed many tales of organizational culture gone wrong. We learned about negative and destructive behaviors in the workplace, mostly perpetrated by powerful leaders, causing serious human, economic and reputational costs for people and organizations. The toxic workplace cultures extended from the pinnacles of political power to the front lines of manufacturing facilities.

Powered by the ubiquity and raw reach of social media, the #MeToo story quickly became universal – told first by the more glamorous denizens of Hollywood and then extending to the most vulnerable hotel, restaurant and factory floor workers. All of them victims of a toxic workplace culture of abuse of power, shame, lies and worse, many of them submitting to terrible work conditions, sidelined from needed jobs or permanently derailed from pursuing desirable careers and professional passions.

The *TIME* Magazine 2017 Person of the Year – the "Silence Breakers" – said it all. Though sparked by the Weinstein expose, this story of the year clearly represented the culmination of decades of pent-up workplace silence, lies, cover-ups, manipulation and anger. The overwhelming impact of the #MeToo phenomenon can only be explained by the explosion and maturation of social media, which has led to the amplification and acceleration of workplace culture-related reputation risk.

Two other relatively recent corporate cultural moments (if we can call them that) come to mind – 2002 and 2008. In 2002 came the downfall of Enron, WorldCom and others, which resulted in an uproar about financial accountability and the adoption of Sarbanes-Oxley. In 2008 came the downfall of financial giants Lehman Brothers and Bear Stearns, and the humiliation of the US financial sector in general for the massive mortgage- and derivative-related scandals, leading to social awakenings such as "Occupy Wall Street" and the adoption of Dodd-Frank.

The most momentous year yet on the topic of culture was arguably 2017. In both the 2002 and 2008 cases, the "cultural" issue revolved mainly around financial malfeasance – greed, negligence, fraud or lack of accountability. The cultural issue of 2017 is qualitatively different – it was about toxic personal behaviors in the workplace perpetrated mainly by leaders against their subordinates.

By 2017 (unlike 2002 and 2008) we had also arrived at the convergence of two other significant developments not fully present or developed before: (1) the rise of the importance to business of environmental, social and governance ("ESG") issues (especially in the US, as Europe has long focused on ESG), and (2) the acceleration and amplified impact of reputation risk associated with ESG risk (which includes workplace cultural issues) because of the age of social media and hyper-transparency.

With all the doom and gloom that these toxic workplace culture issues raise, I would also underscore a hopeful note to boards and executives struggling to deal with the organizational cultural issues so clearly brought to the fore in 2017. Unlike the regulatory responses to the excesses of 2002 (Sarbanes-Oxley) and 2008 (Dodd-Frank), I would suggest that the appropriate response to the 2017 cultural issues is not new regulation but self-regulation, a voluntary upping of the corporate cultural ante by elevating the importance of ethics, compliance and risk management within organizations, powered and driven by a strong culture of accountability and "walk the talk" from the top. This entails a voluntary, value-creation mind-set at the executive and governance levels of an organization that aligns a strong and resilient culture with sustainable profitability and that likewise recognizes that a toxic culture will in the short and long run lead to value and reputational erosion and possibly destruction.

Thankfully, there are positive tales to be inspired by. A case in point: Microsoft. Under its relatively new CEO, Satya Nadella, who has just written a book (*Hit Refresh: The Quest to Rediscover Microsoft's Soul and Imagine a Better Future for Everyone*), has instigated culture change there that by all accounts has had dramatic and beneficial impacts on all stakeholders – internally (employees) and externally (customers). And lo and behold shareholders as well – when Nadella became CEO in 2014, the share price was around $35; today, in mid-2019, Microsoft's share price hovers around at $135.

With all the negative news, now is the time for savvy management and visionary boards to understand, acknowledge and tackle workplace cultural issues head on and in a more systematic and conscientious way. Culture is the fabric of an organization and that fabric can either be healthy and sustainable, able to contribute to the development of resilience and creation of value, or brittle, weak and/or toxic, leading to financial and reputational vulnerability, value erosion or even ruin. It is the direct responsibility of leaders – both management and board – to make the right choices on workplace culture.

In Tables 2.15 and 2.16 executives and boards – the leaders responsible for the creation and deployment of a highly effective ethical culture in their organizations – can find out why culture has been a challenge in the past (Table 2.15) and can learn the questions they can begin to ask to further the implementation of an ethical culture (Table 2.16).

Table 2.15 LRN survey of CECOs and boards[39]

1. Most chief ethics and compliance officers (CECO) feel that boards do not fully understand the ethics and compliance programs they are supposed to be overseeing.
2. CECOs say boards need a more systematic approach or "game plan" for the oversight of ethics and compliance.
3. CECOs believe many boards do not devote significant time and priority to ethics and compliance.
4. Most CECOs believe that boards do not go into sufficient depth on ethics and compliance programs and outcomes.
5. Most CECOs believe that their boards do not focus on the root causes of behavior or appreciate the competitive advantage company culture can provide.

(*Continued*)

Table 2.15 Continued

6. Most CECOs believe that their boards lack appropriate metrics through which to evaluate ethics and compliance.
7. Most CECOs believe that boards do not ask enough of executive management in ensuring effective ethics and compliance.
8. Many boards do not hold executive sessions with their CECOs.

Source: LRN.

Table 2.16 Questions the board should ask the CECO[40]

1. Are senior management and the board receiving the information they need for appropriate oversight?
2. Are we identifying and prioritizing the company's ethics and compliance risks?
3. Are the company's standards, policies and procedures (e.g., code of conduct, etc.) linked to the prioritized ethics and compliance risks?
4. Are we auditing compliance risk areas?
5. How are we communicating to our employees their specific job-related E&C risks and the importance of E&C, generally?
6. Do we have an effective issue escalation policy, procedure, guideline?
7. Do we have the right resources, systems and controls in place to ensure:
   a. Observed misconduct is reported?
   b. Employees are comfortable raising issues?
   c. Building an ethical culture?
8. Are we responding appropriately to compliance allegations, issues and concerns (i.e., escalation, investigation, remediation)?
9. Can we demonstrate that employees are appropriately vetted upon hiring and promotion into certain positions of substantial authority?
10. How do we use performance management tools and processes to ensure all employees are held accountable for creating/maintaining an ethical, values-based culture?

Source: Jacqueline E. Brevard and Andrea Bonime-Blanc for The Conference Board.

Table 2.16 provides the beginning of a solution by suggesting some of the fundamental questions boards and other oversight bodies should be asking management.

2.6 Putting it all together: the leadership, culture and ethics gestalt – a typology of ESGT leadership

Figure 2.6 provides a sneak preview of something we will be touching upon throughout the book and in greater detail when we come to Chapters 7 and 8 where we lay out the importance of the right kind organizational leadership when it comes to managing and deploying ESGT issues, risks and opportunities.

In the meantime, I leave my readers with Figure 2.6 which for now remains self-explanatory.

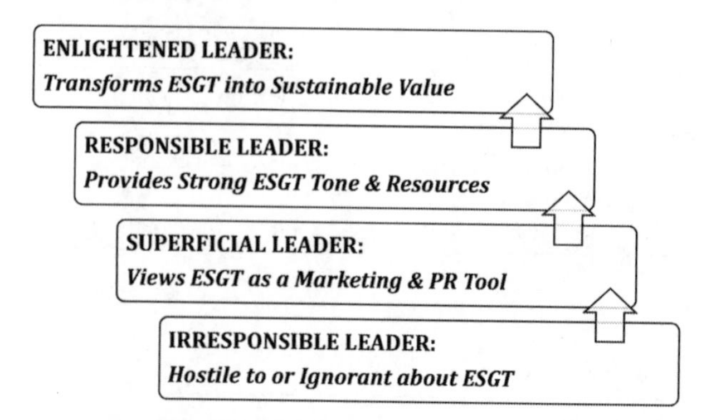

Figure 2.6 An ESGT leadership typology.

Source: GEC Risk Advisory.

Notes

1 https://www.phrases.org.uk/meanings/fish-rot-from-the-head-down.html. Accessed on July 17, 2018.
2 As per Wikipedia: If you lie down with dogs, you get up with fleas, or in Latin, *qui cum canibus concumbunt cum pulicibus surgent.* "He that lieth down with dogs shall rise up with fleas" has been attributed to Benjamin Franklin's *Poor Richard's Almanack*. The Latin has been unreliably attributed to Seneca, but not linked to any specific work. https://en.wikipedia.org/wiki/Wikipedia:If_you_lie_down_with_dogs,_you_get_up_with_fleas. Accessed on August 4, 2018.
3 Machiavelli quotes https://www.brainyquote.com/quotes/niccolo_machiavelli_131418. Accessed on August 30, 2018.
4 Voltaire quote in Quote Investigator. https://quoteinvestigator.com/2015/07/23/great-power/. Accessed on September 3, 2018.
5 Goalcast. Abraham Lincoln Quotes. https://www.goalcast.com/2018/01/05/abraham-lincoln-quotes/. Accessed on November 23, 2018.
6 Andrea Bonime-Blanc, "The Biggest Risks Nobody Talks About". Ethical Corporation. February 2014. http://www.ethicalcorp.com/business-strategy/globalethicist-biggest-risks-nobody-talks-about. Accessed on September 9, 2018.
7 Acton Research. Lord Acton Quote Archive. https://acton.org/research/lord-acton-quote-archive. Accessed on August 20, 2018.
8 *Oxford English Dictionary*. https://en.oxforddictionaries.com/definition/survival. Accessed on June 17, 2018. https://en.oxforddictionaries.com/definition/surviving. Accessed on September 11, 2018.
9 *Oxford English Dictionary*. https://en.oxforddictionaries.com/definition/thriving. Accessed on September 11, 2018.
10 Bandy X. Lee, Editor. "The Dangerous Case of Donald Trump". 2017. https://www.amazon.com/Dangerous-Case-Donald-Trump-Psychiatrists/dp/1250179459/ref=sr_1_1?ie=UTF8&qid=1535062968&sr=8-1&keywords=dangerous+case+of+donald+trump. Accessed on August 23, 2018; and Video of Dr. Kerry Sulkowicz video op-ed "Top Psychiatrist Analyzes Trump's Mental State". https://mic.com/articles/185189/top-psychiatrist-analyzes-trumps-mental-state#.rO6r0Bqpl. Accessed on August 23, 2018.
11 Lea Surugue. "One immoral Act Begets Another, Brain Research Shows". *IB Times*. October 24, 2016. http://www.ibtimes.co.uk/one-immoral-act-begets-another-brain-research-finds-1587966. Accessed on August 23, 2018.

12 Fast Company Staff. "Leadership Quiz: Is Your Boss a Psychopath?" *Fast Company*. July 1, 2005. https://www.fastcompany.com/53265/quiz-your-boss-psychopath. Accessed on August 22, 2018.

13 https://www.ncbi.nlm.nih.gov/pubmed/20422644. Accessed on August 23, 2018.

14 Zhu, David and Chen, Guoli. "Narcissism, Director Selection, and Risk-Taking Spending" (2014). Available at SSRN: https://ssrn.com/abstract=2470037. Accessed on August 23, 2018.

15 James B. Stewart. "Threats and Deception: Why CBS' Board Turned Against Les Moonves". *The New York Times*. September 12, 2018. https://www.nytimes.com/2018/09/12/business/cbs-les-moonves-board.html. Accessed on September 13, 2018.

16 John Carreyrou. *Bad Blood: Secrets and Lies of a Silicon Valley Start-up*. Knopf 2018. Available here: https://www.amazon.com/Bad-Blood-Secrets-Silicon-Startup/dp/152473165X. Accessed July 8, 2018.

17 Jerry Useem. "Power Causes Brain Damage". *The Atlantic*. June 25, 2017. https://www.theatlantic.com/magazine/archive/2017/07/power-causes-brain-damage/528711/. Accessed on August 23, 2018.

18 Daedalus Trust. Hubris Syndrome. http://www.daedalustrust.com/about-hubris/the-14-symptoms-in-full/. Accessed on September 12, 2018.

19 Editorial Board. "Hubris Is an Ever-Present Risk for High Flying Executives". November 20, 2018. *The Financial Times*. https://www.ft.com/content/69343192-ebf0-11e8-8180-9cf212677a57. Accessed on November 22, 2018.

20 Michael Jarrett. "Six Reasons CEOs Fail". INSEAD Leadership and Organizations. July 26, 2018. https://knowledge.insead.edu/leadership-organisations/six-reasons-ceos-fail-9806. Accessed on August 23, 2018.

21 Harry McCracken. "Satya Nadella Rewrites Microsoft's Code". September 18, 2017. *Fast Company*. https://www.fastcompany.com/40457458/satya-nadella-rewrites-microsofts-code?utm_source=postup&utm_medium=email&utm_campaign=Fast%20Company%20. Accessed on September 13, 2018.

22 Harry McCracken. "Satya Nadella Rewrites Microsoft's Code". September 18, 2017. *Fast Company*. https://www.fastcompany.com/40457458/satya-nadella-rewrites-microsofts-code?utm_source=postup&utm_medium=email&utm_campaign=Fast%20Company%20. Accessed on September 13, 2018.

23 Harry McCracken. "Satya Nadella Rewrites Microsoft's Code". September 18, 2017. *Fast Company*. https://www.fastcompany.com/40457458/satya-nadella-rewrites-microsofts-code?utm_source=postup&utm_medium=email&utm_campaign=Fast%20Company%20. Accessed on September 13, 2018.

24 Information gathered from RepRisk proprietary database. https://www.reprisk.com/. Accessed on April 12, 2019.

25 Sunie Giles. "The Most Important Leadership Competencies, According to Leaders Around the World". *Harvard Business Review*. March 15, 2016. https://hbr.org/2016/03/the-most-important-leadership-competencies-according-to-leaders-around-the-world. Accessed on August 23, 2018.

26 Harvard Business Review Staff. "The Best Performing CEOs in the World 2018". *Harvard Business Review*. November/December 2018. https://hbr.org/2018/11/the-best-performing-ceos-in-the-world-2018. Accessed on April 12, 2019.

27 Lou Solomon. "Becoming Powerful Makes You Less Empathetic". *Harvard Business Review*. April 2015. https://hbr.org/2015/04/becoming-powerful-makes-you-less-empathetic 4/7. Accessed on September 12, 2018.

28 Boris Groysberg, Jeremiah Lee, Jesse Price and J. Yo-Jud Cheng. "The Leader's Guide to Corporate Culture". January/February 2018. *Harvard Business Review*. https://hbr.org/2018/01/the-culture-factor. Accessed on September 13, 2018.

29 Boris Groysberg, Jeremiah Lee, Jesse Price and J. Yo-Jud Cheng. "The Leader's Guide to Corporate Culture". January/February 2018. *Harvard Business Review*. https://hbr.org/2018/01/the-culture-factor. Accessed on September 13, 2018.

30 Boris Groysberg, Jeremiah Lee, Jesse Price and J. Yo-Jud Cheng. "The Leader's Guide to Corporate Culture". January/February 2018. *Harvard Business Review*. https://hbr.org/2018/01/the-culture-factor. Accessed on September 13, 2018.
31 NACD. "Report of the Blue Ribbon Commission on Culture as a Strategic Asset". 2017.
32 Ethics and Compliance Initiative. https://www.ethics.org/. Accessed on August 21, 2018; Ethical Systems. https://www.ethicalsystems.org/. Accessed on August 21, 2018; National Association of Corporate Directors (NACD). https://www.nacdonline.org/. Accessed on August 21, 2018.
33 Ethics and Compliance Initiative Blue Ribbon Panel Report. *High-Quality E&C Programs (HQP) Standards*. 2016. https://www.ethics.org/resources/high-quality-ec-programs-hqp-standards/. Accessed on August 21, 2018.
34 Ethical Systems. *Behavioral Science One Sheets*. https://www.ethicalsystems.org/content/behavioral-science-one-sheets. Accessed on August 21, 2018.
35 Ethical Systems. https://www.ethicalsystems.org/. Accessed on August 21, 2018.
36 Ethical Systems. *Behavioral Ethics One Sheets*. https://www.ethicalsystems.org/content/behavioral-science-one-sheets. Accessed on September 14, 2018.
37 InstituteofBusinessEthics."TheIBE9StepModelforDevelopingandImplementingaCode". https://www.ibe.org.uk/nine-steps-for-preparing-a-new-code/103/52. Accessed on September 14, 2018.
38 NACD Blue Ribbon Commission. "Culture as a Corporate Asset". 2017. https://www.nacdonline.org/files/NACD%20BRC%20Culture%20as%20Corporate%20Asset.pdf. Accessed on August 23, 2018; and NACD "Director Essentials: Strengthening Compliance and Ethics Oversight". March 12, 2018. https://www.nacdonline.org/insights/publications.cfm?itemnumber=21600. Accessed on August 23, 2018.
39 LRN. "What's the Tone at the Very Top? He Role of the Board in Overseeing Corporate Ethics and Compliance". August 2018. http://content.lrn.com/research-insights/whats-the-tone-at-the-very-top-the-role-of-boards-in-overseeing-corporate-ethics-compliance. Accessed on August 21, 2018.
40 A. Bonime-Blanc and J. Brevard. "Ethics and the Board". The Conference Board. 2009.

Part II

Navigating the Scylla and Charybdis of environment, society, governance and technology (ESGT)

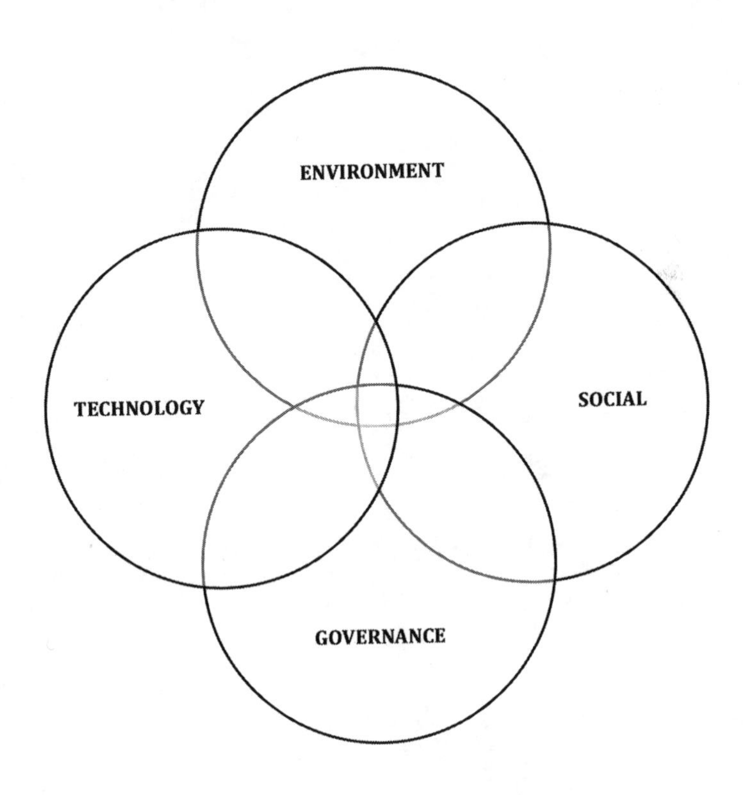

3 Environment

Transforming environmental risk into opportunity from global warming to sustainable resilience

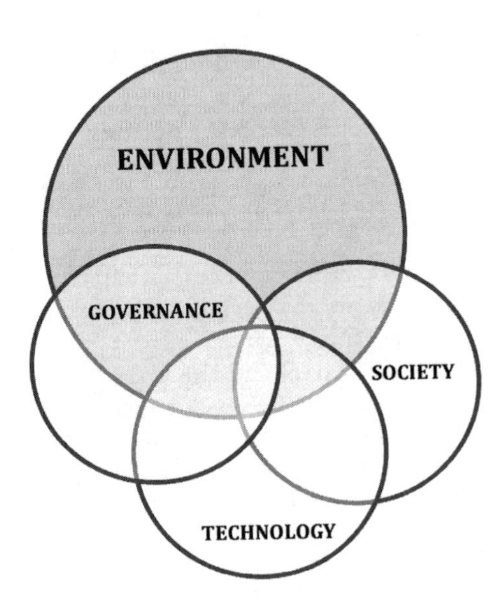

Environment: risk and opportunity

When it comes to environmental risks, Professor Lord Martin Rees has put it in chilling and important terms by stating: "There is no plan B for the world. Our planet is getting more crowded and our climate is warming. Climate change is not under-discussed but it's dismayingly under-acted-upon".[1]

A new global "rock star" of climate-awareness, Greta Thunberg, the amazing 16-year-old Swedish schoolgirl who started a one-person school strike that has lit up the world, put it forcefully and unapologetically to the world's rich and powerful: "I don't want you to be hopeful. I want you to panic. I want you to feel the fear I feel every day. And then I want you to act".[2]

Alarmingly, and in contrast, a climate naysayer, US President Trump, has, since being a candidate, pushed a highly irresponsible, discredited, unscientific and dangerous narrative by making statements like this: "I believe in clean air, immaculate air. But I don't believe in climate change".[3]

On the other side of the coin are the opportunities, encapsulated in this statement by 20th-century cultural anthropologist Margaret Mead: "Never doubt that a small group of thoughtful committed citizens can change the world: indeed, it's the only thing that ever has".[4]

This sentiment is brought home in the statement by impact investor and founder of Domini Impact Investments Amy Domini: "We must continue to stand together to demand that the search for monetary profits not come at the detriment of universal human dignity nor the undermining of ecological sustainability".[5]

Chapter 3: Environment – summary overview

3.1 Introduction and overview

In this chapter, we tackle the "E" in ESGT – the environment. We begin with some fundamentals – definitions of "environment" and "environmental" and a typology of environmental issues derived from several knowledgeable and useful resources.

A common understanding of nomenclature allows us to then place the environment into a clearer context to understand issues, risks, crises and opportunities relating to the environment for leaders of all types of organization. This affords savvy leaders the tools to begin the transformation of their organization's environmental and other ESGT risks into value for stakeholders. Throughout this chapter (and the other three that follow in Part II) we talk constantly about an organization's stakeholders and the need for leaders to know who the most important stakeholders are around their key ESGT issues.

In this chapter we will examine issues and cases and agreements and other things relating to the environment ranging from the Paris Climate Agreement and coal-fired power plants to the Great Pacific Ocean Garbage Patch and plastic-eating bugs and building long-term climate resilience through financial innovation and incentive creation.

The lessons we learn in this chapter and Chapters 4 through 6 – whether good, bad or ugly – will be leveraged in Part III where we will lay out a road map for leaders at every type of organization and at every level of leadership to work together to transform their key ESGT risks into resilience and value for key stakeholders.

3.2 What is "environment"? Definitions and typologies

3.2.a Definitions

To understand the issues embedded in what we call "environmental" risk and opportunity, let's start with some background information on what we mean by "environment", "environmental" and "environmental factors" as defined in various leading dictionaries – moving from the most general to those more specifically relevant to the business world.

What is the "environment"?

From the *Oxford Dictionary*:[6]

> The surroundings or conditions in which a person, animal, or plant lives or operates. The natural world, as a whole or in a particular geographical area, especially as affected by human activity.

From the *Business Dictionary*:[7]

> The sum total of all surroundings of a living organism, including natural forces and other living things, which provide conditions for development and growth as well as of danger and damage.

What is "environmental"?

From the *Oxford Dictionary*:[8]

> relating to the natural world and the impact of human activity in its condition.
>
> acid rain may have caused major environmental damage
>
> relating to or arising from a person's surroundings.
>
> environmental noise.

What is "environmentalism"?

From the *Business Dictionary*:[9]

> A philosophy adhered to by ecologically sensitive advocates who wish to protect the planet from pollution or damage.

What are "environmental factors"?

From an ESG investor analytics perspective:[10]

> Environmental factors include the contribution a company or government makes to climate change through greenhouse gas emissions, along with waste management and energy efficiency. Given renewed efforts to combat

global warming, cutting emissions and decarbonizing is [sic] become more important.

3.2.b Typologies, classifications and categories

Next, let's drill down a little further to understand some of the more important categories of issues that fall under the rubrics of "environment" and "environmental". We do this deeper dive with a specific purpose in mind: to come up with a meaningful survey of some of the key environmental issues that should be on any responsible leader's environmental due diligence list.

What follows is a review of some of the typologies and categorizations of some excellent sources that will help us narrow down a few meaty issues for our own examination later in this chapter. Table 3.1 provides the method to the madness that we are deploying in this chapter to start with some of the broader, more encompassing sources (a planet-wide view as it were) and slowly but surely work our way toward a more focused and granular set of issues and tags.

Table 3.1 On the environment: sources and resources

Level of detail	*Resource*
FROM BIG PICTURE ...	i. Centre for the Study of Existential Risk[11]
	ii. Paris Climate Agreement[12]
	iii. World Economic Forum[13]
	iv. UN Sustainable Development Goals[14]
... TO MORE FOCUSED ...	v. UN Environment Programme[15]
	vi. UN Global Compact[16]
	vii. Allianz Risk Barometer[17]
	viii. MSCI[18]
... TO MORE GRANULAR	ix. RepRisk[19]

Source: Author.

3.2.b.i Centre for the Study of Existential Risk

The Centre for the Study of Existential Risk at Cambridge University (CSER) produces superb research about the end of the world. Literally. This is their mission: "We are dedicated to the study and mitigation of risks that could lead to human extinction or civilizational collapse."[20] And while we are not looking to preach gloom and doom (though there are good reasons to feel trepidation on several of the fronts studied at the CSER), the issues and risks that they identify and document are very much worth examining from the standpoint of the very big picture (the biggest and most all-encompassing) of risks the

Table 3.2 Centre for the Study of Existential Risks, University of Cambridge: Environment-related work[21]

Category of risk	*The challenge*	*Examples of possible extreme risk*	*The opportunities*
Extreme risks and the global environment	"human activity is placing even more pressure on natural processes. Pushing past tipping points might lead to sudden, catastrophic ecosystem collapses or runaway, catastrophic climate change – with dire consequences for human society".	• 10% chance of global warming even higher than projected to 6 degrees centigrade • Biodiversity loss • Ecological collapse – ecosystem collapse • Mass migrations	• Living in the age of the "anthropocene" • Divestment and investment strategies for major investors • Solar radiation management

Source: Centre for the Study of Existential Risks.

planet currently confronts. Table 3.2 provides a summary of some of CSER's environmentally focused research.

CSER's research on "Extreme Risks and the Global Environment" distinguishes several large environmental risks, including the risk of ecosystem collapse, mass migrations and biodiversity loss, and they cite the following research concerning global warming:

> the economists Wagner and Weitzman have used IPCC figures to produce estimates – which are of course highly uncertain – of possible extreme temperature rises. If we continue to pursue a medium-high emissions pathway, they estimate the probability of eventual warming of 6°C is around 10%, and of 10°C is around 3%.[22]

Needless to say, while such probabilities are not extraordinarily high, the mere fact that there is a 10% chance of a 6°C warming makes some of the projections that are already dire and devastating all the amore potentially catastrophic for life on earth.

3.2.b.ii Paris Climate Agreement

Although the 2015 Paris Climate Agreement is more specifically covered later in this chapter, no discussion of environmental typologies, categories and issues would be complete without a high-level summary of the key topics covered in this Agreement.

In essence, the Paris Climate Agreement attempts to incentivize the world to hold global average temperature below 2 degrees centigrade above pre-industrial levels, as soon as possible.

Table 3.3 provides a short summary of the key take-aways of the Agreement.

Table 3.3 Summary of key topics in the Paris Climate Agreement[23]

- **Temperature goals.** "It calls for "holding the increase in the global average temperature to well below 2°c above pre-industrial levels and to pursue efforts to limit the temperature increase to 1.5°c above pre-industrial levels, recognizing that this would significantly reduce the risks and impacts of climate change".
- **Timing.** "Countries should "reach global peaking of greenhouse gas emissions as soon as possible, recognizing that peaking will take longer for developing country parties, and to undertake rapid reductions thereafter".
- **Adverse effects protection**. "The agreement acknowledges "the importance of averting, minimizing and addressing loss and damage associated with the adverse effects of climate change." This was deemed crucial by poor and small-island countries that suffer the most from extreme weather and from long-term impacts like droughts".
- **Continuous progress updates**. "Ahead of the agreement, 186 countries submitted plans detailing how they will reduce their greenhouse gas pollution through 2025 or 2030. The agreement requires all countries to submit updated plans that would ratchet up the stringency of emissions by 2020 and every five years thereafter, a time frame that the United States and the European Union urged. India had initially sought a 10-year review cycle".
- **Global stock-taking 2023 and beyond.** "The deal requires a global "stock-take" – an overall assessment of how countries are doing in cutting their emissions compared to their national plans – starting in 2023, every five years".
- **Greenhouse gas emissions reporting**. "The deal requires countries to monitor, verify and report their greenhouse gas emissions using the same global system".
- **Capacity-building initiative for transparency**. "The agreement sets up something called a "capacity-building initiative for transparency" to help developing countries meet a new requirement that they regularly provide a national "inventory report" of human-caused emissions, by source, and track their progress in meeting their national goals".
- **US$100B collective quantified goal 2020**. "The agreement, which takes effect in 2020, calls on nations to establish'a new collective quantified goal' of at least $100 billion a year in climate-related financing by 2020".
- **Commit to highest possible ambition**. "When countries update their commitments, they will commit to the 'highest possible ambition,' but the agreement does not set a numeric target". It acknowledges "common but differentiated responsibilities and respective capabilities, in the light of different national circumstances".

Source: The Paris Climate Agreement 2015.

The Paris Climate Agreement becomes effective in 2020. Since its signing in late 2015, the world has faced a new geopolitical elephant in the room – the fact that in 2017 the new US President Donald Trump withdrew the US from the Agreement and continues to this day to deny climate change and man's role in it.

The obstacles to the US eventually rejoining can still be removed, as there are plenty of stakeholders in the US and globally that are working toward the Agreement's goals. The commitment to the Paris Climate Agreement goals is especially visible in some of the larger states in the US (including California and New York, which account for a total of about 50 million of the 330 million people living in the US) as well as with forward-thinking businesses that have continued to operate on the longer-term assumption that the US will at some point return to the Agreement.

Plus, smart businesses and investors know that a lot of money is to be made from innovation associated with combating climate change. According to the Global Opportunity Report of 2018, no less than US$12 trillion of new business is associated with new products and services relating to achieving the Sustainable Development Goals (SDGs) (which of course include a number of climate, environmental and energy related goals).[24]

One can only hope that cooler heads will prevail soon in the US, given our currently geopolitically turbulent and literally hotter planet. And there is hope – in late 2018, a broad cross section of US federal agencies published a mandated report to the US Congress and the American people about climate change, and it is a devastating report – despite the climate denials coming from a highly uninformed or deliberately naysaying US president. Moreover, while the US may have a president who is oblivious, even antagonistic, to the challenge, 72% of Americans – the most ever – now believe that climate change is man-made and the cause of global warming.[25]

3.2.b.iii World Economic Forum

Also, from a big-picture but less existential and more practical standpoint, the World Economic Forum (WEF) has been producing an annual Global Risks Report (GRR) for over a dozen years. It always provides riveting information and a truly global road map of five categories of global risk that may have direct or indirect relevance and/or impact on a given organization's – whether company, government agency or NGO's – activity and footprint.

The WEF GRRs serve as an excellent big-picture framework for leaders to map out the larger risks and start the process of zooming into the more relevant, specific and granular issues and risks that their specific organization may be facing. The WEF divides its big categories of risk into the following five:

1. Economic
2. Environmental

3. Geopolitical
4. Societal
5. Technological

In this book, we are examining ESGT issues, risks and opportunities, which leaves out the "economic" and, ostensibly, the "geopolitical" categories (but in the latter case, not really so). Let me explain. The reason we are not covering the field of economics is hopefully obvious: it is an already saturated field full of great content and many experts dating back not just decades but centuries. More humorously, however, economics is above this author's paygrade.

In the case of the WEF's "Geopolitical Risk" category, this book actually covers this category of risk but within the larger category of "governance". Why? Because geopolitical risks by definition have something to do with politics and political decision-making and something to do with geography:

- The "politics" part of the equation has everything to do with governance at various levels of government – global intergovernmental institutions like the World Bank or the United Nations (UN) and its agencies, regional governmental entities like the European Union (EU) in Europe or the Association of Southeast Asian Nations (ASEAN) in Asia or the Organization of American States (OAS) in the Americas, national-level governments for each and every one of the 200 plus nations that currently exist, as well as the full array of state, regional, provincial, local and municipal governments we all live under.
- And the "geography" part of "geopolitics" has to do with the geographic location or locations in which international, national and local governance (or the lack thereof) may or may not take place, including a variety of other national, transnational or extra-national geopolitical actors (terrorist groups, international crime and non-nation state cyber-actors). We will explore this theme further in Chapter 5 on governance.

Like many risk assessments, the WEF GRRs identify risks in terms of greatest likelihood as well as greatest impact. Very significantly, in the two most recent reports available at the time of this writing (the 2018 and 2019 GRRs issued in January 2018 and 2019, respectively), the WEF has identified the most ever environmental risks among the top-ten most likely and highest impact risks of all types in its 11 years of producing GRRs (2008–2019).

Table 3.4 says it all – in 2008, the top-five perceived greatest risks to mankind were dominated by economic and geopolitical risks (in fact, no environmental risk made the top five of either most likely or highest impact global risks). In 2013, two environmental issues were identified as most likely and only one as highest impact.

Table 3.4 World Economic Forum – evolving risk landscape 2008–2019[26]

	2008	*2013*	*2018*	*2019*
TOP-FIVE GLOBAL RISKS IN TERMS OF LIKELIHOOD	ASSET PRICE COLLAPSE	SEVERE INCOME DISPARITY	EXTREME WEATHER EVENTS	EXTREME WEATHER EVENTS
	MIDDLE EAST INSTABILITY	CHRONIC FISCAL IMBALANCES	NATURAL DISASTERS	FAILURE OF CLIMATE-CHANGE MITIGATION AND ADAPTATION
	FAILED AND FAILING STATES	RISING GREENHOUSE GAS EMISSIONS	CYBERATTACKS	NATURAL DISASTERS
	OIL AND GAS PRICE SPIKE	WATER CRISES	DATA FRAUD OR THEFT	DATA FRAUD OR THEFT
	CHRONIC DISEASE, DEVELOPED WORLD	MISMANAGEMENT OF POPULATION AGEING	FAILURE OF CLIMATE CHANGE MITIGATION AND ADAPTATION	CYBERATTACKS
TOP-FIVE GLOBAL RISKS IN TERMS OF IMPACT	ASSET PRICE COLLAPSE	MAJOR SYSTEMIC FINANCIAL FAILURE	WEAPONS OF MASS DESTRUCTION	WEAPONS OF MASS DESTRUCTION
	RETRENCHMENT FROM GLOBALIZATION (DEVELOPED)	WATER CRISES	EXTREME WEATHER EVENTS	FAILURE OF CLIMATE-CHANGE MITIGATION AND ADAPTATION
	SLOWING CHINESE ECONOMY	CHRONIC FISCAL IMBALANCES	NATURAL DISASTERS	EXTREME WEATHER EVENTS
	OIL AND GAS PRICE SPIKE	WEAPONS OF MASS DESTRUCTION	FAILURE OF CLIMATE CHANGE MITIGATION AND ADAPTATION	WATER CRISES
	PANDEMICS	FAILURE OF CLIMATE CHANGE MITIGATION AND ADAPTATION	WATER CRISES	NATURAL DISASTERS

WEF RISK CATEGORY KEY: white = environmental; lightest gray = economic; light gray = geopolitical; medium gray = societal; dark gray = technological.

Source: World Economic Forum.

In 2018, however, the picture changed dramatically: environmental risks became by far the most numerous and most significant risks facing humanity, with three of the top-five most likely and three of the top-five most impactful risks as follows:[27]

- In terms of *Likelihood*:
  - #1 Extreme weather events.
  - #2 Natural disasters.
  - #5 Failure of climate-change mitigation or adaptation.

- In terms of *Impact*:
    - #2 Extreme weather events.
    - #3 Natural disasters.
    - #4 Failure of climate change mitigation or adaptation.

This 2018 picture became what might be a trend when one looks at what happened in 2019: just like in 2018, the top-five most likely risks included three environmental and two cyber-related, and among the top most high-impact risks – all of the risks other than the top one in both years (weapons of mass destruction) had to do with the environment one way or another, including water crises (though categorized by WEF as social).

Moreover, in both 2018 and 2019, the top-three (four, if you count water crises) environmental risks identified in terms of greatest likelihood and greatest impact are the same, meaning that these are super-charged risks, as they are as likely to happen and have the greatest impact in terms of downsides and risk to humanity.

And that's not all. Among the next five most impactful global risks identified in 2018 are two more environmental risks:

- In terms of *Likelihood*:
    - #7 Man-made environmental disasters.
- In terms of *Impact*:
    - #8 Biodiversity loss.

And finally, there are several additional risks the WEF 2018 and 2019 GRRs identify which, while not strictly categorized as "environmental" and instead fall under the "societal" category, are nevertheless greatly driven by environmental conditions and factors as well. Among them in 2018 were the following:

- In terms of *Likelihood*:
    - #6 Large-scale involuntary migration.
- In terms of *Impact*:
    - #5 Water crises.
    - #7 Food crises.
    - #9 Large-scale involuntary migration.
    - #10 Spread of infectious diseases.

If the actual extreme weather events, natural disasters and failures of climate-change mitigation or adaptation that have been showing up in the past couple of years are any indicator, not only is the WEF right on the money of what is going down for the planet from a risk standpoint but these alarm bells should be loudly and rapidly heeded by government, business and NGOs worldwide.

3.2.b.iv UN Sustainable Development Goals

If we turn to the United Nations Sustainable Development Goals (SDGs), we can also glean a preponderance of environmental issues as well as a number of social issues deeply interrelated with environmental issues and risks.

In Table 3.5 we highlight which of the SDGs are primarily environmental, though as any informed observer would note, one could make the argument that there are environmental aspects to almost every SDG.

Table 3.5 United Nations Sustainable Development Goals: environmental issues

SDG – primarily environment-focused	*E*	*S*	*G*
1. No poverty		X	X
2. *Zero hunger*	X	X	X
3. Good health and well-being		X	X
4. Quality education		X	X
5. Gender equality		X	X
6. *Clean water and sanitation*	X	X	X
7. *Affordable and clean energy*	X	X	X
8. Decent work and economic growth		X	X
9. *Industry, innovation and infrastructure*	X	X	X
10. Reduced inequalities		X	X
11. *Sustainable cities and communities*	X	X	X
12. *Responsible consumption and production*	X	X	X
13. *Life below water*	X		X
14. *Life on land*	X		X
15. *Climate action*	X	X	X
16. Peace and justice – strong institutions		X	X
17. *Partnerships for the goals*	X	X	X

Source: United Nations.

3.2.b.v UN Environment Programme

Looking more closely at strictly environmental issues and risks, we can learn something from the UN Environment Programme typology of issues in Table 3.6, which provides a useful view of the key global environmental issues that they are watching and researching.[28]

Table 3.6 UN Environment Programme: list of key environmental issues

- Air
- Biosafety
- Chemicals and waste
- Cities and lifestyles
- Climate change
- Ecosystems
- Education & training
- Energy
- Environment under review
- Environmental right and governance
- Extractives
- Forests

(*Continued*)

Table 3.6 Continued

- Gender
- Green economy
- Oceans and seas
- Resource efficiency
- Sustainable development goals
- Technology
- Transport
- Water

Source: United Nations Environment Programme.

To illustrate some of these issues more concretely, the UN Environment Programme suggests that some of the key environmental stories to watch are the following:[29]

- **Coral Reefs**, with three-quarters of the world's coral reefs at risk, the International Coral Reef Initiative called 2018 "The International Year of the Reef".
- **Plastic Pollution**, with a major emphasis being globally to reduce the use of plastic bags, straws, etc.
- **Greening the World of Sport**, with a focus on sustainability programs within the world of sport.
- **Environment and Migration**, with "climate change and environmental degradation … already officially recognized as drivers of migration – a fact that's driven home by the climate-related disasters", such as drought-plagued parts of Africa, war-torn Yemen and post-Hurricane Maria Puerto Rico, Dominica and the US and British Virgin Islands.

3.2.b.vi UN Global Compact

Turning to a more business-focused environmental issue lens (which can easily apply to any other type of organization where its leaders care about environmental issues they may have an impact on), let's turn in Table 3.7 to the UN Global Compact's three basic environment-related principles, which are primarily aspirational exhortations for business to exercise restraint, precaution, risk management and corporate responsibility and find innovative solutions through environmentally friendly technologies.[31]

Table 3.7 United Nations Global Compact[30]: three principles related to environmental issues

Principle 7	Businesses should uphold a precautionary approach to environmental challenges.
Principle 8	Businesses should undertake initiatives to promote greater environmental responsibility.
Principle 9	Businesses should encourage the development and diffusion of environmentally friendly technologies.

Source: United Nations Global Compact.

The UN Global Compact has developed a number of useful resources that, while mostly targeted at the business world, are equally useful to other types of organizations that have organizational environmental issues (governmental agencies, universities, nonprofits, etc.). Among some of these useful resources are guidance on developing an environmental stewardship strategy, guidance for corporate action on biodiversity and ecosystems and a framework for developing sustainable energy solutions.[32] Finally, they also have a useful guide for organizations desirous of aligning their internal policies with either or both the UN Global Compact Principles and the SDGs that help to tie each SDG to actual measurable goals and objectives.[33]

3.2.b.vii Allianz Risk Barometer

At a more granular level focused mostly on the business world (but again in my opinion useful to any type of organization that is environmentally conscious and/or has environmental issues), the annual Allianz Risk Barometer provides another useful tool to get the pulse of the top risks and challenges facing global organizations each year. Table 3.8 is a distillation of environmentally related risks reported in the Allianz Risk Barometer top-ten risks from 2017 to 2019.

Table 3.8 Allianz Risk Barometer top-ten business risks 2017–2019[34]: focus on environmental issues

Top-ten risks (w/environmental issues)	*2019 Ranking*	*2018 Ranking*	*2017 Ranking*	*ESGT Issue*
Business interruption (including supply chain disruption)	#1	#1	#1	E, S, G, T
Natural catastrophes (e.g., storm, flood, earthquake)	#3	#3	#4	E, S
Fire, explosion	#6	#6	#7	E, S
Climate change/ increasing volatility of weather	#8	#10	N/A	E, S

Source: Allianz.

Very telling from this particular risk ranking is the fact that a whole new category of environmentally related risks broke into the top ten for the first time as #10 for 2018: "climate change, increasing volatility of weather" and moved up two notches to #8 in 2019.

The likelihood of the climate change risk staying put or even moving up the risk ladder is high, especially in light of the records being broken for heat in the summer of 2018 in the northern hemisphere. Witness the following headlines from the summer of 2018:[35]

- In North America: Los Angeles set an all-time high temperature record of 111°F on July 6. Montreal, Canada also set its *all-time high temperature record*, during a deadly Quebec heat wave in early July. This week, Death Valley, California, has broken three straight daily records with a high of 127°F.
- In Europe: Unprecedented heat led to a wildfire outbreak in Scandinavia, and record highs have been set all the way above the Arctic Circle this month. According to the U.N., Sodankyla, Finland hit 89.2°F, or 31.8°C, on July 17, which was an all-time record for that location.
- Friday was the *hottest temperature on record* in Amsterdam, at 34.8°C, or 94.6°F.
- Remarkably, in northern Norway, Makkaur, set a new record high overnight low temperature of 25.2°C, or 77°F, on July 18.
- Heat records have also fallen in the U.K., Ireland and France. In London, high temperatures hit 35°C on Thursday, and were forecast to potentially eclipse that on Friday. The U.K. is suffering through one of its driest years on record.
- In the Middle East: Quriyat, Oman, which likely set *the world's hottest low temperature ever recorded* on June 28, when the temperature failed to drop below 109°F, or 42.8°C.
- In Africa: Ouargla, Algeria, may have set Africa's all-time highest temperature on July 5, with a reading of 124.3°F, or 51.3°C.
- In Asia: *Japan set a national temperature record* of 106°F, or 41.1°C, in a heat wave that followed deadly floods.

And, sadly, those were not the only climate-related disasters that occurred later in 2018 and into 2019. There were also some of the biggest typhoons ever during this time and the most devastating and largest fires ever in California. At the time of final production of this book in the summer of 2019, the 2018 temperature breaking records had been broken once again with July 2019 being declared the hottest month globally ever recorded.[36]

3.2.b.viii MSCI

Table 3.9 offers ESG issue information from the ESG analytics firm, MSCI. It is important to note that the perspective of MSCI is that of an analytics firm that serves the interests and needs of a particular stakeholder – the investor – and as such should be considered with that lens in mind, in other words, the lens of a stock buyer, institutional investor or asset manager.

A significant aspect of this particular typology is that we see not only the issues and the risks but also the opportunities. At the end of the day, the perspective of an investor will always be: why should I buy or sell the stock of this company? Why should I include this company in my asset portfolio? And thus, the idea of opportunity is just as significant as the idea of risk.

Indeed, as we will discuss later in this chapter (as well as in Chapters 4, 5 and 6), any stakeholder (whether an investor, an employee, a customer or a regulator) in any kind of entity (whether for-profit, nonprofit or governmental) should always deploy a risk-to-opportunity lens and ask questions about not only the risks involved in being associated with (or having a stake in) an organization but also the opportunities and potential value associated with that risk or stake and the role the stakeholder might be able to play in moving that risk through the value chain to create opportunity and value.

Table 3.9 MSCI categories of ESG issues[37]: environmental issues

Climate change:
- Carbon emissions
- Product carbon footprint
- Financing environmental impact
- Climate change vulnerability

Natural resources:
- Water stress
- Biodiversity and land use
- Raw material sourcing

Pollution and waste:
- Toxic emissions & waste
- Packaging material & waste
- Electronic waste

Environmental opportunities:
- Opportunities in clean tech
- Opportunities in green building
- Opportunities in renewable energy

Source: MSCI.

3.2.b.ix RepRisk

On a more granular level, RepRisk is a Swiss-based company and a global leader and pioneer in data science, specializing in ESG and business conduct risk research and quantitative solutions. They have developed very useful ESG risk metrics and analytics that dynamically capture risk information related to ESG issues from a broad range of sources. They screen over 80,000 media and stakeholder sources every day in 20 languages. The methodology is issues- and event-driven rather than company-driven – i.e., RepRisk screens sources and stakeholders external to a company for ESG risk incidents in accordance with the RepRisk research scope of 28 ESG Issues.[38]

Among other proprietary risk metrics developed by RepRisk, the RepRisk Index (RRI) is a quantitative measure (0 to 100) of a company's risk exposure to ESG reputational and business conduct risk.[39] In their research scope, derived from key international standards related to ESG and mapping to the ten principles of the UN Global Compact, RepRisk identifies the environmental issues highlighted in Table 3.10.[40]

Table 3.10 RepRisk ESG research scope: environmental issues

Environmental footprint

- Climate change, GHG emissions, and global pollution
- Local pollution
- Impacts on landscapes, ecosystems and biodiversity
- Overuse and wasting of resources
- Waste issues
- Animal mistreatment

Cross-cutting issues (ESG)

- Controversial products and services
- Products (health and environmental issues)
- Violation of international standards
- Violation of national legislation
- Supply chain issues

Source: RepRisk.

Drilling down a little deeper, Table 3.11 shows some of the RepRisk Topic Tags, ESG "hot topics" that are an extension of RepRisk's core research scope and that can be used to apply a finer layer of granularity.

Table 3.11 Sampling of RepRisk ESG topic tags with environmental implications[41]

Abusive/illegal fishing
Alcohol
Animal transportation
Arctic drilling
Asbestos
Automatic and semi-automatic weapons
Biological weapons
Chemical weapons
Cluster munitions
Coal-fired power plants
Coral reefs
Deep sea drilling
Depleted uranium munitions
Drones
Endangered species
Forest burning
Fracking
Fur and exotic animal skins
Gambling

(*Continued*)

Table 3.11 Continued

Genetically modified organisms (GMO)
High conservation value forests
Hydropower (dams)
Illegal logging
Land mines
Marijuana/cannabis
Monocultures
Mountaintop removal mining
Nuclear power
Nuclear weapons
Offshore drilling
Oil sands
Palm oil
Pornography
Predatory lending
Protected areas
Rare earths
Sand mining/dredging
Seabed mining
Ship breaking and scraping
Soy
Tobacco
Water scarcity

Source: RepRisk.

To illustrate what we mean by reputation risk associated with environmental issues or environmental reputation risk, Table 3.12 presents an overview of the "Most Controversial Companies" compiled by RepRisk for 2017 based on their Peak RRI value in the given year involving primarily environmental issues.

Table 3.12 RepRisk most controversial companies 2017[42]: environment-related

MCC 2017 rank	*Company name*	*Peak Reprisk index 2017 (1–100)*[43]	*Sector*	*Country of head-quarters*	*Primary ESGT issue(s)*
#2	Kobe Steel	87	Industrial metals	Japan	E & G
#5	Stalreiniging Barneveld (Chickfriend)	82	Support services (industrial goods and services)	Netherlands	E S & G
#8	Odebrecht Sa	74	Construction and materials	Brazil	E S & G
#9	Petroleos De Venezuela Sa	73	Oil and gas	Venezuela	E S & G

Source: RepRisk.

3.2.c The way forward

This review of definitions, types, typologies and classifications of all things "environmental" now equips us to turn to several big environmental developments and issues, including case studies to deploy some of this knowledge to help leaders begin to think about how to select and frame the key environmental issues, risks, opportunities and value relating to their organizations.

Up next are the following environmental topics:

- We first turn to the topic of climate change – perhaps the single most existential and important topic in the world today. In our review of this topic we deal with some of the big geopolitical risks that are currently intertwined with climate action.
- Next we look at one of the critical environmental issues of our day – the exponential growth in plastic waste and how it is invading our oceans and what some countries, businesses and NGOs are doing to meet this enormous global challenge.
- Finally, we close this chapter with some of the good news stories about what a cross section of society, government and business are doing to build the resilience and mitigation strategies necessary to protect our urban and coastal areas and how to incentivize resilient behaviors in the marketplace.

3.3 Leaders navigating environmental risks, crises and opportunities

In this chapter and the three chapters that follow, we review examples from the global marketplace that illustrate the issues at the center of each chapter – here we are looking at environmental issues. The aim is to get the reader to start thinking about these issues as both manageable and important.

More specifically, for each of the four big categories of ESGT issues, risks and opportunities that we examine in this book, we wish to constantly underscore the fact that leaders have the responsibility to step up to the plate, first, to understand these issues as they relate to their organizations and, second, to lead in addressing and making decisions regarding these issues that are in the best interests of their constituents and stakeholders. That's what it means to lead, not for personal, individual, limited gain or greed, but for the betterment and longer-term good of those who hold a stake in the development and outcomes of these issues – whether they are shareholders, beneficiaries, employees, partners, suppliers, regulators, communities, etc.

3.3.a Transforming climate change and global warming into a sustainable future

3.3.a.i The state of global environmental geopolitics

From *The Economist*, August 2, 2018, cover story:

> In the Line of Fire: The World is Losing the Fight Against Climate Change: Rising energy demand means use of fossil fuels is heading in the wrong direction.[44]

> The Earth is smoldering. From Seattle to Siberia this summer, flames have consumed swathes of the northern hemisphere. One of 18 wildfires sweeping through California, among the worst in the state's history, is generating such heat that it created its own weather. Fires that raged through a coastal area near Athens last week killed 91. Elsewhere people are suffocating in the heat. Roughly 125 have died in Japan as the result of a heatwave that pushed temperatures in Tokyo above 40°C for the first time.[45]

And that was before the deadliest and largest California wildfires to ever consume California and the US that took place in November 2018.

Also ripped from the headlines of *The New York Times* during the summer of 2018:

- "UK 'Heat Wave'? Irish 'Drought'? Unfamiliar Words for Unfamiliar Times". July 4, 2018.
- "Sweden's Tallest Peak Shrinks in Record Heat". August 2, 2018.
- "'Furnace Friday:' Ill-Equipped for Heat, Britain has a Meltdown". July 27, 2018.
- "Europe Swelters Under a Heat Wave Called 'Lucifer'". August 6, 2018.
- "Heat Wave Scorches Sweden as Wildfires Rage in the Arctic Circle". July 19, 2018.
- "Amid Europe's Heat Wave Rare Flamingos Lay First Eggs in 15 Years". August 11, 2018.

In case there was any doubt that climate change is a topic of deep and broad concern and disagreement (even among those who agree that it's happening) about how it should be tackled, witness the following exchange of views and counterviews within the space of one week in August 2018: on the one hand, a provocative *New York Times Magazine* 30,000 word piece titled "Losing Earth: The Decade We Almost Stopped Climate Change" argues that we had the opportunity to address and reverse climate change in the 1980s but squandered that opportunity, which has placed us at the edge of climate disaster today. On the other hand, a *Fast Company* article "A Guide To What *The Times* Bold, Flawed Climate Change Story Left Out" subtitled appropriately:

> *The New York Times Magazine* devoted 30,000 words to why there's been so little progress on climate change. Many environmental writers saw something was missing: blame.

The *Fast Company* article goes on to summarize the views of several environmental activists and experts who disagree for one reason or another with the author of the *NYT Magazine* piece. Their critiques range from the case that the

1980s wasn't a perfect time to implement change because the energy industry's political lobbying made it impossible to the observation that "Capitalism not 'human nature' is what killed our momentum on climate change" and the 'deregulatory ethos of neoliberalism'".[46]

Despite these internal disagreements within the climate change expert community, if there ever was a broad and hard-won consensus achieved by cross-sectoral global leaders on a critical global risk, the Paris Climate Agreement of 2015 can be said to have achieved that consensus. And this, despite the fact that in 2017 the leader of the most powerful and largest economy in the world, US President Donald J. Trump, unilaterally (and many would say irrationally and to the detriment of America and the rest of the world), withdrew the US from the Climate Agreement.

The Climate Agreement focuses on the following key objectives:

(a) Holding the increase in the global average temperature to well below 2°C above pre-industrial levels and to pursue efforts to limit the temperature increase to 1.5°C above pre-industrial levels, recognizing that this would significantly reduce the risks and impacts of climate change;
(b) Increasing the ability to adapt to the adverse impacts of climate change and foster climate resilience and low greenhouse gas emissions development, in a manner that does not threaten food production;
(c) Making finance flows consistent with a pathway towards low greenhouse gas emissions and climate-resilient development.

This was and continues to be a truly global, combined effort of leaders and organizations from government, business and society to address what many consider to be the ultimate existential risk to the planet that humans may still have an ability to influence (both negatively and positively) – unimpeded global warming, threatening the lives, livelihoods and well-being of peoples and life everywhere.

As Jeffrey Sachs succinctly put it in his Project Syndicate article of August 2, 2018: "We Are All Climate Refugees Now", followed by this paragraph:[47]

> Modern humans, born into one climate era, called the Holocene, have crossed the border into another, the Anthropocene. But instead of a Moses guiding humanity in this new and dangerous wilderness, a gang of science deniers and polluters currently misguides humanity to ever-greater danger. We are all climate refugees now and must chart a path to safety.

And therein lies the crux of the matter – we know the risk, and the broad consensus is in about global warming and climate change and the fact that humans can still make a difference. And yet there are challenges, threats, misbehaviors and worse blocking the way of resilient and conscientious leaders trying to transform climate change risk into value – i.e., saving the planet for continued life on earth.

Among the consequences of global warming are other impacts such as the growing desertification of parts of Africa affecting the well-being of large land-locked populations and involving additional severe risks of famine, warfare and large involuntary migrations. Another consequence of global warming is the melting of the planet's icecaps and related higher water levels that may be slowly but surely surging along global coasts, directly affecting and potentially endangering the well-being, economies, survivability and existence of small island nations in the Pacific and the sustainability of some of the largest urban coastal communities around the world.[48]

Indeed, studies show that because of increased temperatures, the countries most at risk for climate change will be those that are least able to manage some of the gargantuan challenges. And "companies operating in or sourcing from" those regions "face escalating risks, new research has revealed".[49]

The Paris Climate Agreement was reached by 194 countries after a long and winding road of blood, sweat and tears by many, including countries that are rapidly growing and still have an incentive to use dirty and relatively cheap sources of energy like coal to fuel their economies, and which are available in their respective backyards – among them are China (the largest emitter of carbon dioxide by far) and India. Though it should be noted that there is a difference between both countries' current approaches to this issue:

> According to a November 2017 study, India is set to surpass China to become the world's largest emitter of sulfur dioxide, a toxic air pollutant that has been blamed for India's current haze problem.[50]

Why would the leadership of these two countries sign the Paris Climate Agreement and look to transform their pollution and related health risks to greater social value? One powerful explanation – especially in the case of China – is that they are listening to their principal stakeholders – their populations, their citizens – who, while benefiting from economic growth, do not benefit from increasing health concerns and reduced life expectancies occasioned by heavy air pollution causing early death, issues that are mainly affecting these countries' most populated areas such as New Delhi, Shanghai and Beijing.

Part of the incentive for countries like India and China to sign an agreement that could be seen to curb their economic freedom to grow unfettered, is that their leaders are gauging the externalities of uncontrolled growth (e.g., pollution) and are making policy choices that balance the interests of economic growth, on the one hand, against an increasingly unhealthy, unhappy and potentially rebellious population, on the other. The tension and the dilemma are clear, as this study signals:

> While it has been claimed that 300 million Indians will be lifted out of energy poverty if India carries through with its plans to build nearly 370 coal-fired power plants, environmentalists warn that hundreds of millions

> of people will be subjected to harmful levels of airborne pollutants, constituting a public health crisis.[51]

In contrast to the longer-term, more strategic policy-making of leaders in China and India, there is no reasonable explanation or rationale behind US President Trump's decision to abandon the Paris Climate Agreement, as arguments in its favor are powerful across the board – including for the business community that is poised to innovate and intent on making a killing on new products and services if they're only allowed to.

To explain this decision-making at the US government level, the analysis of risk and opportunity and the effect on stakeholders has to be made through a different lens. That lens is the very particular lens of the style of leadership and culture that President Trump represents – one that is driven by short term goals and often zigzags between policies, is focused on pleasing large donors and the anti-environmental lobby and doesn't understand either the past or the future of the impact on key stakeholders and policy planning. This short-sighted attitude ignores the economic well-being and health of citizens, on the one hand, and the fact that tackling climate change proactively can bring enormous opportunities for business innovation and growth, on the other.

In late 2018, the Trump administration canceled much of President Obama's legacy of stricter environmental emission standards for the auto industry, as reflected in this headline: "Trump Unveils His Plans to Weaken Car Pollution Rules".[52] And my hope (and I'm pretty sure that of most stakeholders – from shareholders and employees to customers and communities) is that the responsible (and I would argue business savvy) auto companies see beyond the short-term gains such a reversal might allow and understand that, in the longer term, most key stakeholders will support an environmentally more conscientious set of government standards and policies.

Also in late 2018, despite the climate denial taking place at the highest levels of US federal power, facts, reason and science continue to emerge: the *US 2018 Climate Change Report*, mandated by the US Congress and co-written by a dozen federal government agencies and their scientists and peer reviewed by the best of the best in climate science under the auspices of the US Global Change Research Program, issued the Fourth National Climate Assessment – Volume II Impacts, Risks, and Adaptation in the US (*US 2018 Climate Report*).

And this *US 2018 Climate Report* paints a dire and dark picture, as follows:

> Earth's climate is now changing faster than at any point in the history of modern civilization, primarily as a result of human activities. The impacts of global climate change are already being felt in the United States and are projected to intensify in the future—but the severity of future impacts will depend largely on actions taken to reduce greenhouse gas emissions and to adapt to the changes that will occur.

> Climate-related risks will continue to grow without additional action. Decisions made today determine risk exposure for current and future generations and will either broaden or limit options to reduce the negative consequences of climate change. While Americans are responding in ways that can bolster resilience and improve livelihoods, neither global efforts to mitigate the causes of climate change nor regional efforts to adapt to the impacts currently approach the scales needed to avoid substantial damages to the U.S. economy, environment, and human health and well-being over the coming decades.

Table 3.13 contains a summary of the top findings of the *US 2018 Climate Report*. While this is the latest US government report on this topic, a reading of the main findings might as well apply to almost any nation on earth.

Table 3.13 US 2018 Climate Report: principal findings

1. **Communities**
   Climate change creates new risks and exacerbates existing vulnerabilities in communities across the US, presenting growing challenges to human health and safety, quality of life, and the rate of economic growth.
2. **Economy**
   Without substantial and sustained global mitigation and regional adaptation efforts, climate change is expected to cause growing losses to American infrastructure and property and impede the rate of economic growth over this century.
3. **Interconnected impacts**
   Climate change affects the natural, built, and social systems we rely on individually and through their connections to one another. These interconnected systems are increasingly vulnerable to cascading impacts that are often difficult to predict, threatening essential services within and beyond the Nation's borders.
4. **Actions to reduce risks**
   Communities, governments, and businesses are working to reduce risks from and costs associated with climate change by taking action to lower greenhouse gas emissions and implement adaptation strategies. While mitigation and adaptation efforts have expanded substantially in the last four years, they do not yet approach the scale considered necessary to avoid substantial damages to the economy, environment, and human health over the coming decades.
5. **Water**
   The quality and quantity of water available for use by people and ecosystems across the country are being affected by climate change, increasing risks and costs to agriculture, energy production, industry, recreation, and the environment.
6. **Health**
   Impacts from climate change on extreme weather and climate-related events, air quality, and the transmission of disease through insects and pests, food, and water increasingly threaten the health and well-being of the American people, particularly populations that are already vulnerable.

(*Continued*)

Table 3.13 Continued

7. **Indigenous peoples**
Climate change increasingly threatens indigenous communities' livelihoods, economies, health, and cultural identities by disrupting interconnected social, physical, and ecological systems.
8. **Ecosystems and ecosystem services**
Ecosystems and the benefits they provide to society are being altered by climate change, and these impacts are projected to continue. Without substantial and sustained reductions in global greenhouse gas emissions, transformative impacts on some ecosystems will occur; some coral reef and sea ice ecosystems are already experiencing such transformational changes.
9. **Agriculture and food**
Rising temperatures, extreme heat, drought, wildfire on rangelands, and heavy downpours are expected to increasingly disrupt agricultural productivity in the US. Expected increases in challenges to livestock health, declines in crop yields and quality, and changes in extreme events in the US and abroad threaten rural livelihoods, sustainable food security, and price stability.
10. **Infrastructure**
Our Nation's aging and deteriorating infrastructure is further stressed by increases in heavy precipitation events, coastal flooding, heat, wildfires, and other extreme events, as well as changes to average precipitation and temperature. Without adaptation, climate change will continue to degrade infrastructure performance over the rest of the century, with the potential for cascading impacts that threaten our economy, national security, essential services, and health and well-being.
11. **Oceans and coasts**
Coastal communities and the ecosystems that support them are increasingly threatened by the impacts of climate change. Without significant reductions in global greenhouse gas emissions and regional adaptation measures, many coastal regions will be transformed by the latter part of this century, with impacts affecting other regions and sectors. Even in a future with lower greenhouse gas emissions, many communities are expected to suffer financial impacts as chronic high-tide flooding leads to higher costs and lower property values.
12. **Tourism and recreation**
Outdoor recreation, tourist economies, and quality of life are reliant on benefits provided by our natural environment that will be degraded by the impacts of climate change in many ways.

Source: *US 2018 Climate Report.*

Put this *US 2018 Climate Report* together with the other dire warnings issued by the United Nations and leading scientific institutions and scientists, and it is more crystal-clear than ever that the clock is ticking loudly for the world to forcefully address and take action to reverse the worst trends and consequences of climate change now – time is running out.

3.3.a.ii Spotlight on the coal-fired power industry

Now let's take a look at what happens within a business sector – the coal-fired power industry – in terms of leadership, risk, crisis and opportunity on the issue of climate change.

According to data quoted in the RepRisk Special Report on Coal-Fired Power Plants (2018),[53] coal-fired plants account for 40% of the electricity produced worldwide. Additionally, the report states that since 2015, it has "recorded a steady increase in the number of coal-fired power plants facing the threat of divestment and retirement, which is indicative of a worldwide trend toward cleaner energy and a shift from coal-based power generation".

While the overall trend is good, there are still pockets of concern, especially in the three leading countries most exposed to ESG risks with regard to coal-fired power plants – Bangladesh, India and the US. Table 3.14 summarizes the principal actors and some of the key stakeholders (beyond citizens and communities) that have made it to the top of various reputation risk categories regarding coal-fired power plants in 2018.

Table 3.14 RepRisk Special Report on Coal-Fired Power Plants 2018: summary overview of most associated countries, companies, projects and NGOs

Top-Five Most associated countries	*Top-Five Most associated companies*	*Top-Five Most associated projects*	*Top-Five Most associated NGOs*
1. United States	1. NTPC Ltd (India)	1. Rampal Thermal Power Plant	1. Sierra Club
2. Bangladesh	2. Bangladesh India Friendship Power Co. (Bangladesh)	2. Samcheonpo Thermal Power Plant	2. Greenpeace International
3. India	3. Enel Spa (Italy)	3. Hemweg Power Plant	3. Banktrack
4. South Korea	4. Eskom Holdings (South Africa)	4. Badarpur Coal-Based Thermal Power Station	4. Friends Of The Earth
5. South Africa	5. RWE AG (Germany)	5. Dangjin Thermal Power Plant	5. Centre for Environmental Rights

Source: RepRisk.

With respect to the largest coal-fired power developer in the world, a company from India, the study observes the following:

> The world's largest coal-plant developer, National Thermal Power Corporation of India (NTPC), recently announced plans to invest US$10 billion in new coal-fired power stations until 2022, and it has been estimated that India's energy sector will remain dominated by coal over the next decade. While it has been claimed that 300 million Indians will be lifted out of energy poverty if India carries through with its plans to build nearly 370 coal-fired power plants, environmentalists warn that hundreds of millions of people will be subjected to harmful levels of airborne pollutants, constituting a public health crisis.[54]

Finally, regarding the US, the RepRisk study underscores the following:

> the US is not only the leading contributor to coal power development in the world today, but also the only member state within the Group of Seven (G7), representing the seven largest advanced economies in the world, which has not yet committed to a plan to phase out coal within a set timeframe. Domestically, US coal-fired power plants are facing mounting pressure from local communities whose water has been contaminated by coal ash waste, the second largest source of industrial waste in the country. According to the US Environmental Protection Agency (EPA), toxic coal ash waste has already contaminated water sources at 200 sites in 37 states.

And yet where there is smoke there is fire in a good way too – the state of Karnataka in India is the leader in renewable energy in India,[55] once again proving that where there is risk – risk to environment, risk to health, risk to overall citizen well-being – there is also opportunity – that of a cleaner environment, healthier citizens and a more resilient community.

3.3.b *Transforming plastic waste into anti-plastic innovation*

3.3.b.i *The macro picture: oceans full of plastic*

> The modern world had produced around 8.3 billion metric tons of virgin—newly manufactured—plastic. By 2015, 6.3 billion metric tons of that had become plastic waste, but just 9 percent had been recycled, the researchers found.
>
> 322 million tons of plastic, which amounted to more than 900 Empire State Buildings in mass, was produced in 2015, according to the United Nations Environment Programme.[56]

Waste disposal is one of the most critical issues confronting a quickly growing world population that is disposing of waste at an unprecedented level. Especially when it comes to non-biodegradable materials, the world faces an eventual existential crisis which some countries – of course the poorest of the poor where a lot of this waste is shipped to as well as emerging economies not yet properly equipped to manage their own garbage disposal – are already facing.

One of the most critical waste disposal issues surrounds plastic in all of its non-biodegradable forms.

> More than 78 million tons of plastic packaging is produced worldwide every year by an industry worth nearly $198 billion. Just a fraction of that is recycled while the vast majority is thrown away. Plastic litter now clutters every part of our planet, from remote parts of the Antarctic to the deepest ocean trenches.[57]

Who hasn't seen (or at least heard of) the awful images of the massive plastic waste "islands" floating in the Pacific Ocean that some say are as large as the state of Texas?[58]

According to the *Financial Times*, which tracks the issue of waste on its website[59] 80% of the world's plastic waste originates in Asia, though much of it is exported there by the wealthy industrialized nations:

> "Sweeping waste under someone else's carpet" ... environmental advocates say the latest export data illustrates the disconnect between the anti-plastic headlines and the fact that consumption of single-use plastic is still increasing. "Sweeping our waste under someone else's carpet is not the solution to Britain's plastic problem," said Fiona Nicholls, a çampaigner at Greenpeace UK. "Instead of just moving our plastic scrap around the globe, we should turn off the tap at the source." Ocean plastics activists have also warned that plastic waste is more likely to end up as marine litter in Southeast Asian countries with poor waste management infrastructure than if it were handled in the UK.[60]

3.3.b.ii The national lens: anti-plastic initiatives

Meanwhile, the global community of decision-makers at the governmental, corporate and NGO levels have not been able to forge an agreement similar to that of the Paris Climate Agreement. However, a number of governmental entities – at the national, state/provincial and local levels – are implementing programs to combat the scourge of plastic waste. The following is an example from the European Union.

The European Commission (EC) announced in January 2018 a European-wide initiative to "protect the planet, defend our citizens and empower our industries" to move towards a more circular economy when it comes to plastics. As one of the EC's leaders in charge of this strategy stated: [61]

> First Vice-President Frans Timmermans, responsible for sustainable development, said: "If we don't change the way we produce and use plastics, there will be more plastics than fish in our oceans by 2050. We must stop plastics getting into our water, our food, and even our bodies. The only long-term solution is to reduce plastic waste by recycling and reusing more. This is a challenge that citizens, industry and governments must tackle together. With the EU Plastics Strategy we are also driving a new and more circular business model. We need to invest in innovative new technologies that keep our citizens and our environment safe whilst keeping our industry competitive".

The key components of this initiative include:

- Make recycling profitable for business.
- Curb plastic waste.

- Stop littering in the sea.
- Drive investment and innovation.
- Spur change across the world.

Significantly, the first bullet points to a very important concept that we are dealing with in this book: how business (and others) can transform risk into value. The EC plans to accomplish such a feat as follows:

> New rules on packaging will be developed to improve the recyclability of plastics used in the market and increase the demand for recycled plastic content. With more plastic being collected, improved and scaled up recycling facilities should be set up, alongside a better and standardized system for the separate collection and sorting of waste across the EU. This will save around a hundred Euros per ton collected. It will also deliver greater added value for a more competitive, resilient plastics industry.[62]

Around the world at different levels of government – from the national or federal levels to the purely local – we can find a number of initiatives, many of them pushed hard by local stakeholders. In many cases, such initiatives come from waste-minded urban locations looking to "green" their environments. These initiatives reflect the positive attitude of local government as well as corporate and nonprofit leaders and a variety of key stakeholders – from customers to citizens (who are often both) along with innovative organizations. Table 3.15 shows a sampling from around the world.

Table 3.15 A global sampling of national and local environmental initiatives

Location	*Entity sponsoring*	*Program*	*Stakeholders and beneficiaries*
India[63]	New Delhi municipality	India has banned all forms of disposable plastic in its capital	New Delhi residents, visitors, businesses
India[64]	Fishermen in the state of Kerala	Fisherman remove plastic from the sea to reuse it to build roads	Local citizens, residents, businesses, fisherman, municipalities
UK[65]	UK national government	UK plans to end all plastic waste in 25 years	UK residents, citizens, businesses, visitors
Costa Rica[66]	Costa Rica national government	Costa Rica plans to ban single-use plastics, such as bags, straws, bottles, cutlery and cups, by 2021	Costa Rican residents, citizens, businesses, visitors

Source: World Economic Forum.

3.3.b.iii Corporate anti-plastic efforts

3.3.b.iii.A STARBUCKS

A company that already has a long track record of corporate social responsibility on a variety of fronts, Starbucks announced in the summer of 2018 a plan to phase out plastic straws as reported by the *Financial Times* as follows:

> Starbucks jolted higher after plastic straw phase-out plan – It pays to be green. Starbucks shares received a much-needed jolt on Monday after the company revealed plans to phase out plastic straws from its 28,000 stores worldwide by 2020 in a move that will eliminate more than 1bn of the drinking devices each year. The stock rose 1.9 per cent in morning trade versus a 0.7 per cent rise for the broader S&P 500.[67]

What is clear from this headline and from the strategic planning that Starbucks has undertaken is that it pays to go green – to dump plastic in favor of biodegradable waste, and the benefits are not just financial – they are reputational, with positive reverberations from a wide variety of stakeholders.

3.3.b.iii.B ADIDAS

Adidas (and others) have undertaken to support the "Race Against Virgin Plastic". German-based global sports goods powerhouse, Adidas, announced in 2018 that by 2024 it would no longer use "virgin" plastic in any of its products and that all plastic used would be recycled. More specifically, they stated:

> Adidas also said it would stop using virgin plastic in its offices, retail outlets, warehouses and distribution centers, a move that would save an estimated 40 tons of plastic per year, starting in 2018.

3.3.b.iii.C OTHER CORPORATE ANTI-PLASTIC INITIATIVES

Also announcing similar programs in 2018 were the following major companies:

- McDonald's is performing trials in a similar plastic straw phase-out program in the United Kingdom and Ireland.
- IKEA is phasing out single use plastic from its stores and restaurants.

3.3.b.iv Plastic innovation

Leading the pack on all of these initiatives – whether governmental or corporate, global or local – are a variety of inventors and innovators trying to find ways to combat the global scourge that everybody acknowledges plastic to be. Among some of the great opportunities for value creation are the following recent or still-under-development products and services.

3.3.b.iv.A AARDVARK: THE LAST PLASTIC STRAW

Because of the skyrocketing demand for non-plastic straws, Aardvark was founded in Indiana in 2007 to develop alternatives to the plastic straw. The innovative company is the only paper straw maker in the US so far and was recently acquired by a larger company that manufacturers food-service items.[68] The demand for paper straws is expected to grow exponentially as more and more urban areas and restaurant chains and other users of plastic straws move to an environmentally friendly alternative.

3.3.b.iv.B PLASTIC EATING MOTHS[69]

And then there is this in the department of "you can't make this stuff up":

> Moth-eaten: Plastic-eating caterpillars could save the planet.
>
> An escape from a shopping bag triggers an idea.
>
> The experiment behind the paper was inspired when Federica Bertocchini, an amateur beekeeper who is also a biologist at Cantabria University, in Spain, noticed caterpillars chewing holes through the wax in some of her hives and lapping up the honey. To identify them, she took some home in a plastic shopping bag. But when, a few hours later, she got around to looking at her captives she found the bag was full of holes and the caterpillars were roaming around her house.

To make a long story short:

> the discovery that wax-moth larvae can eat plastic is intriguing. Even if the moths themselves are not the answer to the problem of plastic waste, some other animal out there might be.

3.3.c Business transforming environmental risk into opportunity

3.3.c.i General Motors: "moving humanity forward"

There are positive stories to tell as well – stories of risk transforming into opportunity and value for stakeholders. Witness this piece by Mary Barra, CEO of General Motors (GM), titled "Our Vision for Moving Humanity Forward", written in late 2017. In it she extols the benefits of a responsible future together with stakeholders. In her piece, Ms. Barra, addressing the next 25 years of current and future customers, provides the following list of plans:

- More Zero-Emissions Vehicles including "convergence of electric, connected, shared and autonomous vehicles".
- New battery technologies – the "Bolt EV gets an EPA-estimated 238 miles per charge".

- "Other clean-energy technologies such as hydrogen fuel cells".
- Mobility for everyone including a fleet of self-driving Bolt EVs.
- "Driving a Circular, Clean-Energy Economy – Our goal is to produce zero waste everywhere we operate".

She adds:

> We believe that climate change is both a social imperative and economic opportunity, so we have developed a plan to accelerate our use of clean power … our intent (is) to power all of our global operations' electricity with 100 percent renewable energy by 2050.

The State of California – the largest economy within the US and 5th largest economy in the world – has insisted it will retain its stricter standards, so there are likely to be multiple legal challenges to the federal rollback on auto emissions under the Trump administration. We will see who wins out in the end – a majority of stakeholders aligned with science and health or a minority of special interests aligned with a temporary federal administration.

Given the track record so far of the Trump Administration on environmental issues, endangered species and health issues, the likelihood of improvements on carbon footprint and global warming rather than regression on some of the real progress the US had achieved over the past 40 or so years is unfortunately much greater than just two years ago. That can of course change with a new, environmentally friendlier administration in the future that might be able to put a stop to these regressive and socially damaging policies.

3.3.c.ii Apple goes green

As the world faces the very true and difficult obstacles to implementing the goals of the Paris Climate Agreement – specifically addressing #13 of the SDGs "Climate Action" – there are significant opportunities for both nations and companies to implement climate change combating programs and products.

A stellar example of a company putting its carbon emissions commitments where its mouth is has been Apple. Witness the following:

> You have to see Apple's Reno, Nevada, data center from the inside to truly understand how huge it is. It's made up of five long white buildings sitting side by side on a dry scrubby landscape just off I-80, and the corridor that connects them through the middle is a quarter-mile long. On either side are big, dark rooms–more than 50 of them–filled with more than 200,000 identical servers, tiny lights winking in the dark from their front panels. This is where Siri lives. And iCloud. And Apple Music. And Apple Pay. Powering all these machines, and keeping them cool, takes a

> lot of power – constant, uninterrupted, redundant power. At the Reno data center, that means 100% green power from three different Apple solar farms ... Now Apple says it's finished getting the rest of its facilities running on 100% green power – from its new Apple Park headquarters, which has one of the largest solar roofs on the planet, to its distribution centers and retail stores around the world. Though the 100% figure covers only Apple's own operations – not those of the suppliers and contract manufacturers which do much of the work of bringing its ideas to life – it's also convinced 23 companies in its supply chain to sign a pledge to get to 100% renewable energy for the portion of their business relating to Apple products.

While not every company has the power of Apple (after all it was the first company to achieve over $1 trillion in market cap ever) there are powerful lessons to be learned and policies and benefits to be leveraged from its example.[70]

3.3.c.iii Munich Re aspires higher

On August 5, 2018, the largest reinsurance company in the world, Berlin-based Munich Re, announced that it would:

> stop investing in bonds and shares of companies that generate more than 30 percent of their sales with coal-related business, its chief executive said, caving to pressure from investors.

Moreover, the reinsurer's CEO underscored that going forward:

> In the individual risk business, where we can see the risks exactly, we will in future in principle no longer insure new coal-fired power plants or mines in industrial countries.[71]

When insurance companies and the reinsurance industry make commitments of this scale, it has reverberations beyond borders. Companies of all shapes and sizes – with all kinds of products and services – will stand and note the possible absence of insurance coverage for their various risks. This is just another way to transform risk into opportunity for value creation – be it financial, reputational or stakeholder.

3.3.c.iv Green bonds

In another category altogether is the explosion of what are called "corporate green bonds" in the investment marketplace. Only five years ago, the total global investment in these bonds amounted to $3 billion. In 2017 the amount invested catapulted to $49 billion.

A recent study has shown through an analysis of 217 corporate green bonds issued by public companies around the world between January 1, 2013, through December 31, 2017, that:

> [T]hey yield a positive stock market reaction, improvements in financial and environmental performance, an increase in green innovations, and an increase in stock ownership by long-term and green investors.

Who has been issuing these corporate bonds and what for? Table 3.16 shows some of this information.

Table 3.16 Corporate green bond investment 2013–2017: analysis results[72]

- 69% of corporate green bonds were certified by independent third parties to establish that the proceeds are funding projects that generate environmental benefits.
- The stock price increase is larger for companies operating in industries where the natural environment is financially material to the firms' operations.
- Returns are larger for first-time issuers, compared to seasoned issuers.
- Green bond offerings are associated with a 2.4% increase in long-term value.
- In the long run (two years after the green bond issue), ROA increases by 0.6 percentage points.
- Green bonds companies improve their environmental performance:

  Their environmental score rose 6.1 percentage points on the Thomson Reuters' asset scale, which is based on more than 250 key performance indicators such as CO_2 emissions, hazardous waste, recycling, and so on.

- The long-term index (a measure of long-term orientation based on a textual analysis of the firms' annual reports) of green bond issuers increases by 3.9 percentage points.
- The share of long-term investors increases from 7.1% to 8.6% (a 21% increase), and the share of green investors from 3% to 7% (a 75% increase).

Source: *Harvard Business Review*.

Examples of companies that are engaging in these practices include Apple and Microsoft, as summarized in Table 3.17.

Table 3.17 Two examples of leading company green bonds[73]

Apple

Apple issued green bonds on two occasions. The first was $1.5 billion in green bonds and other debt. A Bank of America analysis found that more investors were interested in the green bond tranche, and less so in the other bonds.

In June 2017, Apple issued $1 billion of green bonds to fund renewable energy, in a demonstration of "how a tech company can make huge commitments to clean energy and make the front page of *The Wall Street Journal*, and signal that this is something they're involved in and interested in".

(*Continued*)

Table 3.17 Continued

Microsoft
A global leader in reducing emissions and buying renewable energy, [Microsoft] has an internal carbon tax that pays for many efficiency and renewable energy projects, said Liz Willmott, carbon program lead at Microsoft.
The company levies carbon taxes on its energy-heavy operations, at a rate of roughly $5 to $10 a ton of carbon dioxide equivalent, and sustainability teams can use that money to invest in efficiency and clean energy projects. Microsoft's carbon tax applies to electricity consumption and onsite emissions such as diesel generation and onsite usage, and air travel.

Source: GreenBiz.

Why is this activity important or notable? The fact is that more and more companies and investors are focusing on environmentally friendly practices and investments as they not only promise financial benefits for investors and bond-holders but also benefits for other stakeholders in that these bonds are designed for environmentally friendly purposes and go a long way to satisfying expectations about integrating ESGT issues, risks and opportunities into strategy for broad set value creation.

3.3.d Transforming environmental risk into sustainable resilience and value

Finally, no discussion of environmental challenges, especially those involving the overarching topic of climate change, would be complete without a brief discussion of a few resilience-building efforts that are taking place around the world today – from the governmental and international agency sectors to the business and NGO sectors. These efforts seek to find solutions and innovative thinking, build awareness and construct resilience against climate events from the smallest local efforts to the largest international ones.

Below is a brief overview of several such efforts (a mere drop in the bucket of actual activities worldwide) which deserve our attention and support. Indeed, I would urge readers to think about how you can contribute your time, effort, expertise or even treasure (should you have it) to exercise resilience-building measures from the smallest micro-level in your daily life (in your own home regarding waste management, the use of environmentally safer and greener products and services, building your own family's crisis readiness kit) to supporting local and national government initiatives, donating time and effort to an NGO of your choice, working within your organization or business to make sure they have appropriate crisis management and resilience preparedness. Indeed, don't assume that your employer has a readiness program, ask about it, and if it doesn't exist or is on paper only, volunteer to be part of the solution.

Below are examples of resilience-building efforts – a global NGO effort called 100 Resilient Cities, a post-Hurricane Maria Puerto Rico resilience initiative and several efforts under the general rubric of insurance and financial industry initiatives designed to encourage and incentivize business and government generally to adopt more positive, preventative, early-stage, proactive and resilient approaches to climate change risks and threats.

But let's start this discussion (as we do many in this book) with a couple of definitions of resilience.[74]

From the Rockefeller Foundation:

> Helping cities, organizations, and communities better prepare for, respond to, and transform from disruption.

From the Stockholm Resilience Centre:

> Resilience is the capacity of a system, be it an individual, a forest, a city or an economy, to deal with change and continue to develop. It is about how humans and nature can use shocks and disturbances, like a financial crisis or climate change, to spur renewal and innovative thinking.

3.3.d.i 100 resilient cities

A New York City-based nonprofit, 100 Resilient Cities, was founded in 2013 by the Rockefeller Foundation. Its mission was the following:

> We provide funding, capacity building, and technical assistance to help cities change the way they understand their risks and plan for their futures. We partner closely with cities as they hire a Chief Resilience Officer, develop a holistic resilience strategy, and most importantly, implement the projects identified in their strategy.

They started with 32 cities in 2014. They added their second cohort of an additional 35 cities by the end of 2014 and a third and final cohort in November 2015.

The selection criteria included looking for "innovative mayors, a recent catalyst for change, a history of building partnerships, and an ability to work with a wide range of stakeholders."

Their cities are all over the world, ranging from enormous urban centers like New York City, Mexico City and Seoul to cities known for their livability like Vancouver, Toronto and Sidney to emerging world cities like Tbilisi, Accra and Pune.

They focus on urban resilience, which they define as:

> the capacity of individuals, communities, institutions, businesses, and systems within a city to survive, adapt, and grow no matter what kinds of chronic stresses and acute shocks they experience.

Finally, they have looked at two major categories of resilience building:

- Chronic Stresses that weaken the fabric of a city on a day-to-day or cyclical basis.

  Examples include high unemployment, inefficient public transportation systems, endemic violence and chronic food and water shortages.

- Acute Shocks which are sudden, sharp events that threaten a city.
  Examples include earthquakes, floods, disease outbreaks, and terrorist attacks.

This clearly public/private local and international effort is designed to increase the preparedness and resilience of urban areas to confront and resolve the increasingly complex and interconnected risks and opportunities of city life in the 21st century.

As I finish writing this book, 100 Resilient Cities has just announced that it is closing in July, 2019, but it is transitioning to new initiatives and the Rockefeller Foundation announced "1 billion resilient people" and made this statement:

> Rockefeller Foundation has made the decision to transition the work of 100 Resilient Cities into at least three separate pathways: a new Resilience Office within the Foundation, supporting place-based resilience work within new economic mobility efforts at the Foundation in the US, and funding a resilience effort at the Atlantic Council.

While this particular initiative is morphing into three different ones – which is actually a sign of great resilience in that it is changing and adapting as the needs and demands change – 100 Resilient Cities serves as an inspiration to other initiatives everywhere to build urban and other public/private resilience mechanisms and ventures.

3.3.d.ii ReImagina Puerto Rico

The Resilient Puerto Rico Advisory Commission was formed in November, 2017, as an independent body in Puerto Rico led by locals. Somewhat related but independent of 100 Resilient Cities is the effort that is a combined effort funded partly by the Rockefeller Foundation, the Ford Foundation and the Open Society Foundation.

Their stated mission is to:

> Produce an actionable and timely set of recommendations for how to use philanthropic, local government and federal recovery dollars to not only repair and rebuild the critical systems devastated by Hurricane Maria, but to build back in a way the makes the island stronger – physically, economically, and socially – over the long term.[75]

As they develop their resilience-building recommendations, ReImagina Puerto Rico has identified the acute shocks and chronic stresses that need to be incorporated into their strategic planning process. Table 3.18 summarizes some of the key themes under the rubrics of "Acute Shocks" and "Stresses" and

Table 3.18 ReImagina Puerto Rico/Reimagine Puerto Rico: Strategic considerations[76]

Acute shocks	*Stresses and chronic stresses*
Sudden, sharp events that threaten society, including: • Hurricanes and tropical storms • Earthquakes • Floods • Disease outbreaks • Heatwaves • Landslides • Earthquakes • Tsunamis • Failure of health, communication, energy, fuel, water, and food distribution systems	**Slow moving disasters that weaken society such as:** • High unemployment • Inefficient energy and public transportation system • Endemic violence • Food and water shortage • Urban deforestation • Improper use of the land • Vulnerable populations in high-risk areas • Discrimination (gender, race, sexual orientation, homelessness) • Poor access to mental health and preventive and medical treatment • Lack of available safe and affordable housing • Violence and crime • Migration of health professionals

Source: ReImagina Puerto Rico.

"Chronic Stresses" that Puerto Rico must confront as they rebuild and create new resilience.

To date, ReImagina Puerto Rico has issued a variety of in-depth reports on important resilience subjects and has made a wide variety of recommendations in the following areas:[77]

- ReImagina Puerto Rico Report
- Economic Development Sector Report
- Energy Sector Report
- Health, Education and Social Services Sector Report
- Housing Sector Report
- Physical Infrastructure Sector Report
- Natural Infrastructure Sector Report

Sometimes it takes a disaster or a major setback for us to understand what is at risk and has been lost, and perhaps that's how we learn how best to protect against the severity of another loss. The ReImagina Puerto Rico effort seems to address the idea of transforming climate risk into social opportunity where the island eventually builds itself back but in a stronger, more resilient and more beneficial way to all of its stakeholders, most especially the population of Puerto Ricans and others who call Puerto Rico home.

3.3.d.iii Lloyd's City Risk Index

Now let's take a look at resilience from a somewhat different but certainly powerful and important angle – that of the insurance community. Lloyd's produces something called the Lloyd's City Risk Index, which does the following:

> The Lloyd's City Risk Index, based on original research by the Centre for Risk Studies at Cambridge University's Judge Business School, measure GDP @ Risk of 279 cities around the world from 22 threats in five categories:
>
> - Finance, economics and trade
> - Geopolitics and security
> - Health and humanity
> - Natural catastrophe and climate
> - Technology and space
>
> The cities in the index are some of the world's leading cities, which together generate 41% of global GDP.[78]

The stated purpose of the Lloyd's City Risk Index is to help policy-makers, government decision-makers and insurers achieve greater resilience.

Table 3.19 provides a summary overview of the top-ten cities at risk and the top-ten threats to cities – both with estimated dollars at risk – as compiled by Lloyd's for their City Risk Index.

Table 3.19 Lloyd's City Risk Index: Global Overview: 279 cities – 22 threats

Top-ten cities at risk	*Top-ten threats*
1. Tokyo – $24.31 Billion	1. Market crash – $103.33 Billion
2. New York – $14.83 Billion	2. Interstate conflict – $80.00 Billion
3. Manila – $13.27 Billion	3. Tropical windstorm – $62.59 Billion
4. Taipei – $12.88 Billion	4. Human pandemic – $47.13 Billion
5. Istanbul – $12.74 Billion	5. Flood – $42.91 Billion
6. Osaka – $12.42 Billion	6. Civil conflict – $37.15 Billion
7. Los Angeles – $11.56 Billion	7. Cyberattack – $36.54 Billion
8. Shanghai – $8.48 Billion	8. Earthquake – $33.96 Billion
9. London – $8.43 Billion	9. Commodity price shock – $20.29 Billion
10. Baghdad – $7.91 Billion	10. Sovereign default – $17.97 Billion

Source: Lloyd's.

The following are the key findings from Lloyd's City Risk Index – they are illuminating and speak directly to the point reiterated in this book over and over again (most especially in Part III – Boom: Achieving Sustainable Resilience and Value), that preparedness and resilience will lead to value in the form of financial gain, reputational upside and stakeholder loyalty and trust. Here are some of their key findings in this regard – pay attention to the last two bullet points in particular:

- Across all 279 cities, $546.50bn is at risk from all 22 threats.
- Man-made threats account for 59% of the total GDP@Risk – with market crash the largest single threat, with cities exposed to losses of $103.33bn on an annual basis.
- The ten cities with the highest exposure have a combined $126.82bn of GDP@Risk, almost a quarter of the global total, with Tokyo standing to lose more than any other city.
- Climate-related risks account for $122.98bn of GDP under threat, and this sum will grow as extreme weather events grow in frequency and severity.
- If cities were to improve their resilience, global GDP exposure to loss would drop by $73.4bn.[79]

3.3.d.iv *Financial services sector examples in transforming climate risk to sustainable value: The Task Force on Climate-Related Financial Disclosures (TCFD) and ClimateWise*

One of the more challenging arenas in which to contribute to sustainability and environmental responsibility directly is in the services sectors – as a lawyer, banker, accountant, analyst, etc. Other than minding your own store literally (i.e., ensuring that your offices and similar infrastructure comply with environmental greenhouse, waste and water standards), the services sector arguably does not have as direct a pro-environmental contribution to make as does a manufacturing facility, construction company or chemical plant given their very different and potentially damaging physical footprint.

Or do they?

Indeed, they do – some of the more innovative initiatives with potentially broad and positive social implications are being developed in the financial sector – the TCFD and ClimateWise offer just two of such examples.

3.3.d.iv.A THE FINANCIAL STABILITY BOARD TASK FORCE ON CLIMATE-RELATED FINANCIAL DISCLOSURES (TCFD)

The TCFD is part of the market-based Financial Stability Board, an international entity created in 2009 after the G20 meeting in London whose mission is to monitor the global financial system and make recommendations for its improvement. The TCFD in turn is one of its task forces and is focused on the creation of common reporting standards and mechanisms for environmental metrics. As its chair, Mayor Michael Bloomberg, a huge supporter of climate change action globally as well as one of the most successful business people ever, put it:

> Increasing transparency makes markets more efficient, and economies more stable and resilient.[80]

It's all about transforming climate risk into sustainable value for entities, cities, communities and nations. As of February 2019, more than 580 organizations were supporting the work of the TCFD.

The TCFD is working on a number of recommendations, and in their 2018 TCFD Status Report, one of their clearly stated goals is to improve:[81]

> practices and techniques [that] would further improve the quality of climate-related financial disclosures and, ultimately, support more appropriate pricing of risks and allocation of capital in the global economy.

In doing so, they are developing environmental metrics and reporting that would help in building a common set of standards and language around environmental and climate reporting by business and non-business entities. In Table 1.6 of Chapter 1, we provided a snapshot of some of the issues they are grappling with, and in Table 3.20 is a summary of the key findings from their 2018 status report.

Table 3.20 TCFD 2018 Status Report: key take-aways[82]

- The majority disclose some climate-related information.
- The majority of companies reviewed disclosed information aligned with at least one recommended disclosure, usually in sustainability reports.
- Financial implications are often not disclosed.
- While many companies disclose climate-related information, few disclose the financial impact of climate change on the company.
- Information on strategy resilience under different climate-related scenarios is limited.
- Few companies describe the resilience of their strategies under different climate-related scenarios, including a 2°C or lower scenario, which is a key area of focus for the Task Force.
- Disclosures vary across industries and regions.
- Companies' areas of focus in terms of climate-related financial disclosures vary significantly. For example, a higher percentage of non-financial companies reported information on their climate-related metrics and targets compared to financial companies; but a higher percentage of financial companies indicated their enterprise risk management processes included climate-related risks.
- In terms of regional differences, a higher percentage of companies in Europe disclosed information aligned with the recommendations compared to companies in other regions. Disclosures are often made in multiple reports. Companies often provided information aligned with the TCFD recommendations in multiple reports—financial filings, annual reports, and sustainability reports.

Source: Financial Stability Board – Task Force on Climate-Related Financial Disclosures.

While the challenges are great this is a hopeful and powerful example of what business, government and society can do together to move the needle forward on transforming climate risk to sustainable stakeholder value by providing concrete practices and metrics that will eventually put environmental disclosure on a footing similar to that of financial disclosure, or at least one can hope.

3.3.d.iv.B CLIMATEWISE

Another hopeful and potentially highly leverageable area in which the services sector is capable of making a major, maybe even enormous, contribution to the fight against global warming is the insurance sector. Innovative companies in this space, such as Beazley, are developing new insurance products that incentivize businesses and government and other potential insureds to build their products and services in an environmentally sound way from the beginning, making it financially attractive to them to build resilience up front by providing reduced premiums over time and the like.

One of the leading cross-sectoral initiatives in this space is ClimateWise, part of the Cambridge University Institute for Sustainability Leadership which focuses on building insurance-based climate solutions.[83] Their stated mission:

> ClimateWise supports the insurance industry to better communicate, disclose and respond to the risks and opportunities associated with the climate-risk protection gap. This is the growing divide between total economic and insured losses attributed to climate change.
>
> Representing a growing global network of leading insurance industry organizations, ClimateWise helps to align its members' expertise to directly support society as it responds to the risks and opportunities of climate change.[84]

ClimateWise's principles are summarized in Table 3.21.

Table 3.21 The ClimateWise Principles[85]

Principle 1	Be accountable
Principle 2	Incorporate climate-related issues into our strategies and investments
Principle 3	Lead in the identification, understanding and management of climate risk
Principle 4	Reduce the environmental impact of our business
Principle 5	Inform public policy making
Principle 6	Support climate awareness amongst our customers/clients
Principle 7	Enhance reporting

Source: ClimateWise.

Table 3.22 summarizes actions that ClimateWise recommends that insurers consider when it comes to climate change-related incentivization.

Table 3.22 ClimateWise: actions for insurers[86]

- Initiate a virtuous circle reinforcing the desirability of resilience by adding resilience as an appropriate feature of investments.
- Offer options for resilience investing to policyholders.
- Develop new types of insurance cover or adapt existing ones to support the monetization of returns on investing in resilience.
- Support platforms to package, market and sell investments in resilience projects.
- Adopt indemnity bases that include resilience reinstatement.
- Provide long-term incentives to policyholders through multi-year insurance policies.
- Provide long-term incentives to policyholders through profit-sharing insurance pools.
- Participate in stakeholder partnerships with municipalities and government agencies.
- Second staff to local and national government departments and agencies.
- Increase the focus on resilience in CSR activities.

Source: ClimateWise.

These initiatives should serve as inspiration to other services sectors that are looking for ways to contribute to this common global threat and opportunity for improvement. Two more examples are presented from ClimateWise – one described in Table 3.23, focused on how resilience may be improved regarding flood risk, and the other, quite outside of the box (described in Table 3.24), in which insurers and others are collaborating on addressing the loss of pollinators (i.e., bees and butterflies).

Table 3.23 ClimateWise: example – improving resilience for flood risk[87]

Resilience can be improved by avoiding adverse events, limiting their impact, or enhancing recovery. For flood risk, this could mean:

Avoiding adverse events	**Limiting impact**	**Enhancing recovery**
• Relocating out of a flood plain; • Switching to a supplier that is not located in a flood plain; • Disinvesting from companies located in a flood plain.	• Creating upstream flooding and retention areas; • Building flood defenses to prevent the incursion of water; • Designing floatable buildings or buildings on stilts; • Retrofitting buildings to be more resistant to flood damage, such as by raising electrical systems above ground level; • Having an effective business continuity plan in place; • Avoiding an over-dependence on a single supplier or group of suppliers.	• Having effective evacuation plans in place; • Having sufficient back-up resources to effect rapid repairs.

Source: ClimateWise.

Table 3.24 The Partnership for Pollinators[88]

Pollinator populations are declining rapidly, with 9 per cent or more of many wild bee and butterfly species facing local extinction. This has serious implications for companies and their supply chains – a recent global assessment of the status of pollinators highlighted that around three-quarters of food crops depend on pollination, making pollinators worth up to US$577 billion annually.
The Cambridge University Institute for Sustainability Leadership is working with a group of pollination experts, including the UN Environment World Conservation Monitoring Centre, Fauna & Flora International (FFI) and the University of East Anglia (UEA), to help business take action. The Partnership for Pollinators aims to help companies understand and manage the risks posed by pollinator decline – to improve their supply chain resilience and protect the natural ecosystems on which they rely.

Source: ClimateWise.

3.4 Endnote: an environmental issue dashboard sample

Table 3.25 shows a practical dashboard of big-picture environmental issues for a hypothetical company providing some issues that may be relevant, questions leaders should ask, outlining the key organizational players that should participate in issue discussions and some examples of potential opportunities relating to the overall issue/risk.

Table 3.25 Big-picture environmental issues typology for organizational leaders to consider

Overall issue category	*Leadership threshold questions*	*Cross-functional organizational actors*	*Potential opportunities*
Climate Impact/ Global Warming/ Carbon Footprint	• What is the climate/carbon emissions impact of our operations, assets, research & development, products & services? • Does our business model currently include an analysis of opportunities relating to this category of issues and risks?	• EH&S • Legal & compliance • Risk management • CSR • Quality • Operations • Finance • Crisis management	• Opportunities in clean tech • Opportunities in green building • Opportunities in renewable energy • Product carbon footprint • Financing options

(*Continued*)

Table 3.25 Continued

Overall issue category	*Leadership threshold questions*	*Cross-functional organizational actors*	*Potential opportunities*
Natural Resources	• What resources does our entity use to implement our business plan and strategy? Are they related to water, air, natural resources, other? • Does our business model currently include an analysis of opportunities relating to this category of issues and risks?	• EH&S • Legal & compliance • Risk management • CSR • Quality • Operations • Finance • Crisis management	• Opportunity to change supplies, supply chain to greener, more environmentally friendly, potentially cheaper products • Reputational opportunity • New products or services
Waste Management	• What types of waste do our operations create and where, and what are the applicable legal regulations and desirable policies to deal with the waste? • Does our business model currently include an analysis of opportunities relating to this category of issues and risks?	• EH&S • Legal & compliance • Risk management • CSR • Quality • Operations • Finance • Crisis management • Business continuity	• Opportunity to convert waste into recycled products • Possible revenue stream from recycling • Cheaper resources from recycled materials • Reputational opportunity
Biodiversity/ Animal Well-Being	• Does any of our work impact biodiversity, and the well-being of human or other animal life and how are we prepared to mitigate risk and assure humane and proper treatment? • Does our business model currently include an analysis of opportunities relating to this category of issues and risks?	• EH&S • Legal & compliance • Risk management • CSR • Quality • Operations • Finance • Crisis management • Business continuity	• Appeal to certain potentially larger segment of customers that are anti-animal testing • Reputational opportunity • Development of new testing products and services

(*Continued*)

Table 3.25 Continued

Overall issue category	*Leadership threshold questions*	*Cross-functional organizational actors*	*Potential opportunities*
Natural and Man-Made Disasters	• What is the preparedness of our organization vis-a-vis a natural and/or man-made disaster (hurricanes, tidal waves, earthquakes, severe weather) in terms of protecting both people and property? • Does our business model currently include an analysis of opportunities relating to this category of issues and risks?	• EH&S • Legal & compliance • Risk management • CSR • Quality • Operations • Finance • Crisis management • Business continuity	• Stakeholder care and reputation building • Opportunity to demonstrate care for affected stakeholders such as employees and communities who might be affected by such a disaster

Source: GEC Risk Advisory.

Notes

1 "The Eminent Astronomer Who Says There Is no Plan B for the World". *Medium*. June 14, 2018. https://medium.com/this-cambridge-life/the-eminent-astronomer-who-says-there-is-no-plan-b-for-the-world-dc70dcbe83b6. Accessed on July 4, 2018.
2 Jonathan Watts. "Interview: Greta Thunberg, Schoolgirl Climate Change Warrior: "Some People Can Let Things Go. I Can't." *The Guardian*. March 11, 2019. https://www.theguardian.com/world/2019/mar/11/greta-thunberg-schoolgirl-climate-change-warrior-some-people-can-let-things-go-i-cant. Accessed on April 23, 2019.
3 "Donald Trump: 'I Don't Believe in Climate Change'". *The Wall Street Journal*. September 24, 2015. https://blogs.wsj.com/washwire/2015/09/24/donald-trump-i-dont-believe-in-climate-change/. Accessed on July 4, 2018.
4 Quoted in Amy Domini. "The Next 25 Years: Big Picture Thinking". *GreenMoney*. July/August 2017. https://greenmoneyjournal.com/next-25-years-big-picture-thinking-2/. Accessed on August 8, 2018.
5 Amy Domini. "The Next 25 Years: Big Picture Thinking". *GreenMoney*. July/August 2017.https://greenmoneyjournal.com/next-25-years-big-picture-thinking-2/. Accessed on August 8, 2018.
6 OxfordDictionaries.com. https://en.oxforddictionaries.com/definition/environment. Accessed on July 4, 2018.
7 *The Business Dictionary*. http://www.businessdictionary.com/definition/environment.html. Accessed on July 19, 2018.
8 OxfordDictionaries.com. https://en.oxforddictionaries.com/definition/environmental. Accessed on July 19, 2018.
9 *The Business Dictionary*. http://www.businessdictionary.com/definition/environmentalism.html. Accessed on July 19, 2018.

10 Robecosam. https://www.robeco.com/me/key-strengths/sustainability-investing/glossary/esg-definition.html. Accessed on July 19, 2018.
11 Cambridge Centre for Existential Risk. https://www.cser.ac.uk/. Accessed on July 29, 2018.
12 The Paris Climate Agreement of 2015. https://unfccc.int/sites/default/files/english_paris_agreement.pdf. Accessed on August 8, 2018.
13 World Economic Forum. *Global Risks Report 2019*. http://www3.weforum.org/docs/WEF_Global_Risks_Report_2019.pdf. Accessed on April 12, 2019.
14 UN Sustainable Development Goals. https://sustainabledevelopment.un.org/?menu=1300. Accessed on July 29, 2018.
15 UN Environment Programme. https://www.unenvironment.org/explore-topics. Accessed on July 21, 2018.
16 UN Global Compact. https://www.unglobalcompact.org/. Accessed on July 29, 2018.
17 Allianz. *The Allianz Risk Barometer 2019*. https://www.agcs.allianz.com/content/dam/onemarketing/agcs/agcs/reports/Allianz-Risk-Barometer-2019.pdf. Accessed on April 12, 2019.
18 MSCI. https://www.msci.com/esg-investing. Accessed on July 18, 2018.
19 RepRisk. https://www.reprisk.com/our-approach. Accessed July 17, 2018.
20 Centre for the Study of Existential Risk at Cambridge University. https://www.cser.ac.uk/. Accessed on July 29, 2018.
21 CSER. https://www.cser.ac.uk/. Accessed on July 29, 2018.
22 CSER. https://www.cser.ac.uk/. Accessed on August 4, 2018.
23 Summary gleaned from: The Paris Climate Agreement of 2015. https://unfccc.int/sites/default/files/english_paris_agreement.pdf. Accessed on August 8, 2018; and *The New York Times*. "Key Points of the Paris Climate Pact". December 12, 2015. https://www.nytimes.com/interactive/projects/cp/climate/2015-paris-climate-talks. Accessed on August 8, 2018.
24 DNVVL, UN Global Compact and Sustainia. *The Global Opportunity Report 2018*. https://www.unglobalcompact.org/docs/publications/Global_Opportunity_Report_2018.pdf. Accessed on August 8, 2018.
25 Maria Caspani. "More Americans View Climate Change as Imminent Threat". Reuters. https://www.reuters.com/article/us-climate-change-usa-poll/more-americans-view-climate-change-as-imminent-threat-reuters-ipsos-poll-idUSKBN1OC1FX. Accessed on April 12, 2019.
26 World Economic Forum. *Global Risks Report 2019*. http://www3.weforum.org/docs/WEF_Global_Risks_Report_2019.pdf. Accessed on April 12, 2019.
27 World Economic Forum. *Global Risks Report 2018*. https://www.weforum.org/reports/the-global-risks-report-2018 Accessed on August 7, 2018; World Economic Forum. *Global Risks Report 2019*. http://www3.weforum.org/docs/WEF_Global_Risks_Report_2019.pdf. Accessed on April 12, 2019; and https://www.nytimes.com/interactive/projects/cp/climate/2015-paris-climate-talks. Accessed on August 8, 2018.
28 UN Environment Programme. https://www.unenvironment.org/explore-topics. Accessed on July 21, 2018.
29 UN Environment Programme. https://www.unenvironment.org/explore-topics. Accessed on July 21, 2018.
30 United Nations Global Compact. https://www.unglobalcompact.org/what-is-gc/mission/principles. Accessed on August 7, 2018.
31 UN Global Compact Principles. https://www.unglobalcompact.org/what-is-gc/mission/principles accessed July 17, 2018.
32 UN Global Compact. Environment Resources. https://www.unglobalcompact.org/what-is-gc/our-work/environment. Accessed on August 4, 2018.
33 UN Global Compact. https://www.unglobalcompact.org/sdgs/17-global-goals accessed July 17, 2018.

34 Allianz. *The Allianz Risk Barometer 2019*. https://www.agcs.allianz.com/content/dam/onemarketing/agcs/agcs/reports/Allianz-Risk-Barometer-2019.pdf. Accessed on April 12, 2019.

35 Andrew Freedman. "Global Heat Wave Is so Pervasive, It's Surprising Scientists". Axios. July 27, 2018. https://www.axios.com/global-heat-wave-stuns-scientists-as-records-fall-4cad71d2-8567-411e-a3f6-0febaa19a847.html?utm_source=newsletter&utm_medium=email&utm_campaign=newsletter_axiosam&stream=top-stories. Accessed on July 29, 2018.

36 Andrew Freedman. "July Was Earth's Hottest Month Since Records Began, with the Globe Missing 1 Million Square Miles of Sea Ice". *The Washington Post*. August 15, 2019. https://www.washingtonpost.com/weather/2019/08/15/independent-data-confirms-july-was-earths-hottest-month-since-records-began/. Accessed on August 17, 2019.

37 MSCI. https://www.msci.com/esg-investing. Accessed on July 18, 2018.

38 The results of the screening are then curated by analysts and quantified to provide users with a systematic way to benchmark risk exposure and track risk trends over time and enable them to identify, assess, and monitor ESG and business conduct risks in business and investments. RepRisk's ESG Risk Platform covers more than 115,000 companies and more than 28,000 projects such as mines, pipelines, and power plants globally, covering both private and publicly listed companies and emerging and frontier markets. RepRisk. https://www.reprisk.com/our-approach. Accessed July 17, 2018.

39 According to RepRisk, it is expected that most large multinationals have an RRI between 26–49, due to their global footprint and salience vis-à-vis media and stakeholders. RepRisk. https://www.reprisk.com/our-approach. Accessed July 17, 2018.

40 RepRisk. https://www.reprisk.com/our-approach. Accessed July 17, 2018.

41 RepRisk. https://www.reprisk.com/our-approach. Accessed July 17, 2018.

42 RepRisk. *Most Controversial Companies 2017*. https://www.reprisk.com/content/5-publications/1-special-reports/53-most-controversial-companies-of-2017/mcc-2017.pdf. Accessed on August 5, 2018.

43 According to RepRisk, "Peak RRI: equal to the highest level of the RRI over the last two years – a proxy for overall reputational exposure related to ESG and business conduct risk." https://www.reprisk.com/our-approach. Accessed on August 5, 2018.

44 "The World Is Losing the War Against Climate Change". *The Economist*. August 2, 2018. https://www.economist.com/leaders/2018/08/02/the-world-is-losing-the-war-against-climate-change. Accessed on August 2, 2018.

45 "The World Is Losing the War Against Climate Change". *The Economist*. August 2, 2018. https://www.economist.com/leaders/2018/08/02/the-world-is-losing-the-war-against-climate-change. Accessed on August 2, 2018.

46 Eillie Anzilotti. "A Guide to What the Times Bold, Flawed Climate Story Left Out". *Fast Company*. August 6, 2018. https://www.fastcompany.com/90214347/a-guide-to-what-the-times-bold-flawed-climate-story-left-out?utm_source=postup&utm_medium=email&utm_campaign=Fast%20Company%20Daily&position=7&partner=newsletter&campaign_date=08072018. Accessed on August 7, 2018. Referencing this story in *The New York Times Magazine*: "Nathaniel Rich. 'Losing Earth: The Decade We Almost Stopped Climate Change'". *The New York Times Magazine*. August 1, 2018. https://www.nytimes.com/interactive/2018/08/01/magazine/climate-change-losing-earth.html#part-one. Accessed on August 7, 2018.

47 Jeffrey Sachs. "We Are All Climate Refugees Now". Project Syndicate. https://www.project-syndicate.org/commentary/climate-change-disaster-in-the-making-by-jeffrey-d-sachs-2018-08. Accessed on August 2, 2018.

48 Damien Cave. "His Pacific Island Was Swallowed by Rising Seas. So He Moved to a New One". *The New York Times*.
https://www.nytimes.com/2018/07/26/world/asia/solomon-islands-south-pacific.html?smtyp=cur&smid=tw-nytimesworld. Accessed on August 11, 2018.

49 Lucy Hook. "The Markets That Will Be Hardest Hit by Rising Global Temperatures". Corporate Risk and Insurance. August 6, 2018. https://www.corporateriskandinsurance.com/news/geo-political/the-markets-that-will-be-hardest-hit-by-rising-global-temperatures/108096. Accessed on August 7, 2018.
50 RepRisk. *Special Report Coal-Fired Power Plants.* June 2018. https://www.reprisk.com/content/home/reprisk-special-report-coal-fired-power-plants.pdf. Accessed on July 23, 2018.
51 RepRisk. *Special Report Coal-Fired Power Plants.*
52 Coral Davenport. "Trump Unveils His Plans to Weaken Car Pollution Rules". *The New York Times.* August 2, 2018. https://www.nytimes.com/2018/08/02/climate/trump-auto-emissions-california.html?hp&action=click&pgtype=Homepage&clickSource=story-heading&module=first-column-region®ion=top-news&WT.nav=top-news. Accessed on August 2, 2018.
53 RepRisk. *Special Report Coal-Fired Power Plants.*
54 RepRisk. *Special Report Coal Fired Power Plants.*
55 World Economic Forum. Costly Coal Has Pushed Karnataka to Become India's Renewables Leader. https://www.weforum.org/agenda/2018/07/costly-coal-has-pushed-karnataka-to-become-india-s-renewables-leader/. Accessed on August 5, 2018.
56 CNBC. "Plastic Pollution: Firms and Governments Are Combating Millions of Tons of Waste". April 22, 2018. https://www.cnbc.com/2018/04/22/plastic-pollution-firms-and-governments-fight-waste.html. Accessed on August 8, 2018.
57 Richard Grey. "What's the Real Price of Getting Rid of Plastic Packaging?" BBC.com. July 6, 2018. http://www.bbc.com/capital/story/20180705-whats-the-real-price-of-getting-rid-of-plastic-packaging. Accessed on August 2, 2018.
58 "The Great Pacific Ocean Pacific 'Garbage Patch' Even Worse Than Feared, New Research Shows".
https://www.youtube.com/watch?v=0uU1ZyQ1OwA; "The Great Pacific Ocean Patch Explained". https://www.youtube.com/watch?v=0EyaTqezSzs. Accessed on August 2, 2018.
59 *Financial Times.* Waste Management and Recycling. https://www.ft.com/stream/5c3cfe00-2de5-41b6-aaaf-c7cf0f81a8ce. Accessed on August 5, 2018.
60 Leslie Hook. "Plastic Waste Export Tide Turns to Southeast Asia after China Ban." June 13, 2018. *Financial Times.* https://www.ft.com/content/94ee72d0-6f26-11e8-852d-d8b934ff5ffa. Accessed on August 5, 2018.
61 European Commission. "Plastic Waste: A European Strategy to Protect the Planet, Defend Our Citizens and Empower our Industries". January 16, 2018. http://europa.eu/rapid/press-release_IP-18-5_en.htm. Accessed on August 8, 2018.
62 European Commission. "Plastic Waste".
63 World Economic Forum, "India Has Banned all Forms of Disposable Plastic in Its Capital". March 13, 2017. https://www.weforum.org/agenda/2017/03/india-bans-disposable-plastic-in-delhi. Accessed on August 11, 2018.
64 John McKenna. "These Indian Fisherman Take Plastic out of the Sea and Use It to Build Roads". June 28, 2018. World Economic Forum. https://www.weforum.org/agenda/2018/06/these-indian-fishermen-take-plastic-out-of-the-sea-and-use-it-to-build-roads/. Accessed August 11, 2018.
65 https://www.weforum.org/agenda/2018/01/Britain-to-end-plastic-waste-in-25-years. Accessed on August 11, 2018.
66 "Costa Rica Ants to Be the First Country to Ban all Single Use Plastics". World Economic Forum. https://www.weforum.org/agenda/2017/08/costa-rica-plastic-ban-2021. Accessed on August 11, 2018.
67 Peter Wells. "Starbucks Jolted Higher after Plastic Straw Phase-Out Plan". *Financial Times.* July 9, 2018.
68 https://www.usatoday.com/story/money/2018/08/06/aardvark-only-u-s-producer-paper-straws-acquired-meet-demand/916755002/. Accessed on August 11, 2018.

69 *The Economist.* "Plastic Eating Caterpillars Could Save the Planet". April 29, 2017. https://www.economist.com/science-and-technology/2017/04/29/plastic-eating-caterpillars-could-save-the-planet. Accessed on August 11, 2018.
70 Mark Sullivan. "Apple Now Runs on 100% Green Energy, and Here's How It Got There". *Fast Company.* April 9, 2018. https://www.fastcompany.com/40554151/how-apple-got-to-100-renewable-energy-the-right-way?utm_source=twitter.com&utm_medium=social. Accessed August 6, 2018.
71 Reuters. "Munich Re to Back Away from Coal-Related Business – CEO." https://in.reuters.com/article/munich-re-group-coal/munich-re-to-back-away-from-coal-related-business-ceo-idINL5N1UW0M2. Accessed August 6, 2018.
72 Caroline Flammer. "Green Bonds Benefit Companies, Investors and the Planet". *Harvard Business Review.* November 22, 2018. https://hbr.org/2018/11/green-bonds-benefit-companies-investors-and-the-planet?utm_medium=email&utm_source=newsletter_daily&utm_campaign=dailyalert_activesubs&utm_content=signinnudge&referral=00563&deliveryName=DM19598. Accessed on November 23, 2018.
73 Cassandra Sweet. "More Companies Prefer Green Bonds, Carbon Taxes". February 7, 2018. GreenBiz. https://www.greenbiz.com/article/more-companies-prefer-green-bonds-carbon-taxes. Accessed on November 23, 2018.
74 ClimateWise. University of Cambridge Institute for Sustainability Leadership. "Investing for Resilience". December 2016. https://www.cisl.cam.ac.uk/resources/publication-pdfs/Investing-for-resilience.pdf. Accessed on April 23, 2019.
75 ReImagina Puerto Rico. http://www.resilientpuertorico.org/en/. Accessed on November 23, 2018.
76 ReImagina Puerto Rico. http://www.resilientpuertorico.org/en/resilience-2/. Accessed on November 23, 2018.
77 ReImagina Puerto Rico. http://www.resilientpuertorico.org/en/reports-2/. Accessed on November 23, 2018.
78 Lloyds of London. Lloyds City Risk Index. https://cityriskindex.lloyds.com. Accessed on November 23, 2018
79 Lloyds of London. Lloyds City Risk Index. https://cityriskindex.lloyds.com/wp-content/uploads/2018/06/Lloyds_CRI2018_executive%20summary.pdf. Accessed on November 23, 2018.
80 TCFD. https://www.fsb-tcfd.org/. Accessed on April 24, 2019.
81 TCFD 2018 Status Report. https://www.fsb-tcfd.org/wp-content/uploads/2018/08/FINAL-2018-TCFD-Status-Report-092518.pdf. Accessed on April 24, 2019.
82 TCFD 2018 Status Report. https://www.fsb-tcfd.org/wp-content/uploads/2018/08/FINAL-2018-TCFD-Status-Report-092518.pdf. Accessed on April 24, 2019.
83 ClimateWise. https://www.cisl.cam.ac.uk/business-action/sustainable-finance/climatewise/about-membership. Accessed on April 24, 2019.
84 These insurers include ABI, Allianz, Aon, Argo International, Aviva, Beazley, CII, Chubb, Ecclesiastical, Hiscox, Lloyd's, MS Amlin, Navigators, Prudential, QBE, Renaissance Re, RSA, Sanlam, Santam, Swiss Re, Tokio Marine Kiln, Tokio Marine and Nichido, Willis Tower Watson, XL Catlin, Zurich.
85 ClimateWise. The ClimateWise Principles. https://www.cisl.cam.ac.uk/business-action/sustainable-finance/climatewise/principles. Accessed on April 23, 2019.
86 ClimateWise. University of Cambridge Institute for Sustainability Leadership. "Investing for Resilience". December 2016. https://www.cisl.cam.ac.uk/resources/publication-pdfs/Investing-for-resilience.pdf. Accessed on April 23, 2019.
87 ClimateWise. University of Cambridge Institute for Sustainability Leadership. "Investing for Resilience". December 2016. https://www.cisl.cam.ac.uk/resources/publication-pdfs/Investing-for-resilience.pdf. Accessed on April 23, 2019.
88 Partnership for Pollinators. ClimateWise. https://www.cisl.cam.ac.uk/business-action/natural-capital/pollinators/partnership-for-pollinators. Accessed April 24, 2019.

4 Society

Transforming social risk into opportunity from slavery and trafficking to safety and diversity

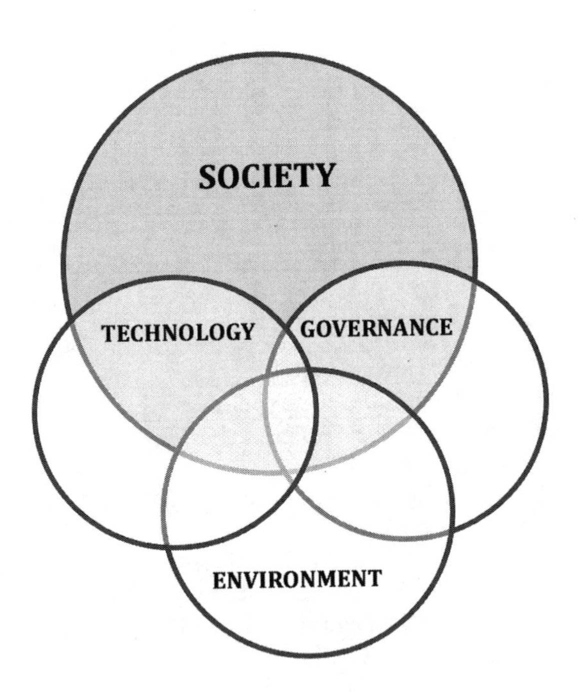

Society: risk and opportunity

Both risk and opportunity exist in the sphere of social issues. The risks range from human trafficking and slavery as one of the uglier realities that put the lives of more than 40 million human beings at risk every day to a wide variety of everyday risks to the health and safety of employees and consumers (among other stakeholders). Opportunities such as achieving mutual respect and real diversity from the front lines of the workplace to the highest levels of an organization – boards of directors and other oversight bodies – are also present.

This quote from a Romanian forced sex worker in Sicily describes the horror of the first type of risk:

> I came to Sicily with my husband. We needed to send money back to support our children in Romania. But the greenhouse farmer where we found work said I had to sleep with him, and if I refused, he wouldn't pay us. My husband said it was the only way we could keep our work. My employer threatened me with a gun, and when he finished, he just walked away. This went on for months. I left both the farm and my husband but found out it is the same wherever you try to find work here in Sicily.[1]

On the other end of the spectrum is the opportunity embedded in some of the social issues we discuss in this chapter, as this quote from Mary Barra, CEO of General Motors, illustrates:

> Ultimately, we want a world of zero crashes, zero emissions and zero congestion ... By living our values and continuing to invest in innovation, I am confident we will get there.[2]

Chapter 4: Society – summary overview

4.1 Introduction and overview

Much as we did in Chapter 3, on environment, in this chapter we tackle the concept of "society", "social", "civil society" and "societal" as it relates to the focus of this book – how leaders transform social/societal risk into opportunity and value.

We begin with the fundamentals – definitions of "social", "civil society" and "societal" – and then explore typologies of social issues, once again making use of several authoritative and knowledgeable resources, some we already deployed in Chapter 3 – the WEF, Allianz, RepRisk, UN Global Compact – and some more specific and directly relevant resources like the UN Guiding Principles on Business and Human Rights and the Global Slavery Index for 2018.

The key social topics we look at more deeply later in this chapter focus on three general categories of risk and opportunity: (1) the state of affairs in global human slavery and efforts to combat it; (2) the topic of organizational responsibility for environment, health, safety and security (EHS) of products and services and what must be done organizationally to mitigate and eliminate EHS risks in the workplace as well as in products and services; and (3) a critically important issue facing the world today at every level – international, national, corporate, governmental and non-profit – diversity and inclusion in the workplace, a set of issues that have become more visible than ever since the rise of the #MeToo events in 2017. These are all key social issues, risks and opportunities that a resilient and properly led organization should tackle in the 21st century.

4.2 What is "society"? Definitions and typologies

4.2.a Definitions

Below are definitions for three words or concepts that help put into context what we mean by the social issues, risks and opportunities. These are society, civil society and social.

"Society:"

- "The aggregate of people living together in a more or less ordered community".
- "The community of people living in a particular country or region and having shared customs, laws, and organizations".
- "A specified section of society".
- "A plant or animal community".
- "An organization or club formed for a particular purpose or activity".
- "The situation of being in the company of other people".[3]

"Civil Society:"

- "Society considered as a community of citizens linked by common interests and collective activity".[4]

"Social:"

- "Relating to society or its organization".
- "Relating to rank and status in society".
- "Needing companionship and therefore best suited to living in communities".
- "Relating to or designed for activities in which people meet each other for pleasure".
- "(of an insect) living together in organized communities, typically with different castes, as ants, bees, wasps, and termites do".
- "(of a mammal) living together in groups, typically in a hierarchical system with complex communication".
- "An informal social gathering, especially one organized by the members of a particular club or group".[5]

Also useful and more specifically from the perspective of the investor community, is the following definition of "Social":[6]

> Social include human rights, labor standards in the supply chain, any exposure to illegal child labor, and more routine issues such as adherence to workplace health and safety. A social score also rises if a company is well integrated with its local community and therefore has a "social license" to operate with consent.

4.2.b Typologies, classifications and categories

The following are a series of increasingly more focused lenses on what we mean by "social" issues, risks and opportunities in terms of issue categories and sub-categories. Table 4.1 lists the resources we have consulted for this discussion.

It is worth noting that, by definition, anything that involves or impacts human life on earth is or could be "social". What we are attempting here, however, is to shine a light on distinct "social" issues that the average leader should think about in managing or overseeing their organization and stakeholders' needs and expectations. Social issues may require attention both from an issue understanding standpoint to assist with better stakeholder relations and from risk and crisis management and opportunity identification perspectives as well.

Table 4.1 On society: sources and resources

Level of detail	*Resource*
FROM BIG PICTURE ...	i. Centre for the Study of Existential Risk[7] ii. World Economic Forum[8] iii. UN Sustainable Development Goals[9] iv. UN Universal Declaration of Human Rights[10]
... TO MORE FOCUSED ...	v. UN Human Rights Office Guiding Principles on Business and Human Rights[11] vi. UN Global Compact[12] vii. Allianz Risk Barometer[13]
... TO MORE GRANULAR	viii. MSCI[14] ix. RepRisk[15]

Source: Author.

4.2.b.i Centre for the Study of Existential Risk

As we saw in the previous chapter on environment, Cambridge-based Centre for the Study of Existential Risk (CSER) is engaged in a set of research projects around existential risks. Table 4.2 summarizes the issues they are tackling from a societal impact standpoint. While most organizations will hopefully not necessarily confront the existential level of risk embodied in CSER's work, we all benefit from the analysis and research at the extremes of risk. I am convinced that understanding extreme, even existential, risk can bring a valuable perspective to our understanding of risk as it relates to each of our particular organizations.

Table 4.2 Centre for the Study of Existential Risks, University of Cambridge: social issue research

Category of risk	*The challenge*
Extreme Risks and the Global Environment	Human activity is placing even more pressure on natural processes. Pushing past tipping points might lead to sudden, catastrophic ecosystem collapses or run-away, catastrophic climate change – with dire consequences for human society.
Global Catastrophic Biological Risks	Pandemics are as old as humanity, but in today's interconnected world we are more vulnerable than ever. The increase in the capability and spread of biotechnology poses new risks, from accidental release to intentional misuse.
Risks from Artificial Intelligence	Recent years have seen dramatic improvements in artificial intelligence, with even more dramatic improvements possible in the coming decades. In both the short-term and the long-term, AI should be developed in a safe and beneficial direction.

Source: Centre for the Study of Existential Risk.

4.2.b.ii World Economic Forum

From a somewhat less existential but nevertheless significant standpoint, we can turn again to the WEF GRR which, as we explored in Chapter 3 regarding the environment, also identified a number of socially impactful global risks with direct or indirect relevance to a given organization's activities and footprint – whether it is a company, government agency or NGO.

The most impactful societal global risks identified by WEF for 2019 in its top-ten category include the following:

- In terms of Likelihood:
  - #7 Large-scale involuntary migration.
  - #9 Water crises.
- In terms of Impact:
  - #4 Water crises.
  - #10 Spread of infectious diseases.

To make clear some of the trendlines, Table 4.3 presents the top-five most likely or impactful societal risks identified by WEF in the prior five years (2013, 2014, 2015, 2016 and 2017).

Table 4.3 Top societal risks identified by WEF GRR in its top-ten most likely/highest impact global risks 2012–2019[16]

2012	**Likelihood**	#1 Severe income disparity
	Impact	#3 Food shortage crisis
2013	**Likelihood**	#1 Severe income disparity
		#5 Mismanagement of population ageing
	Impact	N/A
2014	**Likelihood**	#1 Income disparity
	Impact	N/A
2015	**Likelihood**	N/A
	Impact	#1 Water crisis
		#2 Rapid and massive spread of infectious diseases
2016	**Likelihood**	#1 Large-scale involuntary migration
	Impact	#3 Water crises
		#4 Large-scale involuntary migration
2017	**Likelihood**	#2 Large-scale involuntary migration
	Impact	#3 Water crises
2018	**Likelihood**	#6 Large-scale involuntary migration
	Impact	#5 Water crises
		#7 Food crises
		#9 Large-scale involuntary migration
		#10 Spread of infectious diseases
2019	**Likelihood**	#7 Large-scale involuntary migration
		#9 Water crises
	Impact	#4 Water crises
		#10 Spread of infectious diseases

Source: World Economic Forum.

Societal risks do not appear to dominate in terms of numbers of top-ten risks the WEF has identified over the past six years (in contrast to environmental risks which dominated 2019, as discussed in Chapter 3), though it bears underlining that a couple of very significant social risks have appeared in the top ten repeatedly over the past decade – water crises and large-scale involuntary migration. This points to the fact that the top-ten risks generally can be said to be those that are deeply interconnected with societal risks. This combined approach to risks paints a more alarming societal risk picture than if viewed in isolation. The interconnectedness of these major categories of risk is an important theme we address repeatedly in this book.

4.2.b.iii UN Sustainable Development Goals

It's now time to turn to the Sustainable Development Goals (SDGs) and select the ones that have something to do with society. Once again one could make the case that most (if not all) of these issues are related in some way to society and social issues, even the two I have chosen to omit from this list – #14 Life Under Water and #15 Life on Earth, as the condition and fruits of the sea and land have a direct and impactful connection to society and humans, of course. However, it bears highlighting how socially important all of the SDGs are, so here goes once again (Table 4.4).

Table 4.4 United Nations sustainable development goals: social issues

SDG – Primarily society focused	*E*	*S*	*G*
1. No poverty		X	X
2. Zero hunger	X	X	X
3. Good health and well-being		X	X
4. Quality education		X	X
5. Gender equality		X	X
6. Clean water and sanitation	X	X	X
7. Affordable and clean energy	X	X	X
8. Decent work and economic growth		X	X
9. Industry, innovation and infrastructure	X	X	X
10. Reduced inequalities		X	X
11. Sustainable cities and communities	X	X	X
12. Responsible consumption and production	X	X	X
13. Climate action	X	X	X
14. Life below water	X		X
15. Life on land	X		X
16. Peace and justice – strong institutions		X	X
17. Partnerships for the goals		X	X

Source: United Nations.

4.2.b.iv The Universal Declaration of Human Rights

As is widely known, the Universal Declaration of Human Rights (UDHR) was first issued in 1948 by the then newly established United Nations and at the heels of successive world wars, other atrocities and crimes against humanity. Here's what the UN itself says about the UDHR on its website:

> The Universal Declaration of Human Rights (UDHR) is a milestone document in the history of human rights. Drafted by representatives with different legal and cultural backgrounds from all regions of the world, the Declaration was proclaimed by the United Nations General Assembly in Paris on 10 December 1948 (General Assembly resolution 217 A) as a common standard of achievements for all peoples and all nations. It sets out, for the first time, fundamental human rights to be universally protected and it has been translated into over 500 languages.[17]

Among the key elements contained in the 30 articles that comprise the UDHR – which serve as antecedents to and inspiration for the later development of the human rights field and some of the practical frameworks we work with today such as the UN's Guiding Principles on Business and Human Rights (discussed later) – are the following:

1. Dignity, liberty, equality, and brotherhood.
2. Right to life.
3. Prohibition of slavery.
4. Rights of defense against human rights violations.
5. Rights of the individual towards the community, including freedom of movement.
6. "Constitutional liberties", including spiritual, public and political freedoms, such as freedom of thought, opinion, religion, conscience and peaceful association.
7. Individual economic, social and cultural rights, including healthcare, food, clothing, housing and medical care and necessary social services.
8. Accommodations for security in case of physical debilitation or disability.
9. Care for motherhood or childhood.

4.2.b.v The UN Guiding Principles on Business and Human Rights

As we delve a little more deeply into the issue of "human rights" in the context of leaders and their organizations (whether companies, government agencies or NGOs), it's good to turn to a definition from one of the experts. In her book, *The Business of Human Rights*, Alex Newton, provides the following definition of "Human Rights":

> Human rights are the universal rights and freedoms that apply to everyone, irrespective of their gender, race, religion or country of origin. They aim to secure dignity and equality for all people.[18]

The UN Human Rights Office of the High Commissioner states the following about the UN Guiding Principles on Business and Human Rights:

> These Guiding Principles are grounded in recognition of: (a) States' existing obligations to respect, protect and fulfil human rights and fundamental freedoms; (b) The role of business enterprises as specialized organs of society performing specialized functions, required to comply with all applicable laws and to respect human rights; (c) The need for rights and obligations to be matched to appropriate and effective remedies when breached.[19]

Table 4.5 provides some of the highlights of the 31 Guiding Principles.

Table 4.5 Select UN Guiding Principles on Business and Human Rights[20]

- States must protect against human rights abuse.
- States should set out clearly the expectation that all business enterprises domiciled in their territory and/or jurisdiction respect human rights throughout their operations.
- States should enforce laws, enable human rights enforcement, provide guidance, encourage communication.
- States should take additional steps to protect against human rights abuses by business enterprises that are owned or controlled by the state.
- States should exercise adequate oversight in order to meet their international human rights obligations when they contract with, or legislate for, business enterprises.
- States should promote respect for human rights by business enterprises with which they conduct commercial transactions.
- States should help ensure that business enterprises operating in conflict-ridden areas are not involved with the heightened abuse that can occur in such contexts.
- States should ensure that governmental departments, agencies and other state-based institutions that shape business practices are aware of and observe the state's human rights obligations when fulfilling their respective mandates.
- States should maintain adequate domestic policy space to meet their human rights obligations when pursuing business-related policy objectives with other states or business enterprise.
- Business enterprises should respect human rights.
- The responsibility of business enterprises to respect human rights refers to internationally recognized human rights.
- The responsibility of business enterprises to respect human rights applies to all enterprises regardless of their size, sector, operational context, ownership and structure.
- In order to meet their responsibility to respect human rights, business enterprises should have in place policies and processes appropriate to their size and circumstances.

Source: United Nations Human Rights Office of the High Commissioner.

4.2.b.vi The UN Global Compact

Another angle comes from the UN Global Compact's six basic principles that relate to large buckets of social and individual rights categories applicable to organizations, as summarized in Table 4.6.[22]

Table 4.6 United Nations Global Compact 10 principles:[21] social issues

Human Rights	**Principle 1**	Businesses should support and respect the protection of internationally proclaimed human rights.
	Principle 2	Businesses should make sure that they are not complicit in human rights abuses.
Labor	**Principle 3**	Businesses should uphold the freedom of association and the effective recognition of the right to collective bargaining.
	Principle 4	Businesses should uphold the elimination of all forms of forced and compulsory labor.
	Principle 5	Businesses should uphold the effective abolition of child labor.
	Principle 6	Businesses should uphold the elimination of discrimination in respect of employment and occupation.

Source: United Nations Global Compact.

As noted in Chapter 3, the UN Global Compact also has useful resources for organizations that want to align their internal policies with either or both the UN Global Compact Principles and the SDGs helping to tie each SDG to actual, measurable goals and objectives.[23]

4.2.b.vii Allianz Risk Barometer

If we now turn to the more granular review of top-ten business risks from the Allianz Risk Barometer, we will find quite a few with social issue implications. A couple of observations include that a brand-new business risk with deep social implications entered the ranking in 2019 at #10 – shortage of skilled workforce, a reflection of both positive (almost full employment in some economies) and negative (lack of properly skilled/trained workers) developments (Table 4.7).

Table 4.7 Allianz Risk Barometer Top-Ten Business Risks 2017–2019:[24] focus on social issues

Top-Ten Risks (w/Social Issues)	*2019 Ranking*	*2018 Ranking*	*2017 Ranking*	*ESGT Issue*
Business Interruption (Including Supply Chain Disruption)	#1	#1	#1	E, S, G, T
Natural Catastrophes (e.g., Storm, Flood, Earthquake)	#3	#3	#4	S, E
Changes in Legislation and Regulation (e.g., Government Change, Economic Sanctions, Protectionism, Brexit, Eurozone Disintegration)	#4	#5	#5	G, S

(*Continued*)

Table 4.7 Continued

Top-Ten Risks (w/Social Issues)	*2019 Ranking*	*2018 Ranking*	*2017 Ranking*	*ESGT Issue*
Fire, Explosion	#6	#6	#7	S, E
New Technologies (e.g., Impact Of Increasing Interconnectivity, Nanotechnology, Artificial Intelligence, 3d Printing, Drones)	#7	#7	#10	G, T
Climate Change/Increasing Volatility Of Weather	#8	#10 (*NEW RISK*)	N/A	
Loss of Reputation or Brand Value	#9	#8	#9	G
Shortage of Skilled Workforce	#10 (*NEW RISK*)	(*POLITICAL RISK #9*)	(*POLITICAL RISK #8*)	S, G

Source: Allianz.

4.2.b.viii MSCI

For the investor community ESG perspective, Table 4.8 provides a typology of social issues that MSCI utilizes as part of its analytical framework. As noted in Chapter 3 which dealt with environmental issues, the MSCI framework also includes a category of "social opportunities".

Table 4.8 MSCI categories of ESG issues:[25] "social"

Human capital
- Labor management
- Human capital development
- Health and safety
- Supply chain labor standards

Product liability
- Product safety and quality
- Chemical safety
- Financial product safety
- Privacy and data security
- Responsible investment
- Health and demographic risk

Stakeholder opposition
- Controversial sourcing

Social opportunities
- Access to communications
- Access to finance
- Access to healthcare
- Opportunities in health and nutrition

Source: MSCI.

4.2.b.ix RepRisk

RepRisk provides a more granular and useful data set containing ESG Issues and Topic Tags that are helpful in identifying some of the more immediate issues that leaders and their organizations may confront, depending of course on their purpose, sector, footprint, products, services, etc.

Once again, using the research scope that RepRisk developed for ESG issues, Table 4.9 outlines their list of social issues, which they break down into two major categories – Social – Community Relations and Social – Employee Relations plus several cross-cutting issues.[26]

Table 4.9 RepRisk ESG research scope for social issues

Community relations	*Employee relations*
Human rights abuses, corporate complicity	Forced labor
Impacts on communities	Child labor
Local participation issues	Freedom of association and collective bargaining
Social discrimination	Discrimination in employment
	Occupational health and safety issues
	Poor employment conditions
Cross-Cutting Issues	
Controversial products and services	
Products (health and environmental issues)	
Violation of international standards	
Violation of national legislation	
Supply chain issues	

Source: RepRisk.

Drilling down even deeper, Table 4.10 shows the ESG Topic Tags linked to social issues that RepRisk uses to tag specific content linked to risk incidents across 80,000+ public sources on a daily basis:

Table 4.10 Sampling of RepRisk ESG topic tags for ESG issues with social implications[27]

Abusive/illegal fishing
Agricultural commodity speculation
Asbestos
Automatic and semi-automatic weapons
Biological weapons
Chemical weapons
Cluster munitions
Conflict minerals
Cyberattack

(*Continued*)

Table 4.10 Continued

Depleted uranium munitions
Diamonds
Drones
Gambling
Gender inequality
Genetically modified organisms (GMO)
Genocide/ethnic cleansing
Human trafficking
Hydropower (dams)
Indigenous people
Involuntary resettlement
Land mines
Land grabbing
Marijuana/cannabis
Migrant labor
Nuclear power
Nuclear weapons
Palm oil
Pornography
Predatory lending
Privacy violations
Protected areas
Rare earths
Security devices
Ship breaking and scraping
Soy
Tobacco
Water scarcity

Source: RepRisk.

To illustrate how such ESG issues/risk incidents can translate into real-world cases, Table 4.11 shows RepRisk's Most Controversial Companies for 2017 showing the cases from that year with mostly social issue implications.

Table 4.11 RepRisk Most Controversial Companies 2017:[28] cases with social issues

MCC 2017 rank	*Company name*	*Peak RepRisk index 2017 (1–100)*[29]	*Sector*	*Country of head-quarters*	*Primary ESGT issue(s)*
#1	**The Weinstein Company**	92	Media	USA	S & G
#3	**J&F Invest-mentos SA**	83	Food and beverage, personal and household goods	Brazil	S & G
#5	**Stalreiniging Barneveld (Chickfriend)**	82	Support services (industrial goods and services)	Netherlands	E S & G

(*Continued*)

Table 4.11 Continued

MCC 2017 rank	*Company name*	*Peak RepRisk index 2017 (1–100)*[29]	*Sector*	*Country of head-quarters*	*Primary ESGT issue(s)*
#6	**Equifax Inc.**	79	Financial support services (industrial goods and services)	USA	S G & T
#8	**Odebrecht SA**	74	Construction and materials	Brazil	E S & G
#9	**Petroleos de Venezuela SS**	73	Oil and gas	Venezuela	E S & G

Source: RepRisk.

Finally, in 2017, there were many – in fact a preponderance of – of social issues in RepRisk's Ten Most Controversial Projects as illustrated in Table 4.12.

Table 4.12 RepRisk Most Controversial Projects 2017:[30] cases with social issues

MCC 2017 rank	*Company name*	*Peak RepRisk index 2017 (1–100)*	*Sector*	*Country of head-quarters*	*Primary ESGT issue(s)*
#1	**Grenfell Tower**	93	Media	UK	S
#2	**Ctrip Day Care Center**	73	Industrial metals	China	S
#4	**Brook House Immigration Removal Centre**	68	Support services (industrial goods and services)	UK	S & G
#6	**Jalabiya Cement Works**	59	Financial support services (industrial goods and services)	Syria	S & G
#7	**OPL 245 Oil Block**	59	Aerospace and defense industrial engineering	Nigeria	S & G
#8	**Changwon Jinhae Shipyard**	58	Construction and materials	South Korea	S
#9	**Guangzhou No 7 Thermal Power Plant**	57	Oil and gas	China	S
#10	**Imperial Pacific Resort Hotel**	57	Industrial transportation	USA	S

Source: RepRisk.

4.2.c The way forward

After this review of definitions, types, typologies and classifications of all things "social", we are hopefully better equipped, indeed armed, with concepts that will allow the reader to deploy some of this knowledge in furtherance of your own organization (whatever it might be) to identify and distill some of your key social issues, risks and opportunities.

In the next section, we provide a tour of several of the larger strategic social issues, risks and opportunities organizations of every type and stripe may face at some point in their journey through today's unpredictable, turbulent and choppy waters. They are:

- **Modern-day human slavery.** We first turn to the topic of modern-day human slavery to understand what it means and how pioneering organizations are learning to understand and even lead on the resolution of this complex topic.
- **Workplace environment, health and safety.** Next, we look at workplace environment, health and safety by contrasting a company that has taken it seriously for many years – Public Service Enterprise Group (PSEG) in the US (full disclosure: I used to work there) – with one that did not (at least before its disastrous scandal) – Japanese nuclear power company, TEPCO Fukushima.
- **Workplace conduct and behavioral topics – discrimination, harassment, diversity and inclusion.** We close this chapter by looking at another key social issue: a package of interrelated workplace conduct and behavioral topics – discrimination, harassment, diversity and inclusion.

4.3 Leaders navigating social risks, crises and opportunities

4.3.a From human rights abuse to valuing human beings: eradicating modern-day human slavery

> In 2016, 40.3 million people were living in modern slavery. It exists in every corner of the world yet is seemingly invisible to most people. Unravelling this problem requires sustained vigilance and action. Take this fire in a clandestine textile workshop in Buenos Aires, Argentina. These images are from 2006, yet the fight for justice for the five boys and a pregnant woman who were forced to work at this facility, and died in this fire, is still ongoing. In 2016, a court sentenced the workshop operator to 13 years prison for servitude and destruction of property causing death. This year, the court called for a deposition from the owner of the clothing brands, who also owns the property. The fight to end modern slavery continues. We can, and must, do more.[31]

Sadly, the origin of human slavery probably coincides with the origin of human life on earth and the creation of forms of social order in which the darker angels of our nature began to abuse fellow members of the human race in an organized or systematic way.

This book will not delve into the remote antecedents of this scourge. However, this book fully intends to lay bare the modern-day expressions of this quintessential and ultimate inequality that is part of many societies worldwide. Frankly, as we will see below, modern-day human slavery touches almost everyone in the world today – even those of us who are horrified that this is still taking place and yet indirectly allow it to happen. Whether we are conscious of it or not or willing to admit it or not, human slavery touches all of us through the supply chain of products and services that we all enjoy, especially in the advanced industrialized world.

4.3.a.i What is "modern-day human slavery"?

According to the Global Slavery Index, published by the leading global anti-slavery nonprofit organization The Walk Free Foundation, modern-day slavery consists of three main issues:

- Forced labor.
- Human trafficking.
- Slavery and slavery-like practices (including forced marriage).

Additional violations occur at the intersection of these main topics, as follows (see Figure 4.1):

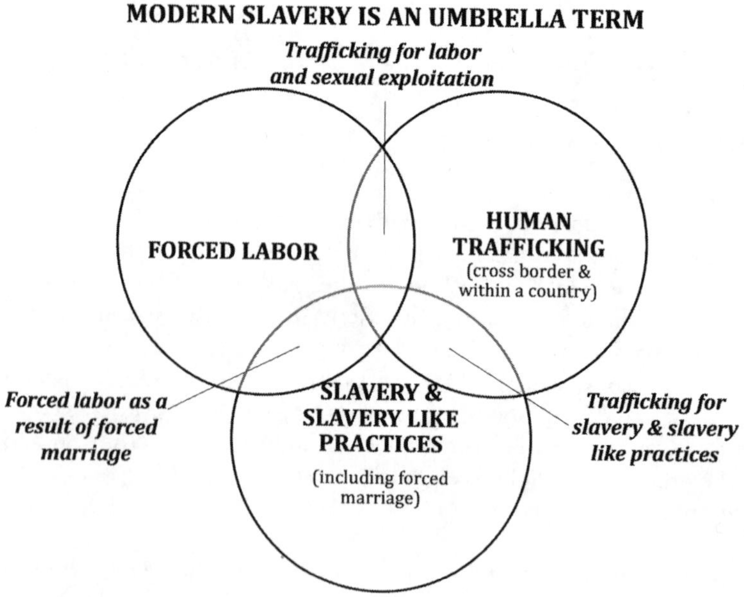

Figure 4.1 What is modern-day slavery?

Source: The Global Slavery Index 2018. https://www.globalslaveryindex.org/. Accessed on July 21, 2018. Graphic reconstructed by the author.

- *Trafficking for labor and sexual exploitation* is at the intersection of forced labor and human trafficking.
- *Trafficking for slavery and slavery-like practices* is at the intersection of human trafficking and slavery or slavery-like practices.
- *Forced labor as a result of forced marriage* is at the intersection of slavery and slavery-like practices and forced labor.[32]

Table 4.13 provides a chronological overview of the developments on a global level via government, business and society on this topic starting in 2000 when the first global agreement was reached to tackle and try to eradicate the modern-day versions of an age-old stain on humanity – the enslavement of people by people.

Table 4.13 A chronology of key developments in global cooperation against human slavery since the year 2000[33]

2000	The United Nations passes the protocol to prevent, suppress, and punish trafficking in persons as part of the convention against transnational organized crime. It is the first global legally binding treaty with an internationally agreed definition of trafficking in persons.
2001	The countries of the economic community of western African states agree on an action plan to tackle slavery and human trafficking in the region.
2002	The international cocoa initiative is established as a joint effort of anti-slavery groups and major chocolate companies to protect children and contribute to the elimination of child labor. It is the first-time members of an entire industry have joined forces to tackle slavery in its supply chain.
2004	Brazil launches a national pact for the eradication of slave labor. It brings together civil organizations, businesses and government to get companies to commit to the prevention and eradication of forced labor in their supply chains. It also includes a provision to create a "dirty list" if companies are found selling products produced by slaves. The United Nations appoints a special rapporteur on human trafficking.
2005	The International Labor Organization's (ILO) first global report on forced labor puts the number of slaves worldwide at 12.3 million. A 2012 update increases the number to 20.9 million.
2008	The Council of Europe convention on action against trafficking in human beings comes into force. The convention is the first international law to define trafficking as a violation of human rights, and it guarantees minimum standards of protection to victims.
2011	The ILO adopts a convention laying down basic rights of domestic workers. California enacts the California Transparency in Supply Chains Act. It requires major manufacturing and retail firms to disclose what efforts they are making to eliminate forced labor and human trafficking from their supply chains.
2012	The US Securities and Exchange Commission passes the conflict minerals rule, requiring major publicly held corporations to disclose if their products contain certain metals mined in areas of conflict in Eastern Congo or neighboring countries and if payment for these minerals supports armed conflict.

(*Continued*)

Table 4.13 Continued

2013	The first Global Slavery Index released by the Walk Free Foundation estimates that there are 29.8 million slaves globally. The 2014 index increases that to 35.8 million, and the 2016 index to 45.8 million.
2014	The ILO adopts a protocol on forced labor, bringing its 1930 convention on forced labor into the modern era to address practices such as human trafficking.
2015	Britain's Modern Slavery Act comes into force. It requires businesses to disclose what action they have taken to ensure their supply chains are free of slave labor. It also increases the maximum jail sentence for traffickers to life from 14 years and allows authorities to force traffickers to pay compensation to their victims. It also brings in measures to protect people feared at risk of being enslaved. The United Nations adopts 17 Sustainable Development Goals, including a target of ending slavery and eradicating forced labor and human trafficking.
2016	In April, businesses in Britain must start reporting steps they are taking to tackle slavery in their supply chains, to comply with the Modern Slavery Act.
2018	The *2018 Global Slavery Report* states that there are a total of 40.3 million people around the world who can be characterized as suffering from one or more types of human slavery.

Sources: ILO, Walk Free Foundation, Anti-Slavery International, Free the Slaves, Anti-Trafficking Review, Reuters.

In their *2018 Global Slavery Report*, the Walk Free Foundation points out the following disturbing numbers (applicable as of 2016):

- In 2016, there were 40.3 million people worldwide in modern slavery.
- 71% or 28.6 million were women or girls.
- 29% or 11.7 million were men or boys.
- Of the total female population, more than half or 15.4 million were in forced marriages.
- Of the total population, over half or 24.9 million were in forced labor situations.[34]

Table 4.14 Governments with best and worst policies regarding modern-day slavery[35]

Best/Most proactive	*Worst/Least proactive*
Netherlands	North Korea
United States	Libya
United Kingdom	Eritrea
Sweden	Central African Republic
Belgium	Iran
Croatia	Equatorial Guinea
Spain	Burundi
Norway	Republic of the Congo
Portugal	Sudan
Montenegro	Mauritania

Source: The Global Slavery Index.

Finally, while the more industrialized and democratic nations generally took greater responsibility for combating modern-day slavery than did some of the less developed or more authoritarian nations globally (see Table 4.14), the more developed world was the greater consumer of products and services tied into supply chain modern-day slavery by far, especially the US (the largest consumer purchasing about 40% of such products), followed by Japan, Germany, the UK, France and Canada.

Table 4.15 illustrates the primary industries or products at risk for supply chain-related modern-day slavery.

Table 4.15 Top five products at risk of modern-day slavery imported into G20 countries[36]

1. Technology sector (laptops, computers, phones)	US$200.1 Billion
2. Garment sector	US$127.7 Billion
3. Fish	US$12.9 Billion
4. Cocoa	US$3.6 Billion
5. Sugarcane	US$2.1 Billion

Source: The Global Slavery Index.

It is clear who the beneficiaries of the modern-day slavery supply chain are. It is for this reason that clear cross-sectoral policies and measures need to be undertaken by the collective action of governments, businesses and non-profits.

In the global fight against modern-day slavery, the UK government has taken a leadership role to try to transform this inhuman risk into an opportunity for a better world. But governments cannot fight this problem alone, and that is why a public/private/NGO partnership to combat slavery in the supply chain, for example, is the only way that we will be able to tackle and reduce (eventually eliminate) this horrific plague on humanity that is still so prevalent in the 21st century.

The *Global Slavery Index Report* for 2018 suggests the following general policies and collaboration that need to take place at the intersection of government, business and nonprofits:[37]

1. Governments and business need to prioritize human rights in decision-making when engaging with repressive regimes.
2. Governments must proactively anticipate and respond to modern-day slavery in conflict situations.
3. Governments must improve modern-day slavery responses at home.
4. G20 governments and businesses must address modern-day slavery in supply chains.
5. Government must prioritize responses to violations against women and girls.

Several examples of human rights abuses are presented below, including human slavery in the supply chain. Also presented is a review of some of what is being done to transform these serious social risks into an opportunity for a better world for so many who have been and continue to be enslaved.

4.3.a.ii Conflict mineral supply chain and human rights

Mini-case study: the conflict minerals supply chain and human rights in the electronics industry[38]

- As supply chains go increasingly global, we rarely stop to consider the human costs of producing advanced electronic devices – like smartphones, tablets and laptops. Yet human rights challenges are particularly pertinent to the industry given its reliance on so-called "conflict minerals" to make capacitors for electronic goods.
- Conflict minerals are tantalum, tin, tungsten and gold (often referred to as "3TG") when sourced from the Democratic Republic of the Congo (DRC). While these minerals are also sourced from various other countries around the world, when derived from mines in the DRC, they have been associated with the decades-long civil war there in which more than 5.4 million people have died.
- The civil war in the DRC has been funded, in large part, by rebel and government-backed militias. The militias often control the country's mineral deposits – either directly or indirectly (through taxing and exploiting artisanal miners and local populations). Egregious human rights violations have been connected to the militias' activities. These include murder, mass-rape as a weapon of war, torture, forced labour, and the conscription of child soldiers. Accordingly, electronics companies purchasing conflict minerals from the DRC have been implicated in the gross human rights violations occurring there.[39]
- For information on what public/private actions can be taken, see the OECD due diligence guidelines for conflict minerals.[40]

Source: Alex Newton.

4.3.a.iii Chinese forced labor/prison labor supply chain

Mini-case study: The Chinese forced /prison labor supply chain[41]

- Prison labor is common in China, where the law states that prisoners able to work must do so—a system known as "reform through labor". China is home to around 2.3m prisoners and pre-trial detainees, according to the Institute for Criminal Policy research, giving it the world's second-largest prison population after the US.
- "Most of the companies set up under prison provincial administration bureaus in China look, from the outside, like ordinary companies," says Joshua Rosenzweig of Amnesty International in Hong Kong. "Foreign corporations are in a pretty tough position to do the kind of due diligence that would be needed to identify whether their supply chains are connected to prison labor".

Source: *Financial Times.*

4.3.a.iv Fixing global fishing industry slavery in the supply chain

An example of the modern-day slavery debacle is also very visible in the fishing industry, which has been in the crosshairs of various media investigations over the past few years, and has received greater attention from governments, non-profits and the business sector.

In 2017, the US Department of State in its annual Human Rights Report made noise with its findings of extensive human rights abuses, including slavery, sexual exploitation and child labor in the Thai fishing industry.[42] A couple of years before this report was issued, in 2015, several leading media, including *The New York Times*, *The Guardian* and the *Associated Press*, produced important reports about the linkage of the fishing industry with human slavery.

Some of the excesses reported include the following aberrations of human behavior:[43]

- people bought and sold between boats.
- "the sick cast overboard".
- "the defiant beheaded".
- "the insubordinate sealed for days below deck in a dark, fetid fishing hold".
- "long-haul fishing, in which vessels stay at sea, sometimes for years, far from the reach of authorities".

And why is this happening?

> While forced labor exists throughout the world, nowhere is the problem more pronounced than here in the South China Sea, especially in the Thai fishing fleet, which faces an annual shortage of about 50,000 mariners, based on United Nations estimates. The shortfall is primarily filled by using migrants, mostly from Cambodia and Myanmar.

Among the findings is this chilling reporting from *The New York Times*:

> Most of Thailand's seafood workers are migrants from neighboring Cambodia or Myanmar; they were brought into Thailand illegally by traffickers, provided fake documents and often sold to boat captains, the report said. On fishing boats, these workers routinely faced limited access to medical care for injuries or infection; worked 16-hour days, seven days a week; endured chronic sleep deprivation; and suffered from an insufficient supply of water for drinking, showering or cooking, the report found.
>
> "Sometimes, the net is too heavy, and workers get pulled into the water and just disappear," one Burmese worker said, according to the report. "When someone dies, he gets thrown into the water".[44]

This awareness-raising by the media, government and non-profit sectors has led to a broad variety of efforts to combat the scourge of slavery in the fishing industry. While this book doesn't claim to know of all of the efforts that are being made, several are worth noting. There are a variety of nonprofits, business and cross-sector collaborative resources and tools that the fishing industry can and is deploying worldwide.

Seafood Slavery Risk has published a variety of business tools. One of them involves a toolkit of key questions businesses should ask, as outlined in Table 4.16.

Table 4.16 Seafood Slavery Risk tool: key questions[45]

Key question	*Explanation*	*Good practice*
1. **Do you know where the seafood you purchase is coming from?**	Knowing basic sourcing details for the seafood you purchase is essential to understanding the risks of forced labor, human trafficking, and hazardous child labor associated with fisheries.	You should look to require and provide updates about where your seafood is coming from at least twice a year, if not more frequently.
2. **Does your company have a clear agreement or policy covering human rights standards in place with suppliers and business partners?**	Legal documents such as a master purchase agreement, code of conduct, or even a purchase order can be used, or a separate stand-alone policy can be created.	The agreement should include key issues such as forced labor and child labor, and all business partners should be required to sign and agree to the terms.
3. **Has your company conducted a risk assessment in the last year?**	At least annually, use the risk tool and other sources to search and review your seafood purchases to identify if any potential risks may exist.	The risk tool isn't a direct indicator of human rights issues in your supply chain, but it can be used to give notice of potentially high-risk areas that should be prioritized for further review.
4. **How will your company further investigate potential high-risk seafood purchases from suppliers or business partners?**	Based on the potential risks identified in your search of the risk tool and other sources, your company should then follow-up with any suppliers or business partners in an area designated as a critical or high risk to better determine if potential human rights violations exist.	While the risk tool is a useful database of information, it's not able to capture all potential instances of human rights abuses. Therefore, it's important that you continue to conduct additional reviews of all purchases and business partners using data requests, audits, or other tools.

(*Continued*)

Table 4.16 Continued

Key question	*Explanation*	*Good practice*
5. Does your company have a plan for how it will respond to a human rights issue or violation, including a corrective action plan?	It's essential that a company discusses and plans for this ahead of time, so as to be prepared before any cases might occur.	If a supplier or business partner is found to be in violation of a human rights agreement, it's important your company stays engaged and works to implement a corrective action plan. If a company immediately walks away, then those human rights violations will continue and possibly be driven further underground. In the event your company identifies human rights violations, be sure to look at engaging with on-the-ground organizations and stakeholders who can assist with implementing a corrective action plan.

Source: Seafood Slavery Risk.

On the heels of much of this reporting and increased government focus on correcting these abuses, corporations have also done their part. Nestlé, in 2015, issued a report titled "Nestle Takes Action to Tackle Seafood Supply Chain Abuses"[46] in which they extensively reported on the deep problems in the seafood supply chain and issued an Action Plan[47] that states the following, as quoted in *The New York Times*:

> The seafood industry in Thailand suffers from widespread labor and human rights abuses, exposing virtually all American and European companies that buy seafood from there to the "endemic risk" of having these problems as part of their supply chain, according to a report released on Monday by the food giant Nestlé.[48]

By working together on the twin serious problems of slavery and environmental depletion, it is possible to create transparency in the supply chain which would end up not only assisting in the elimination of forced labor, including hazardous child labor, but also have other corollary benefits: the elimination of fraud on the consumer (a governance benefit), the reduction of depletion of fisheries (an environmental benefit), and more than anything else, a huge social benefit – eliminating slavery conditions and perhaps creating living wages and conditions for fishing industry workers and their families.

There are so many other examples of areas in which human slavery is taking place and of combined or individual efforts by government, business and the nonprofit sector regarding this scourge that are beyond the scope of this book, but in Table 4.17 are additional examples and resources.

Table 4.17 RepRisk migrant labor findings[49]

Most Associated Countries	Most Associated Companies
1. United States	1. FIFA
2. Russia	2. Walmart
3. Thailand	3. Dhammakaset Co.
4. Brazil	4. Sarbanand Farms LLC
5. Qatar	5. Betagro Group; Industria de Diseno Textil SA; Marks & Spencer Group
Most Associated Projects	**Most Associated NGOs**
1. 2022 FIFA World Cup – Qatar	1. Human Rights Watch
2. 2018 FIFA World Cup – Russia	2. Amnesty International
3. Lop Buri Chicken Farm (Thailand); New Zenit Stadium (Russia)	3. Dutch Federation of Trade Unions
4. Crist Shipyard; Imperial Pacific Resort Hotel (Saipan)	4. Migrant Workers Rights Network
5. Saadiyat Island Campus (New York University Abu Dhabi)	5. Building and Wood Workers International; International Transport Workers Federation

Source: RepRisk.

And, as this synopsis of developments reflects, the corporate world has had to forcibly or voluntarily step up to this plate. Indeed, one of the great revelations of our time is that as supply chains have become increasingly global and complex, we have discovered how deeply embedded, and in some ways intractable, some of the worst human rights violations (which qualify as modern-day slavery) are. That calls for a proactive "call to arms" by business, government and society and not a passive, collective burying of heads in the sand.

4.3.b From health, safety and security irresponsibility to accountability

Most adults have had one or more experiences either in the workplace or personally with a product or service with a health, safety or security issue. The array of such issues is broad and diverse as we explored in the typology section earlier in this chapter.

To underscore the responsibility that leaders have to their stakeholders on these types of issues in the workplace or with regard to the quality and safety of products and services, the spectrum of situations that could go wrong at any given time from the minor to the downright life threatening or life-taking bears highlighting.

Here are some fact-based hypothetical situations you may have come across in your own experience or through the headlines:

Office safety risk. You are an office worker and there are computer cables not properly secured that you can trip on; what should you do? Common sense would say that everyone is responsible for safety awareness and simple corrective actions in any workplace – so just speak up and make sure this apparently minor but potentially significant hazard is removed. Better yet, go to your office manager or HR and ask for help in creating more awareness around this issue so it's not just a one-off situation but a lesson learned that helps to improve office safety and a sense of collaboration and mutual care.

Facility built on earthquake faultline. You work in an industrial plant that is located on an earthquake fault line; does your facility have an emergency plan should there be an event? What should such preparations look like? A range of possible options should exist, but if you work in a nuclear power plant in an earthquake prone area near the sea (like TEPCO Fukushima, explored later), hopefully everything has been done from a crisis preparedness standpoint to protect and secure life and assets and other surrounding properties to avoid the unmitigated disaster that became Fukushima.[50]

Power outage. You work in an office building and the power goes out; what do you do? Does your company or the building have a facilities or crisis management protocol in place? Did they train you in advance on what to do? I lived through just such an experience in New York City in the summer of 2003 when the entire East Coast of the US went dark for two days. See the Crisis Readiness section of Chapter 7 for a mini-case study on what happened next.

Oil rig safety risk. You are an oil rig worker and the health and safety issues you are exposed to are exponential to those of an office worker; what are the safety protocols that your company has in place and have you experienced regular training, including for speaking up and reporting of concerns or hazards without fear of retaliation? Probably not, if you were a worker at BP's Deepwater Horizon in 2010, which led to the death of 11 workers and a multi-billion dollar environmental disaster affecting many stakeholders.

Hurricane zone risk. A hurricane threatens the place you live; what do you do? It's a Category 5 hurricane and you're in Puerto Rico in September 2017. Was the federal government prepared? Was the local government prepared? Were businesses prepared? Was the electric utility, PREPA, prepared? We all now know the answers to these questions, which are neither good nor heartening. The question going forward is: will Puerto Rico and similarly vulnerable geographies be prepared for the next one?

Travel safety risk. You are traveling for business or pleasure and are caught up in a political crisis with potential or actual violence taking place; what do you do? Do you have travel back-up plans, a safety/security kit, some form of travel rescue insurance, access to colleagues and/or resources for help? If you're there for work, did your company equip you with standard procedures, access to information and resources that might be needed under such travel crisis circumstances?

Automobile safety risk. You are the manufacturer of a safety device used in automobiles that appears to be faulty to the point of threatening and even taking lives. What happens next? Ask Takata, the once dominant Japanese-based global manufacturer of automobile airbags that turned out to be defective and in some cases even deadly, exactly the opposite of what they were designed to do – to save lives. After endless investigations, lawsuits, bad press and ascendant reputation risk, they declared bankruptcy in 2017 and were acquired by a Chinese-based global competitor.

Garment industry facility safety. You are in the garment industry, and your supply chain consists of several layers of contractors and subcontractors, and the facilities at which some of these workers are housed violate all manner of building codes. As the retailer of these clothing products, what is your responsibility for policing these somewhat remote subcontractors and their health and safety standards? Ask the global retailers in the supply chain of Rana Plaza in Bangladesh – one of the worst disasters to befall garment workers anywhere, ever, with over 1,000 people losing their lives because of such subcontractors' gross negligence in ignoring building safety standards and allowing a mountain of violations to go unheeded over time.

Different approaches to risk and crisis management provide different outcomes. What follows is an analysis primarily using reputation risk data that provides an interesting lens on what corporate social responsibility versus irresponsibility looks like.

By examining this data, in addition to understanding the relevant recent histories of these companies, one can make a few observations that are valuable to industry generally – mostly to leaders who have oversight, supervising or operational responsibility – on what creates resilience, protects stakeholders and converts social risk into social opportunity and value creation in the process.

Let's start by contrasting the health, safety, security and environmental disaster that was TEPCO Fukushima nuclear plant meltdown in Japan of 2011 to a similar company – a utility in the US, PSEG. Important lessons can be derived from these very different stories. We will conclude this section of Chapter 4 by examining several mini-case studies of positive social issue cases that we can all learn from and be inspired by.

4.3.b.i Spotlight: the TEPCO–Fukushima environmental, health, safety and security catastrophe

Many of us remember waking up the morning of March 11, 2011, to the news and the horrific images of the compounded catastrophe that we now know as "Fukushima" where one natural disaster (an earthquake) led to another natural disaster (a tsunami), which then led to an even more severe man-made disaster (the shutdown and later meltdown of the fuel rods at the TEPCO utility's nuclear power plants on the northeast coast of Japan), which stopped short of being an unprecedented and unmitigated disaster for Japan and the surrounding northern Asian region (radiation exposure could have been much worse).

A report issued on this event relates the details in an objective yet horrifying way:

> March 11, 2011, on the east coast of northern Japan, is believed to be one of the largest earthquakes in recorded history. Following the earthquake on Friday afternoon, the nuclear power plants at the Fukushima Daiichi,

> Fukushima Daini, Higashidori, Onagawa, and Tokai Daini nuclear power stations (NPSs) were affected, and emergency systems were activated. The earthquake caused a tsunami, which hit the east coast of Japan and caused a loss of all on-site and off-site power at the Fukushima Daiichi NPS, leaving it without any emergency power. The resultant damage to fuel, reactor, and containment caused a release of radioactive materials to the region surrounding the NPS.[51]

Years later, the Japanese courts found TEPCO and its management to have been negligent in its pre-accident preparation for predictable risks, finding that:

> the ruling by the Maebashi District Court in Gunma Prefecture, the court said that the disaster, considered the worst nuclear calamity since Chernobyl in 1986, was "predictable" and that it was "possible to prevent the accident.[52]

TEPCO's management issued the following apology years later in 2017:

> "We again apologize from the bottom of our hearts for giving great troubles and concerns to the residents of Fukushima and other people in society by causing the accident of the nuclear power station of our company," Isao Ito, a spokesman, said. "Regarding today's judgment given at the Maebashi local court today, we would like to consider how to respond to this after examining the content of the judgment".[53]

To this day, there are ongoing cases, prosecutions and legal action taking place regarding the after-shocks of this catastrophe. While mother nature wrought particular fury on the east coast of Japan on that terrible day in March 2011, the most important lesson learned from this series of events was that it was the man-made component of this disaster that became the most potentially dangerous and catastrophic event of them all (a nuclear meltdown), affecting many stakeholders (employees; contractors; neighbors; towns; businesses; owners of animal, fishing and agricultural holdings; animals and wildlife; nature and the nation of Japan). Even a megalopolis like Tokyo with a combined population of 13 million people was on high alert for a lengthy period of time.

And all this because proper environmental, health, safety and security preparations on behalf of human and animal life as well as protection of property, risk and crisis management all failed or simply weren't in place. Here is the view of one of TEPCO's stakeholders:

> "The money is not a problem," said Koichi Muramatsu, 66, a former resident of Soma City in Fukushima and the secretary of a victims group representing 4,200 plaintiffs in the suit being handled by Mr. Managi. "Even if it's ¥1,000 or ¥2,000, it's fine. We just want the government to admit their responsibility. Our ultimate goal is to make the government admit their responsibility and remind them not to repeat the same accident".[54]

Figure 4.2 visualizes the alleged breaches of the UN Global Compact Principles most salient to TEPCO in the past two years (2017–2019) and past ten years (2009–2019), respectively, as compiled by RepRisk. An interpretation of these visuals is that the company may have done a better job managing the human rights principle (#1) and environmental principles (7, 8 and 9) from a reputation risk standpoint in recent years in contrast to TEPCO's ten-year track record. The ten-year track record, in turn, reflects the reputation risk exposure against the underlying human rights and environmental risks after the 2011 catastrophe.

While this data reflects the lens of negative media and social media stories, one should also always remember that for an organization – whether TEPCO or any other entity – the absence of negative stories affecting a reputation doesn't mean that something negative and/or potentially explosive isn't brewing underneath the surface.

Indeed, the reason I am sharing this data is to underscore the importance to leaders of any type of organization of building resilience and reputation over time, as a long-term proposition that requires constant tending, building, reforming and innovating to maintain and create sustainable value, as we will dissect in greater detail in Chapter 7, where we explore the eight elements of organizational resilience.

Let's drill a little deeper into the TEPCO Fukushima affair. Looking at several additional visual snapshots of TEPCO's ESG reputation risk profile before 2012, from 2011 to 2016 and most recently from 2016 through 2019 (as displayed in Figures 4.3, 4.4 and 4.5), several additional observations can be made:

- TEPCO's RepRisk Index before their crisis event in 2011 was moderate-to-low. This reflected an absence of bad press over a long period of time but did not necessarily reflect the existence of resilient environment, health, safety and security (EHS) systems or crisis preparedness.
- The moment Fukushima struck, the RepRisk Index grew dramatically.
- TEPCO went through a lengthy period of reputation recovery – about four years from the incident – with the concomitant negative reputational, financial and stakeholder consequences, until they returned to a relatively moderate RepRisk Index in 2015.
- A return to a moderate RepRisk Index, while laudatory, should not lead to a sense of complacency – one need only look at the modest RepRisk TEPCO had prior to the Fukushima disaster for a reminder that these things can change literally overnight.
- However, in the past two years, the RepRisk Index for TEPCO seems to have increased significantly, as Figure 4.5 shows. The critical question for TEPCO's management, board and stakeholders is whether this is a reflection of continuing fallout from 2011 Fukushima-related events or something more recent that TEPCO management and their board should be tuned into and possibly concerned about.
- Overall, one would hope that TEPCO as an organization has learned some serious lessons that might have led to building appropriate (even leading) internal programs and resilience especially in the areas of health, safety, environment, and related risk and crisis management.

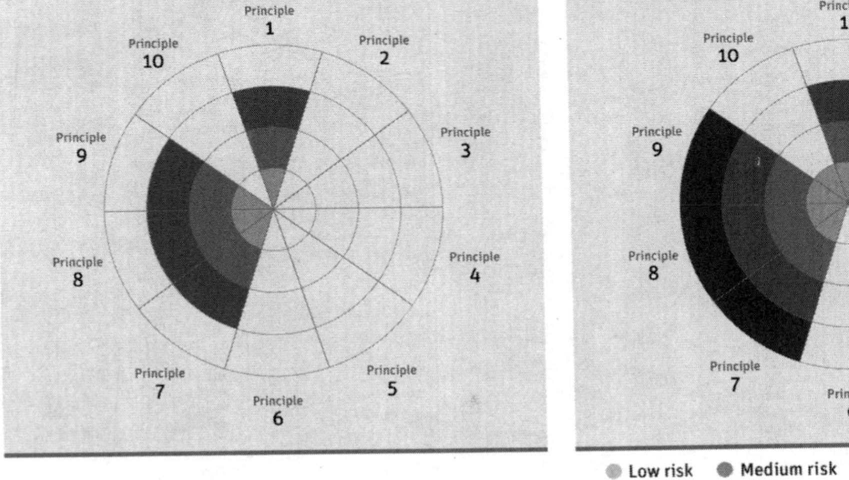

Figure 4.2 TEPCO UN Global Compact principles visualization.

Source: RepRisk.

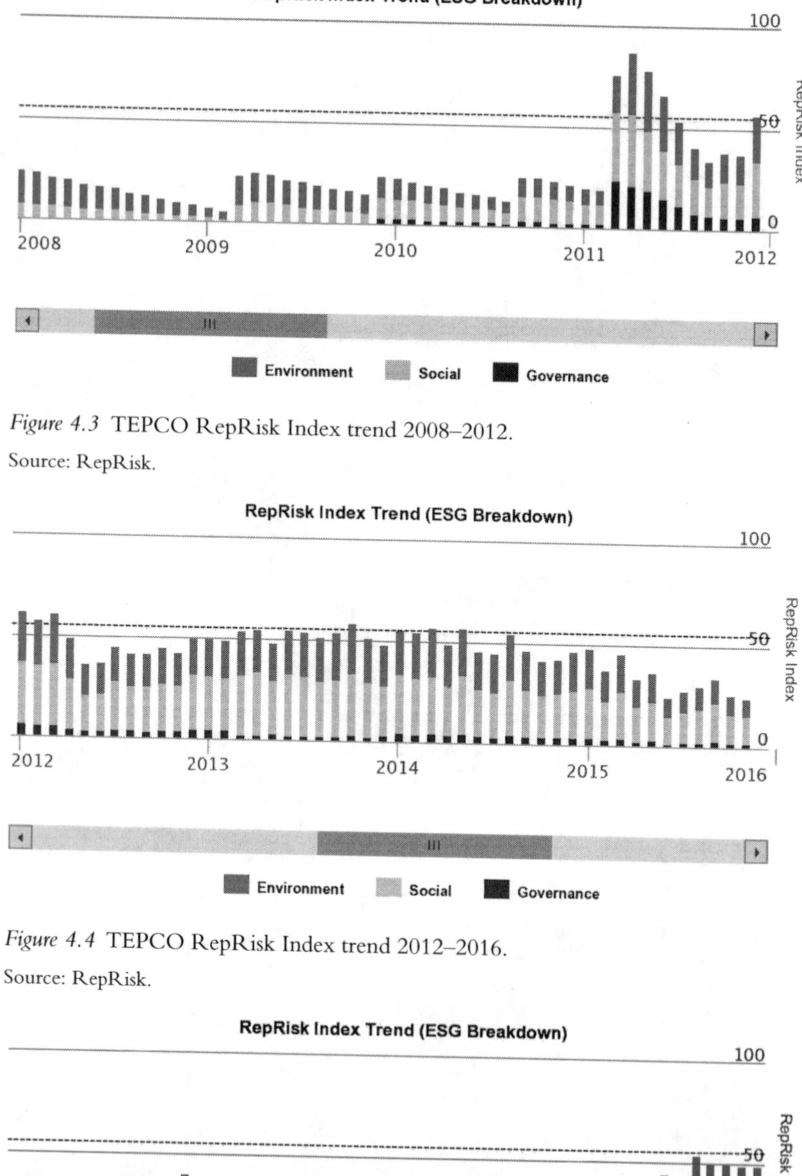

Figure 4.3 TEPCO RepRisk Index trend 2008–2012.

Source: RepRisk.

Figure 4.4 TEPCO RepRisk Index trend 2012–2016.

Source: RepRisk.

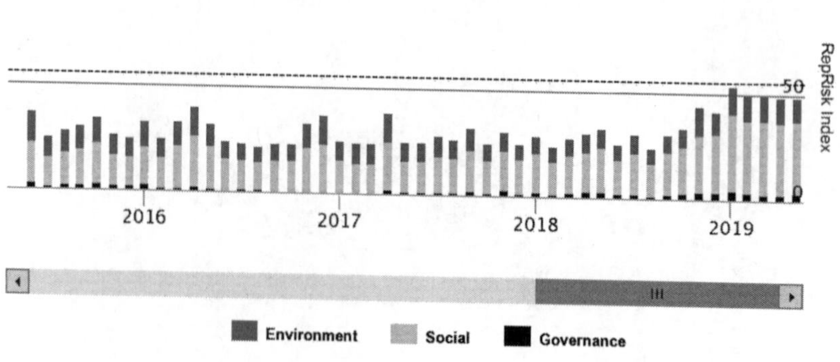

Figure 4.5 TEPCO RepRisk Index trend 2016–2019.

Source: RepRisk.

4.3.b.ii PSEG: sustainable environment, health, safety and security (EHS) responsibility in action

In contrast to what we saw with TEPCO Fukushima, PSEG is an example of a company that has generally taken good care of its environmental, health, safety and security issues over time.[55] No entity is perfect, and in the case of PSEG they did have a variety of environmental, health and safety events in the 1980s involving environmental and health and safety violations. However, instead of papering over those problems or dismissing them as one-offs, this company took their lessons learned seriously and built new programs and long-term resilience into their EHS systems.

Figure 4.6 shows a snapshot of the RepRisk Index for this utility located in the Northeast of the US. The notable fact about PSEG for the ten years of reputation risk data that RepRisk has on them is that they never had a major event during that period and indeed, their average RepRisk Index has pretty much remained at or below 25, considered to be a low reputation risk profile.

Looking at the alleged breaches of one or more of the UN Global Compact Principles for PSEG over a two-year and ten-year period, once again one can interpret a low level of risk or crisis compared to many similarly situated companies (electric utilities) and certainly compared to TEPCO, a very similar company, as further discussed later (Figure 4.7).[56]

TEPCO and PSEG are indeed very similar companies – one based in Japan and the other in the US; both are power generation companies with nuclear power generation, both are heavily regulated by government and the two companies have similar environmental, health and safety challenges and risks. Such government-enabled, highly regulated monopolies like PSEG and TEPCO require a good faith, proactive and engaged board, management and private/public partnerships to protect both the financial and ESGT interests of their most important stakeholders – from shareholders to customers and from employees to regulators.

Additionally, one of these companies – TEPCO – has the geographical burden of being located right on top of the Pacific Ring of Fire – meaning the most volcanic, earthquake- and by dint of that, tsunami-prone area in the world. Given this distinct and alarming feature, one would assume and expect that TEPCO would, if nothing else, have a heightened sense of awareness, responsibility and readiness for the possible consequences of a natural disaster such as an earthquake, volcanic eruption and/or tsunami, as those events might affect their nuclear power plants and by association all proximate stakeholders (employees, contractors, neighbors, pets, livestock, wildlife, homes, towns, businesses, etc.).

But based on everything we have learned from the Fukushima disaster, it would appear that TEPCO did not have that heightened sense of responsibility and readiness to protect the environment, health and safety of key stakeholders and their property.

In contrast, PSEG has proven that it is a learning organization, one that reacted constructively to its EHS risks and crises of the 1980s by building appropriate EHS programs that have proven sustainable and resilient to this day. Indeed, in Chapter 7 we profile how PSEG actually exported their EHS

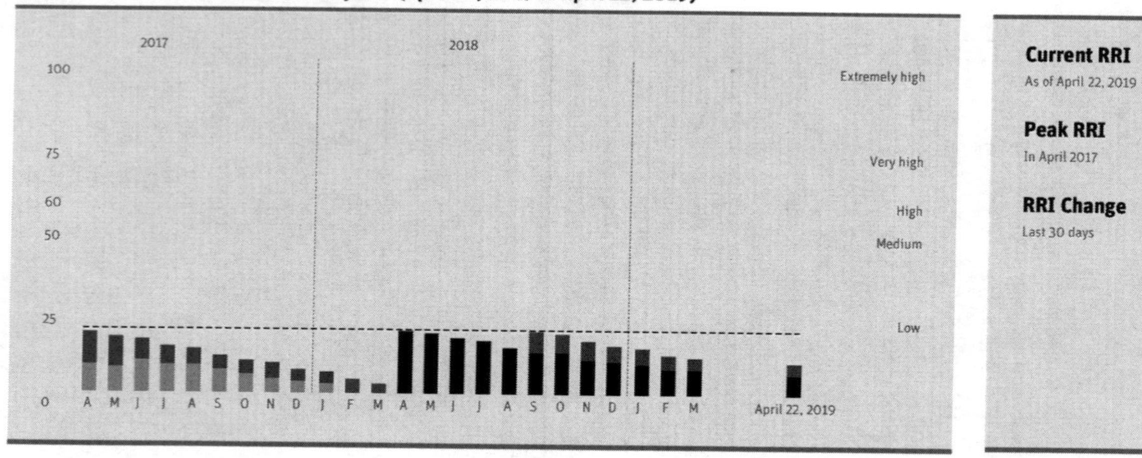

Figure 4.6 PSEG RepRisk Index trend 2017–2019.

Source: RepRisk.

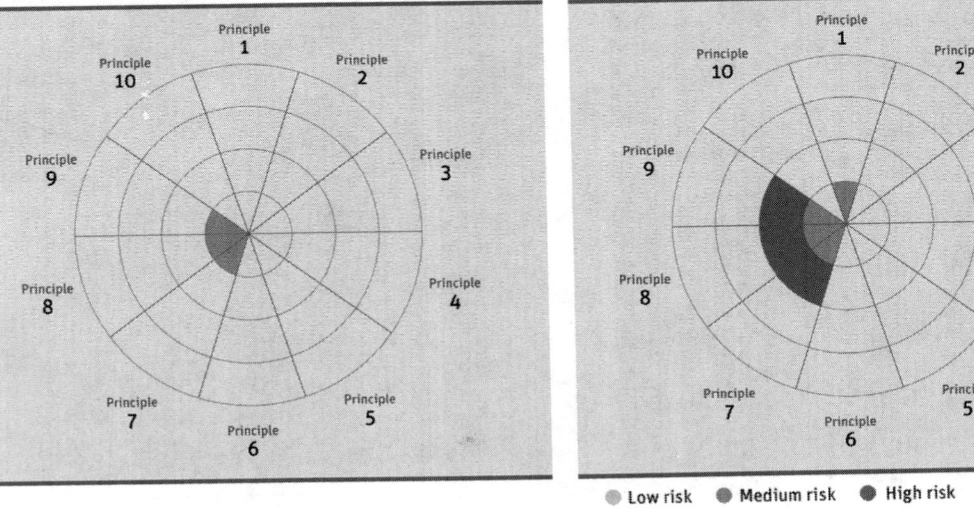

Figure 4.7 PSEG UN Global Compact principles visualization.

Source: RepRisk.

programs to their Latin American power assets in the late 1990s and early 2000s when they had holdings there, in effect extending their corporate social responsibility to places where laws either didn't exist or were largely unenforced.

Today, the PSEG website in addition to providing a lot of information for stakeholders on issues of environment, health and safety, also engages in a dialogue on the SDGs, selecting seven of them as core focuses for the company as well as listing particular programs under each category from supporting quality education (#4) to climate action (#13). Among other things, PSEG also pays attention to ESG issues and was named in 2018 for the 11th year in a row to the Dow Jones Sustainability Index.

A look at PSEG's track record on EHS matters reveals a company that has taken a long-term, sustainable and responsible approach to the environment and the health and safety of employees and communities alike. Next, I share a personal professional story from my days serving as PSEG Global general counsel and head of EHS (among other things).

4.3.b.iii Mini-case study: aligning corporate health and safety performance with individual annual bonus incentive programs

---

The challenge

- When PSEG Global began a major growth strategy to expand into Latin America from 1995 through 2002, one of the key components of the program was ensuring that the chain of command in charge of EHS in that region had their annual bonus potential tied directly to achieving the reduction of accident rates and the elimination of deaths at the assets acquired through government privatizations in the region.
- PSEG Global acquired old state-run assets with very poor environmental and health and safety records.

The solution

- Leadership from the very top of the utility – from the CEO and the board.
- Development and enforcement of metrics to measure progress.
- Development and enforcement of performance management system tying EHS metrics to annual bonus potential up and down the chain of command from the EHS team (100% of their annual bonus) to the general counsel (30%), the regional and the company CEOs (20% and 15% respectively).
- Availability of necessary and desirable resources to audit all assets, undertake inventory of problems and root cause analysis of previous EHS failures.

The results

- Within a year of deploying the program throughout the targeted Latin American assets, PSEG Global's regional accident rate was reduced by 40% and no deaths were recorded in any of the assets.
- Stakeholder and organizational benefits:
  - Worker didn't die
  - Worker accident rates dropped by 40%
  - Worker morale increased
  - Workers from competitor power plants wanted to work for PSEG Global

---

- The company achieved enormous goodwill both internally and externally
  - Fewer/no liabilities
  - Fewer/no fines or regulatory action
  - Local community and government respect
  - The PSEG Global EHS team received 100% of their bonus the following year (and so did everyone up and down the chain of command at PSEG)
  - The PSEG Global EHS team received external recognition with several awards

Source: Author.

4.3.b.iv Mini-case study: Johnson & Johnson Tylenol and Merck Vioxx – Exemplars of product quality and safety management

Pharma product safety risk. As a pharmaceutical company, you produce products that people ingest or apply to their bodies. What could go wrong? Ask Johnson & Johnson about Tylenol in 1984. Ask Merck about Vioxx in the early 2000s. Both of these companies handled their respective crises far better than many others have over the years. What did they do that was different or helpful? They were both led by conscientious leaders (CEOs) who placed a high value on the safety of their products and the welfare of their stakeholders (in addition to being profitable):

- In the case of Johnson & Johnson, they reacted to a terrible event (deliberate criminal poisoning and tampering of Tylenol bottles) in a way that is today considered to be a textbook case of exceptional crisis management – first, by quickly removing Tylenol from the marketplace, and, second, by creating product safety after the event that not only protected stakeholders (in this case consumers) against future possible tampering but turned a crisis into an opportunity to distinguish themselves from competitors in the marketplace in terms of their consumer safety mind-set.
- In the case of Merck (unlike Pfizer who had a similar product), their then CEO, Raymond Gilmartin, decided to voluntarily withdraw the product from the marketplace to protect future patients from the possible downsides of the medication. CEO Gilmartin's principled withdrawal of Vioxx ended costing Merck market share in that particular product line (compared to Pfizer, for example) but they decided that protecting one of their key stakeholders (patients) was more important and in the process also gained the support of other key stakeholders like employees, enhancing their brand and reputational standing in the marketplace for patient safety over the long term.

Source: Author.

4.3.b.v Mini-case study: keeping power plant construction personnel safe in a war zone

- When I worked at PSEG Global in the late 1990s, we had entered into a joint venture with AMOCO (pre-BP) to develop a power plant in the jungles of Colombia close to where AMOCO was developing gas fields.
- At the time, guerilla warfare was a serious health and safety risk for anyone traveling in and to/from Colombia. Additionally, the threat of kidnapping was a very high risk at that time.

- We had employees – mostly engineers – who needed to travel not only to Bogota regularly (which at the time was challenging enough from a travel safety standpoint) but also to the remote locations where the plant was to be built.
- One of the first questions we asked as we negotiated our deal with AMOCO was about the safety of our personnel and of plant construction and development personnel generally (both foreigners and locals).
- We were able to assess and mitigate this danger by (1) having a strong crisis management plan and team in place at PSEG Global and (2) piggybacking off of the superb security and crisis management program that AMOCO already had in place in Colombia and under whose umbrella our employees would operate while in Colombia.
- We never had a safety or security incident thanks to this corporate planning and the deliberate policy to put our most important stakeholders – our employees and contractors – first.

Source: Author.

4.3.c From harassment and discrimination to diversity and inclusion

4.3.c.i Harvey Weinstein and the birth of the #MeToo movement

Having now covered two serious social issues focused on liberty, health, life and death implications as well as serious financial damage potential – human slavery and workplace health and safety – let's tackle another serious social topic. While not literally involving life and death (although in isolated cases it might), there are potentially enormous consequences from this next cluster of issues on the financial and emotional well-being of individuals and their families, the workplace and indirectly, society. We are talking about a collection of social issues that fall under the broad umbrella of discrimination, harassment and bullying, on the one hand, and their antidote or counterpart – diversity and inclusion.

Let's start with a couple of dramatic examples. Here is what the *Financial Times* reported on August 7, 2018:[57]

> A prestigious Japanese medical school has confessed to systematically rigging its entrance exams against women, in a scandal that has highlighted the nation's deep problem with gender discrimination. An internal investigation found that Tokyo Medical University had for more than a decade subtracted marks from female applicants in a deliberate effort to produce more male doctors, and falsified exams to help specific individuals. The revelations have shocked the nation and turned a spotlight on a number of other Japanese universities that admit a suspiciously small number of women.

And here is what *The New York Times* reported in the fall of 2017 that set off the #MeToo movement in the US and eventually worldwide:

> An investigation by *The New York Times* found previously undisclosed allegations against Mr. Weinstein stretching over nearly three decades, documented through interviews with current and former employees and film

> industry workers, as well as legal records, emails and internal documents from the businesses he has run, Miramax and the Weinstein Company.
>
> During that time, after being confronted with allegations including sexual harassment and unwanted physical contact, Mr. Weinstein has reached at least eight settlements with women, according to two company officials speaking on the condition of anonymity.[58]

And here is what investigative journalist Ronan Farrow contemporaneously reported for *The New Yorker*:

> In the course of a ten-month investigation, I was told by thirteen women that, between the nineteen-nineties and 2015, Weinstein sexually harassed or assaulted them. Their allegations corroborate and overlap with the *Times*'s revelations, and also include far more serious claims. [59]

The latest on the Harvey Weinstein story is that the founder and CEO of the company resigned, as did most of his board of directors, and the company subsequently and rapidly went bankrupt. At the time of this writing, Harvey Weinstein is being prosecuted in New York for multiple accusations of rape and other sexual abuses and under investigation for similar charges in Los Angeles.

And thus was born the #MeToo movement which has had a pervasive effect on issues of culture, harassment, discrimination and similar toxic workplace behaviors certainly in the business world but also beyond the business world into other workplaces (like government, nonprofits, universities), not only in the US but in many other countries as well.

What has made the #MeToo movement so intense and widespread is that it has come together with other international trends, including the rise of the Trump presidency, the rise of the latest international women's movement (also partly in response to the Trump presidency), the awakening in the C-Suite and boardroom to the need to understand the importance of #MeToo in the context of their own corporate and organizational culture (and liabilities) and finally, the intensification and amplification of the #MeToo and related hashtags and movements such as Hollywood's #TimesUp initiative through intense social media campaigns and in-depth and widespread investigatory journalism.[60]

These examples are merely the tip of the iceberg of a much more pervasive, usually subtle, series of discrimination, harassment and bullying situations that not only affect women but plague members of other minority or discriminated-against groups. Let's turn to some additional examples.

4.3.c.ii The hospitality industry: transforming exploitation risk to stakeholder safety and value

The hospitality industry for many years got away with suffering no repercussions for some of the most egregious and potentially violent behaviors that took place within their facilities – from human trafficking to prostitution and from harassment to discrimination.

That changed somewhat in the past decade when a number of horrific stories came to light once again thanks to intrepid reporting and investigative journalism around the world. What follows is a mini-case study of developments in the hotel industry that were occasioned mainly by investigative journalism. But it also shows how the industry has actually responded both practically and constructively. Today some of the largest hotel chains are leaders in trying to eradicate these abusive and illegal practices from their facilities worldwide and creating broad awareness of these aberrations so that those of us who may be oblivious to the fact that these things are happening in the hotels we stay in are made aware and even proactive about reporting situations we might witness or suspect.

4.3.c.iii Mini-case study: the hospitality industry and human rights[61]

- There have been many headlines about trafficking, but these three reports illustrate the range of possible risks:
  - *The Telegraph* in June, 2016, reported that the Hilton Hotel in the Chinese city of Chongqing closed over an alleged brothel operating in the basement.
  - A report by the CNN Freedom Project spurred a change.org petition that resulted in Wyndham Worldwide's agreement in 2011 to sign the tourism child protection code of conduct to prevent child sex trafficking.
  - In advance of the World Cup, the interfaith Center of Corporate Responsibility's 2012–2013 annual report pressed more than 300 institutional investors and members of the faith-based community to publicly ask hotels what they do to deter trafficking.
- According to the nonprofit organization *Businesses Ending Slavery and Trafficking*, human trafficking at hotels brings safety risks to guests and staff, as trafficking is often connected to violent assaults by gangs that can put everyone present in jeopardy.
- Hits to reputation are quickly followed by financial risks resulting from reservation cancellations and fewer bookings.
- Additional costs can come in the form of legal fees and property damage associated with trafficking activity.
- Various state and municipal laws could hold hoteliers liable, at least in part, for any trafficking that occurs in their hotels.

Source: Jeff Hoffman and Andrea Bonime-Blanc, *NACD Directorship Magazine*.

Some of the leading hotel chains are now proactively partnering with some NGOs and local governments to maintain safe environments within their properties by adopting a variety of awareness, training and reporting mechanisms, including the following as a result of their collaboration with NGO ECPAT-USA.

For example, this is what Marriott has pledged to implement as part of a code of conduct created with ECPAT and other hotel chains:

1. Establish a corporate policy and procedures against sexual exploitation of children

2. Train employees in children's rights, the prevention of sexual exploitation and how to report suspected cases
3. Include a clause in further partner contracts stating a common repudiation and zero tolerance policy of sexual exploitation of children
4. Provide information to travelers on children's rights, the prevention of sexual exploitation of children and how to report suspected cases
5. Support, collaborate and engage stakeholders in the prevention of sexual exploitation of children
6. Report annually on the company's implementation of Code-related activities.[62]

In a related and also important stakeholder protection policy, several of the major hotel chains – including Marriott, Hilton and Hyatt – have also announced that they will equip their hotel workers – a major, high priority and often underserved and exploited stakeholder – with panic buttons by 2020.[63]

In what could perhaps be characterized as a preview of the #MeToo moment, who can forget the sordid story in 2011 involving the former managing director of the IMF and well-known French politician, Dominique-Strauss-Kahn, and a hotel chambermaid who accused him of sexual assault, with criminal charges later dropped and the finalization of a civil settlement between them. This incident only serves to underscore the often-extreme danger that the very vulnerable hotel worker community (low paid, mostly female, mostly immigrant workers) can suffer at the hands of abusive or otherwise entitled patrons – ranging from harassment to violence.

The fact that leading hotel chains are beginning to equip their workers and other stakeholders with both tools and protections against some of the vile social risks that take place within their four walls is a reason for hope and something that may actually empower and drive transformation of these risks into opportunity and value for all concerned.

4.3.c.iii Diversity and inclusion: the role of boards in transforming social risk into stakeholder value[64]

It is my deepfelt belief that the ultimate responsibility for making sure there is a culture of respect, diversity and inclusion rests first and foremost with the CEO or head of an organization and her/his team as well as with the board of directors or other oversight body who are supposed to hold that leader accountable.

Let us turn to some thoughts about how such an oversight body can and should play a positive and central role in ensuring a better organizational culture. It starts with the board or other oversight body being or becoming diverse and inclusive in the first place and reflecting the demographic of (and expertise needed by) their stakeholders.

Hardly radical thought leaders like McKinsey, for example,[65] are increasingly showing the deep connection between diversity, on the one hand, and greater corporate resilience and profitability, on the other. Likewise, the negative track record at the highest levels of leadership for lack of diversity and

ESGT expertise is also mounting. To put it in a less politically correct way: boards of far too many companies are still too pale, stale and male, populated by current or former CEOs and CFOs, and often devoid of any notable expertise in ESGT and risk governance more generally.

Governance diversity means being "inclusive" of those that are most relevant to the business of a company in the full sense of the word – from both a financial/operational and an ESGT standpoint, whether that diversity is because of gender, ethnicity, race, nationality, age, functional or professional expertise or cultural and leadership qualities.

The figures on board composition say it all. Most boards in the US are 85% dominated by men, have an average age in excess of 62 and a diversity score almost too low to register. The operating reality is that most boards exclude women, people of color, young up-and-coming professionals and global and nonfinancial business experts. In defense of their staid ways, many boards claim to be diverse in thought. While this may be marginally true in some cases, the bottom line is that most boards continue to be almost singularly made up of CEOs and CFOs (i.e., strictly financial C-Suite executives, which also explains why the modus operandi for many boards is growth, profits and maintaining the status quo). Worse still, many boards are composed of "friends and family" of the CEO, appointed by him (mainly) and who are often loath to challenge the status quo and subject to paycheck persuasion.

Table 4.18 summarizes some of the governance characteristics of six companies with major scandals to give the reader a sense of their "diversity" profile or, in their case, the lack thereof. We will come back to why this is important a little later in the chapter.

Table 4.18 Governance diversity characteristics of several scandal-plagued companies at or around the peak of their crisis[66]

Company and crisis date	*Scandal*	*Board members*	*No. of women*	*Other diversity*	*Average age*
Equifax (2017)	Massive data breach of private information on more than 150 million consumers. Poor cyber risk oversight.	10	2 (20%)	Low	63
The Weinstein Company (2017)	Pervasive culture of abuse, sexual harassment and allegations of rape in exchange for access and opportunities.	3 (after resignation of 6)	0 (0%)	Low	62
Uber (2015–2017)	Multiple cumulative scandals, including harassment, regulatory and employment breaches.	11	2 (18%)	Low	50

(*Continued*)

Table 4.18 Continued

Company and crisis date	*Scandal*	*Board members*	*No. of women*	*Other diversity*	*Average age*
Volkswagen (2015 TO DATE)	Emission rigging scandal affecting more than 11 million vehicles underreporting environmental impacts.	23	5 (22%)	Low	59
Wells Fargo (2015 TO DATE)	Among other scandals, account rigging and consumer fraud where millions of fee-bearing accounts were created without customer knowledge or consent.	30	9 (30%)	Low	65
Wynn Resorts (2018)	Allegations that founder and CEO engaged in decades long culture of direct sexual harassment and abuse of employees and others including multi-million-dollar payment to former employee for sexual harassment.	9	1 (11%)	Low	69

Source: Dante A. Disparte and Andrea Bonime-Blanc. *Risk Management Magazine.*

A review of the world of business and governance reveals little progress on diversity despite the apparent attempts, especially in the US, where there are few (though growing on the heels of California's and Illinois's recent state law changes) board gender quotas, compared to most European countries. It is also easy to dismiss the "diversity of thought" claim many companies and their boards often make as a cop-out for having real diversity of background and experience. The more insidious argument used to justify low diversity scores is that there is a "lack of suitable talent" – all too often subtle prejudice masked as meritocracy.

The fact of the matter is that many boards have become parochial in thought, judgment and action, and the lack of meaningful commitments to diversity, reflecting society, markets, investors and other stakeholders is harming private enterprise. The raft of corporate scandals and the often clumsy response or deafening silence from boards is revealing of a more endemic problem.

If we look at the companies singled out in Table 4.18, the following observations can be made:

Equifax. In the case of Equifax – who on the board was minding data privacy and cyber-risk oversight? Indeed, five years' worth of annual reports reveals troublingly low risk awareness when it comes to cyber threats. Did they have a digital native?[67] Did they have a board member well versed in technology, cyber-security and digital transformation expertise? In risk management at all? Indeed, one of the scapegoats to fall on their own sword in the Equifax scandal was the chief security officer, who was derisively dismissed as a music major.[68]

Wells Fargo. In the case of Wells Fargo, were board members asking the right questions and digging deeper on consumer fraud prevention, auto insurance, foreign exchange, expense account and a growing list of additional scandals?[69] According to reports, internal whistle blower calls not only went unanswered, they resulted in punitive actions against Wells Fargo employees who had the temerity to cross the line, showing the danger of a dial tone at the top.[70]

The Weinstein Company. Who on the TWC board did not know about the alleged sexual harassment (or worse)? And who did not know about the reported special "get out of jail free" clause in Harvey Weinstein's contract,[71] which also appears to have applied in Bill O'Reilly's contract with Fox following a $32 million harassment settlement?[72]

Wynn Resorts. Likewise, as to Wynn Resorts, did the nine-member board (with only one female director) know and/or inquire about cultural issues at the company? Did they have an ethics and compliance officer? If so, did that person report regularly to the board about ethics issues or concerns at the company, including possible sexual harassment and/or discrimination issues, let alone involving the founder and CEO? And if they did have such reporting, did the board do anything to inquire further about such issues, or hold the CEO accountable for setting the tone for an ethical culture? The Las Vegas gaming industry could have learned a thing or two from the Wynn episode: according to the *Las Vegas Review Journal*, as of late 2017, of the six largest such companies, two had no female directors and of the 57 combined total of directors, only eight were women.[73]

Uber. In the case of Uber, was anyone on the founder's handpicked board proactively overseeing the many, varied and repeated risk and compliance issues popping up all around the world until they catalyzed in mid-2017 with a major blogpost by a female ex-employee?[74] The example of the new CEO brought in to clean up the many messes left by the founding CEO is certainly a study in contrasts that augurs better for the longer-term sustainability of the company as we discuss in other chapters in this book.

Volkswagen. And who on the 23-person supervisory board at Volkswagen was properly overseeing environmental risks and other forms of possible cheating and consumer fraud on their journey to become the #1 auto company in the world? It appears as if none of the Lower Saxony government, workers' council, and actual board members were minding the ESGT store, certainly before the emissions scandal erupted in late 2015.

These cases have resulted in billions of dollars in lost market value, not to mention ongoing reputational and other financial/liability reverberations. In many such cases, CEOs are gently let go courtesy of golden parachutes, with boards remaining largely intact and in place.

There are two common and deeply interrelated threads running through these cases:

1. The breakdown or non-existence of appropriate risk oversight at the board level, especially relating to the two most important strategic risks the board is uniquely responsible for: leadership and culture (as we discussed in great detail in Chapter 2).
2. The lack of diversity on these boards or the prevailing homogeneity of background, expertise and demographics of individual board members that exacerbates these problems.

Evidence from the fields of psychology and behavioral economics underscores how boards that are too homogenous, complacent or agreeable do a major disservice to shareholders and other stakeholders.[75] There is equally valuable and voluminous research that more diverse boards and executive teams deliver greater financial value and engage in more successful strategic decision-making. In short, we are not only what we preach, we are what we are composed of and increasing board diversity sends a powerful signal to all stakeholders while driving improved governance and business outcomes.

There is an increasing number of companies that get this equation right and understand that good ESGT management, including diversity and inclusion within their employee population, their management and their board is good for business, strategy, resilience and value creation. Companies such as Microsoft, Starbucks and L'Oréal are making their mark. They are diverse (including at their board levels), they are focused on ESGT issues, they understand reputation risk management and they are doing well financially over time.

The bottom line is that whether or not more diverse boards actually have a better financial impact for shareholders, one thing can be said for sure: companies that are diverse in management and board representation are companies that are in tune and better synchronized with their stakeholders, meaning that their decision-making, product and service innovation and creation are going to be in greater synch with their employees, customers and regulators. Whether they actually create more financial value is almost beside the point, because the name of the game in the new era of capitalism is that all stakeholders must be heard from and catered to and thereby shareholders will also do better financially certainly over the longer term.

More recently, as we discuss in greater detail in the "values" section of Chapter 8, we are seeing a raft of additional direct or indirect research and evidence from a variety of fields and from leading corporate, consulting and investor voices that are finally putting two and two together – i.e., that to be attractive for long-term investment and sustainable profitability, companies must pay substantial (if not equal) attention to nonfinancial or ESG (or as I would argue, ESGT) factors as they do to strictly financial ones. The evidence for this (including the need for ethics and compliance programs and more diverse and agile boards) is mounting as fast as is the evidence that a diverse board brings value to a company. Table 4.19 provides a summary of these benefits.

Table 4.19 The benefits of board diversity and inclusion

- Multiple and diverse backgrounds and experiences – from geographical, ethnic and racial to skills, talents and education.
- The benefit of looking at strategy, risk and opportunity from a 360 lens instead of a myopic one.
- Having better oversight of key management roles – not only financial and operational, but also functional and skills-based (ethics, legal, sustainability, regulatory, digital, cyber, risk governance, etc.).
- Knowing what questions to ask of management about emerging and newer issues, risks, opportunities and technologies.
- Diversity also leads to different and complementary personal styles and behaviors that can lead to more discussion, greater insights, better decision-making.
- Better overall risk management and profitability from diversity.
- A greater emphasis on the nonfinancial aspects of doing business, which can have severe financial and reputational consequences on the one hand and lead to unplanned and unexpected financial and reputational opportunity on the other hand.

Source: Dante A. Disparte and Andrea Bonime-Blanc. *Risk Management Magazine*.

4.4 Endnote: diversity and governance – *quo vadis*?

We are living at a time of unprecedented global risk – at no time since WWII has the world been in greater turmoil because of the convergence of serious and rapidly scaling challenges, disruptions and potentially existential threats – from technological innovation and disruption (e.g., artificial intelligence, industrial automation and blockchain), climate change, pandemics, bio-hazards, global geopolitical instability, the severe decline in trust in institutions (including NGOs) and attendant acceleration and amplification of reputation risk in a world of instantaneous social media.[76] In this rapidly moving, disruptive and changing world, companies need to have boards that are agile and able to keep pace with these challenges – we need boards that are moving from a sit-back to a lean-in attitude on actual diversity of people, backgrounds, expertise and skills.

All too often large enterprises have no notable separation of powers when it comes to the chairman and CEO role, which most often is dominated by a single person, typically a man. Additionally, there is no perceivable independence when it comes to risk oversight and corporate governance standards. As a result, many boards have ossified into a decision avoidance body at worst and a status quo rubberstamp at best. Decision avoidance is no more a strategy than the lack of board diversity is a sign of agility. A status quo attitude is dangerous in our complex and fast-moving world, as it creates severe blind sides and, worse, value destruction.

An oft-cited argument against achieving better representation on boards is the scarcity of qualified candidates from diverse backgrounds. How can women expect to crack this glass ceiling if the men on boards are recycled across multiple organizations? What special endowment does this shallow pool of board candidates have that does not exist in other groups? Similarly, the growing ranks of entrepreneurs around the world offer another unique and valuable perspective on board governance, especially as many entrepreneurs are the very players debunking classical notions of the corporation and its role in society. Additionally, many of these digitally native professionals bring with them an untapped skill set when it comes to harnessing technology, whether for managing risk, or driving digital transformation. Indeed, the fact that for many years Equifax annual reports barely mentioned rampant cyber threats, information security or privacy speaks to the potentially extreme dangers of homogeneity.

All too often the role of the board is about maintaining the status quo and a semblance of order. It is thus in challenging times when we realize the true effectiveness of a board. These times, set against a profoundly shifting risk landscape, provide little comfort for organizations large and small. Therefore, challenging conventional board structures with more diverse candidates will invariably produce a different set of questions while breaking the radio silence on the status quo. In short, values matter most when it is least convenient, and corporate boards should act as the third rail that keeps management in check.[77]

Boards have a crucial role to play in guiding not only financial returns but also ESGT impacts. A dramatic shift is taking place in governance for the dual purpose of maximizing financial returns and creating common social good for the broader stakeholder community – employees, customers, communities and third parties. The role of the board must evolve beyond check-the-box compliance and move to embrace a holistic and integrated strategic oversight of ESGT, enterprise risk management, leadership and culture.[78] If management is responsible for an organization's body, the board is responsible for its soul. All boards should focus on diversity, ESGT and profitable sustainability on behalf of their shareholders and other stakeholders.

Table 4.20 provides some ideas to get a board from being un-diverse to diverse, from being focused exclusively on financial issues to including ESGT, from focusing not only on strategy but also on the role of risk and opportunity in achieving that strategy.

Table 4.20 Pathways to board diversity: actions boards can take[79]

- Percentage target for diversity that is customized to your business needs: gender, race, ethnicity, national origin, age.
- Broadening talent pool with several new kinds of experts including – risk, digital, technology, ESGT, sustainability, ethics and compliance.
- Reshuffling and/or recreating committees to represent current market realities and operating norms.
- Separating risk and opportunity oversight from audit, perhaps creating a specialized strategic risk and opportunity committee.
- Bringing third party specialists to conduct crisis simulations, tabletop exercises and across-industry benchmarking to shake things up.
- Separating CEO from chair and strategic risk management oversight.
- Enforcing term limits, as well as capping the total number of concurrent board seats.
- Creating advisory committees of key outside experts (e.g., tech giants are creating AI ethics advisory panels).
- Bringing in truly independent qualified directors and weaning powerful CEOs from the habit of appointing "friends and family" to the board.

Source: Dante A. Disparte and Andrea Bonime-Blanc. *Risk Management Magazine*.

The role of the board is not only to challenge management to achieve financial results but to achieve them in a manner congruent with stakeholder trust and value systems. In a market where trust becomes the most important asset organizations have, the board is the sentinel of institutional trust.

If the boards of Wells Fargo, Uber, Volkswagen, Equifax, Wynn and TWC had been significantly more diverse, perhaps they would have had more risk resilience-building programs in place and fewer or none of these cases would have exploded into the strategic scandals they became. More diverse boards are better at understanding, overseeing and balancing not only financial issues but ESGT issues as well. If you are a member of a board, take a look at the ESGT chapters in this book for a sense of the breadth and depth of such issues and ask yourself: Does your board have the right members in place today who have the insight, experience and leadership qualities required to understand and oversee the breadth and depth and fast-changing nature of the ESGT issues, risks and opportunities that are relevant to your business or other type of organization for today and tomorrow?

Notes

1 https://www.globalslaveryindex.org/. Accessed on July 21, 2018.
2 Mary Barra. "Our Vision for Moving Humanity Forward". *GreenMoney*. July/August 2017. https://greenmoneyjournal.com/next-25-years-big-picture-thinking-2/. Accessed on August 8, 2018.
3 *Oxford Dictionaries*. https://en.oxforddictionaries.com/definition/society. Accessed on August 16, 2018.
4 *Oxford Dictionaries*. https://en.oxforddictionaries.com/definition/civil_society. Accessed on August 16, 2018.
5 *Oxford Dictionaries*. https://en.oxforddictionaries.com/definition/social. Accessed on August 16, 2018.
6 Robeco. https://www.robeco.com/en/key-strengths/sustainability-investing/glossary/esg-definition.html. Accessed on June 20, 2018.
7 Cambridge Centre for Existential Risk. https://www.cser.ac.uk/. Accessed on July 29, 2018.
8 World Economic Forum. *Global Risks Report 2019*. http://www3.weforum.org/docs/WEF_Global_Risks_Report_2019.pdf. Accessed on April 12, 2019.
9 UN Sustainable Development Goals. https://sustainabledevelopment.un.org/?menu=1300. Accessed on July 29, 2018.
10 United Nations. *Universal Declaration of Human Rights*. https://www.ohchr.org/EN/UDHR/Documents/UDHR_Translations/eng.pdf. Accessed on August 17, 2018.
11 United Nations Human Rights Office of the High Commissioner. *The Guiding Principles on Business and Human Rights*. https://www.ohchr.org/Documents/Publications/GuidingPrinciplesBusinessHR_EN.pdf. Accessed on August 16, 2018.
12 UN Global Compact. https://www.unglobalcompact.org/. Accessed on July 29, 2018.
13 Allianz. *The Allianz Risk Barometer 2019*. https://www.agcs.allianz.com/content/dam/onemarketing/agcs/agcs/reports/Allianz-Risk-Barometer-2019.pdf. Accessed on April 12, 2019.
14 MSCI. https://www.msci.com/esg-investing. Accessed on July 18, 2018.
15 RepRisk. https://www.reprisk.com/our-approach. Accessed on July 17, 2018.
16 World Economic Forum. *Global Risks Report 2019*. http://www3.weforum.org/docs/WEF_Global_Risks_Report_2019.pdf. Accessed on April 12, 2019.

17 United Nations. *Universal Declaration of Human Rights*. http://www.un.org/en/universal-declaration-human-rights/. Accessed on August 17, 2018.
18 Alex Newton. *The Business of Human Rights: Best Practice and the UN Guiding Principles*. Routledge 2019.
19 United Nations Human Rights Office of the High Commissioner. *The Guiding Principles on Business and Human Rights.* https://www.ohchr.org/Documents/Publications/GuidingPrinciplesBusinessHR_EN.pdf. Accessed on August 16, 2018.
20 United Nations Human Rights Office of the High Commissioner. *The Guiding Principles on Business and Human Rights.* https://www.ohchr.org/Documents/Publications/GuidingPrinciplesBusinessHR_EN.pdf. Accessed on August 16, 2018.
21 United Nations Global Compact. https://www.unglobalcompact.org/what-is-gc/mission/principles. Accessed on August 7, 2018.
22 UN Global Compact Principles. https://www.unglobalcompact.org/what-is-gc/mission/principles. Accessed on July 17, 2018.
23 UN Global Compact. https://www.unglobalcompact.org/sdgs/17-global-goals accessed July 17, 2018.
24 Allianz. *The Allianz Risk Barometer 2019*. https://www.agcs.allianz.com/content/dam/onemarketing/agcs/agcs/reports/Allianz-Risk-Barometer-2019.pdf. Accessed on April 12, 2019.
25 MSCI. https://www.msci.com/esg-investing. Accessed on July 18, 2018.
26 RepRisk. https://www.reprisk.com/our-approach. Accessed on July 17, 2018.
27 RepRisk. https://www.reprisk.com/our-approach. Accessed on July 17, 2018.
28 RepRisk. *Most Controversial Companies 2017*. https://www.reprisk.com/content/5-publications/1-special-reports/53-most-controversial-companies-of-2017/mcc-2017.pdf. Accessed on August 5, 2018.
29 According to RepRisk, "Peak RRI: equal to the highest level of the RRI over the last two years -a proxy for overall reputational exposure related to ESG and business conduct risk". https://www.reprisk.com/our-approach. Accessed on August 5, 2018.
30 RepRisk. *Most Controversial Projects 2017*. https://www.reprisk.com/content/5-publications/1-special-reports/1-most-controversial-projects-of-2017/reprisk-most-controversial-projects-of-2017-report.pdf. Accessed on August 5, 2018.
31 https://www.globalslaveryindex.org/. Accessed on August 16, 2018.
32 The Global Slavery Index 2018. https://www.globalslaveryindex.org/. Accessed on July 21, 2018.
33 "Milestones in the Fight Against Modern Slavery". Reuters. https://www.reuters.com/article/us-slavery-index-timeline/milestones-in-the-fight-against-modern-slavery-idUSKCN0YM1ZX. Accessed on August 16, 2018. Other sources cited in this article: ILO www.ilo.org/; Walk Free Foundation www.walkfreefoundation.org; Anti-Slavery International www.antislavery.org/english; Free the Slaves www.freetheslaves.net/; and Anti-Trafficking Review http://www.antitraffickingreview.org/index.php/atrjournal.
34 The Global Slavery Index 2018. https://www.globalslaveryindex.org/. Accessed on August 16, 2018.
35 The Global Slavery Index 2018. https://www.globalslaveryindex.org/. Accessed on August 16, 2018.
36 The Global Slavery Index 2018. https://www.globalslaveryindex.org/. Accessed on August 16, 2018.
37 The Global Slavery Index 2018. https://www.globalslaveryindex.org/. Accessed on July 21, 2018.
38 Alex Newton. *The Business of Human Rights*. Sheffield, UK: Greenleaf, 2018.
39 For more information, see Alex Newton and Jude Soundar, "Blood in Your Mobile: Implications for the Electronics Industry of Human Rights Violations in the Democratic Republic of the Congo". *Ethical Corporation Magazine*. 14 August 2014.

40 OECD. "OECD Due Diligence Guidance for Responsible Supply Chains of Minerals from Conflict-Affected and High-Risk Areas". 2016. http://www.oecd.org/daf/inv/mne/mining.htm. Accessed on September 7, 2018.
41 Yuan Yang. "Supply Chains: The Dirty Secret of China's Prisons". The Financial Times. August 29, 2018. https://www.ft.com/content/1416a056-833b-11e7-94e2-c5b903247afd?segmentId=6132a895-e068-7ddc-4cec-a1abfa5c8378. Accessed on August 30, 2018.
42 United States Department of State. *Human Rights Report 2017*. https://www.state.gov/documents/organization/277365.pdf. Accessed on August 18, 2018.
43 Ian Urbina. "Sea Slaves: The Human Misery that Feeds Pets and Livestock". *The New York Times*. July 27, 2015. https://www.nytimes.com/2015/07/27/world/outlaw-ocean-thailand-fishing-sea-slaves-pets.html?_r=0. Accessed on August 18, 2018.
44 Ian Urbina. "The Outlaw Ocean Series". The New York Times. July 25, 2015. http://www.nytimes.com/interactive/2015/07/24/world/the-outlaw-ocean.html. Accessed on August 18, 2018; Felicity Lawrence. "Thai Seafood Industry Censured over Burmese Migrant's Trafficking Ordeal". The Guardian. March 4, 2014. http://www.theguardian.com/global-development/2014/mar/04/thai-seafood-industry-burmese-migrant-trafficking-ordeal. Accessed on August 18, 2018; Robin McDowell, Margie Mason, and Martha Mendoza. "AP Investigation: Slaves May Have Caught the Fish You Bought". Associated Press, March 25, 2015. https://www.ap.org/explore/seafood-from-slaves/ap-investigation-slaves-may-have-caught-the-fish-you-bought.html. Accessed on August 18, 2018.
45 Seafood Slavery Risk. http://www.seafoodslaveryrisk.org/what-business-can-do/. Accessed on August 18, 2018.
46 Nestle. "Nestle Takes Action to Tackle Seafood Supply Chain Abuses". https://www.nestle.com/media/news/nestle-tackles-abuses-seafood-supply-chain. Accessed on August 18, 2018.
47 Nestle. "Responsible Sourcing of Seafood – Thailand Action Plan 2015–2016". https://www.nestle.com/asset-library/documents/library/documents/corporate_social_responsibility/nestle-seafood-action-plan-thailand-2015-2016.pdf. Accessed on August 18, 2018.
48 Ian Urbina. "Nestle Reports on Abuses in Thailand Seafood Industry". *The New York Times*. November 23, 2015. https://www.nytimes.com/2015/11/24/business/nestle-reports-on-abuses-in-thailands-seafood-industry.html. Accessed on August 18, 2018.
49 RepRisk. *Special Report on Migrant Labor*. September 2018. https://www.reprisk.com/. Accessed on August 29, 2018.
50 Here's what PGE, a company that has had a number of health and safety disasters, has on their website regarding the Diablo Canyon Nuclear facility: https://www.pge.com/en_US/safety/how-the-system-works/diablo-canyon-power-plant/diablo-canyon-power-plant.page. Accessed on August 18, 2018.
51 American Nuclear Society Special Committee on Fukushima. *Fukushima Daiichi: ANS Committee Report*. March 2012.
52 Motoko Rich. "Japanese Government and Utility Are Found Negligent in Nuclear Disaster". *The New York Times*. March 17, 2017.
53 Motoko Rich. "Japanese Government and Utility Are Found Negligent in Nuclear Disaster". *The New York Times*. March 17, 2017.
54 Motoko Rich. "Japanese Government and Utility Are Found Negligent in Nuclear Disaster". *The New York Times*. March 17, 2017.
55 PSEG. *PSEG Sustainability Report 2017*. https://investor.pseg.com/sites/pseg.investorhq.businesswire.com/files/doc_library/file/sustainability_report.pdf. Accessed on August 7, 2018.
56 RepRisk. *RepRisk Analytics Report on PSEG Power*. August 2, 2018.
57 Robin Harding. "Japanese Medical School Admits Rigging Exams to Favor Men". *Financial Times*. August 7, 2018. https://www.ft.com/content/5332a832-9a11-11e8-9702-5946bae86e6d. Accessed on August 19, 2018; and "Japan Medical School Scandal

Highlights Deeper Gender Problem". *Financial Times*. August 10, 2018. https://www.ft.com/content/54e98c1e-9c54-11e8-9702-5946bae86e6d. Accessed on August 19, 2018

58 Jodi Kantor and Megan Twohey. "Harvey Weinstein Paid Off Sexual Harassment Accusers for Decades". *The New York Times*. October 5, 2017. https://www.nytimes.com/2017/10/05/us/harvey-weinstein-harassment-allegations.html. Accessed on August 19, 2018.

59 Ronan Farrow. "From Aggressive Overtures to Sexual Assault: Harvey Weinstein's Accusers Tell Their Stories". *The New Yorker*. October 10, 2017. https://www.newyorker.com/news/news-desk/from-aggressive-overtures-to-sexual-assault-harvey-weinsteins-accusers-tell-their-stories. Accessed on August 19, 2018.

60 Somini Sengupta. "The #MeToo Moment: What Happened After Women Broke the Silence Elsewhere?". *The New York Times*. December 22, 2018. https://www.nytimes.com/2017/12/22/us/the-metoo-moment-what-happened-after-women-broke-the-silence-elsewhere.html?rref=collection%2Fseriescollection%2Fmetoo-moment&action=click&contentCollection=us®ion=stream&module=stream_unit&version=latest&contentPlacement=10&pgtype=collection. Accessed on August 19, 2018.

61 Jeff Hoffman and Andrea Bonime-Blanc. "Seeing Opportunity in Reputation Risk". *Directorship Magazine*. March/April 2017. https://gecrisk.com/wp-content/uploads/2017/04/ABonimeBlanc-JHoffman-NACD-Directorship-MarApr17-In-Practice_Reputation-Risk.pdf. Accessed on September 7, 2018.

62 Elliott Mest. "Marriott, ECPAT-USA Partner to Combat Human Trafficking in Hotels". January 30, 2018. Hotel Management.com. https://www.hotelmanagement.net/operate/marriott-ecpat-usa-partner-to-counter-human-trafficking-hotels. Accessed on November 23, 2018.

63 Dee-Ann Durbin. "Major Hotels Will Give Panic Buttons to Staff to Help Protect Them from Harassment and Assault". *TIME Magazine*. September 6, 2018. http://time.com/5389070/hotels-panic-buttons-me-too/. Accessed on September 7, 2018.

64 This section of the chapter is based heavily from a piece cowritten with Dante A. Disparte. "Stale, Pale and Male, Does Board Diversity Matter?". *Risk Management Magazine*. September 2018. http://www.rmmagazine.com/2018/09/04/pale-stale-male/. Accessed on September 9, 2018. Permission obtained from *Risk Management Magazine* to reprint portions of this article in this book.

65 McKinsey. "Delivering Through Diversity". January 2018. https://www.mckinsey.com/~/media/McKinsey/Business%20Functions/Organization/Our%20Insights/Delivering%20through%20diversity/Delivering-through-diversity_full-report.ashx. Accessed on September 9, 2018.

66 Andrea Bonime-Blanc and Dante A. Disparte. "Stale, Pale and Male, Does Board Diversity Matter?". *Risk Management Magazine*. September 2018. http://www.rmmagazine.com/2018/09/04/pale-stale-male/. Accessed on September 9, 2018.

67 Dante A. Disparte. "The Equifax Breach and Five Years of Missed Warnings". September 17, 2017. *HuffPost*. https://www.huffingtonpost.com/entry/the-equifax-breach-and-5-years-of-missed-warning-signs_us_59bf2480e4b06b71800c3b07. Accessed on September 9, 2018.

68 Brian Fung. "Equifax Security Chief Had Some Big Problems – Being a Music Major Wasn't One of Them". *The Washington Post*. September 19, 2017. https://www.washingtonpost.com/news/the-switch/wp/2017/09/19/equifaxs-top-security-exec-made-some-big-mistakes-studying-music-wasnt-one-of-them/?noredirect=on&utm_term=.b843787d7dfa. Accessed on September 9, 2018.

69 Matt Egan. "The Two Year Wells Fargo Horror Show Just Won't End". CNN Money. September 7, 2018. https://money.cnn.com/2018/09/07/news/companies/wells-fargo-scandal-two-years/index.html. Accessed on September 9, 2018.

70 Matt Egan. "Wells Fargo's Whistleblower Problem Worsens". CNN Money. April 6, 2017. https://money.cnn.com/2017/04/06/investing/wells-fargo-whistleblower-retaliation-osha/index.html. Accessed on September 9, 2018.

71 TMZ. "Harvey Weinstein Contract with TWC Allowed for Sexual Harassment". October 12, 2017. http://www.tmz.com/2017/10/12/weinstein-contract-the-weinstein-company-sexual-harassment-firing-illegal/. Accessed on September 9, 2018.
72 BBC News. "Fox Renewed Bill O'Reilly Deal Despite Harassment Suit". October 22, 2017. https://www.bbc.com/news/world-us-canada-41712527. Accessed on September 9, 2018.
73 Nicole Raz. "Las Vegas Gaming Company Boards Below Average in Women Members". *Las Vegas Review Journal*. February 13, 2018. https://www.reviewjournal.com/business/casinos-gaming/las-vegas-gaming-company-boards-below-average-in-women-members/. Accessed on September 9, 2018.
74 Susan Fowler. "Reflecting on One Very, Very Strange Year at Uber". SusanJFowler.com. February 19, 2017. https://www.susanjfowler.com/blog/2017/2/19/reflecting-on-one-very-strange-year-at-uber. Accessed on September 9, 2018.
75 David Rock, Heidi Grant and Jacqui Grey. "Diverse Teams Feel Less Comfortable – and That's Why They Perform Better". *Harvard Business Review*. September 22, 2016. https://hbr.org/2016/09/diverse-teams-feel-less-comfortable-and-thats-why-they-perform-better. Accessed on September 9, 2018.
76 Edelman. Edelman Trust Barometer 2017. January 2018. https://www.edelman.com/trust2017/. Accessed on September 9, 2018; and A. Bonime-Blanc. *The Reputation Risk Handbook*. Greenleaf, 2014.
77 Dante A. Disparte. "Simple Ethics Rules for Better Risk Management". *Harvard Business Review*. November 8, 2016. https://hbr.org/2016/11/simple-ethics-rules-for-better-risk-management. Accessed on September 9, 2018.
78 Andrea Bonime-Blanc. "Implementing a Holistic Governance, Risk and Reputation Strategy for Multinationals: Guidelines for Boards". *Ethical Boardroom*. 2014. https://ethicalboardroom.com/implementing-a-holistic-governance-risk-and-reputation-strategy-for-multinationals-guidelines-for-boards/. Accessed on September 9, 2018.
79 Andrea Bonime-Blanc and Dante A. Disparte. "Stale, Pale and Male, Does Board Diversity Matter?". *Risk Management Magazine*. September 2018. http://www.rmmagazine.com/2018/09/04/pale-stale-male/. Accessed on September 9, 2018.

5 Governance

Transforming governance risk into opportunity from authoritarianism to stakeholder centricity

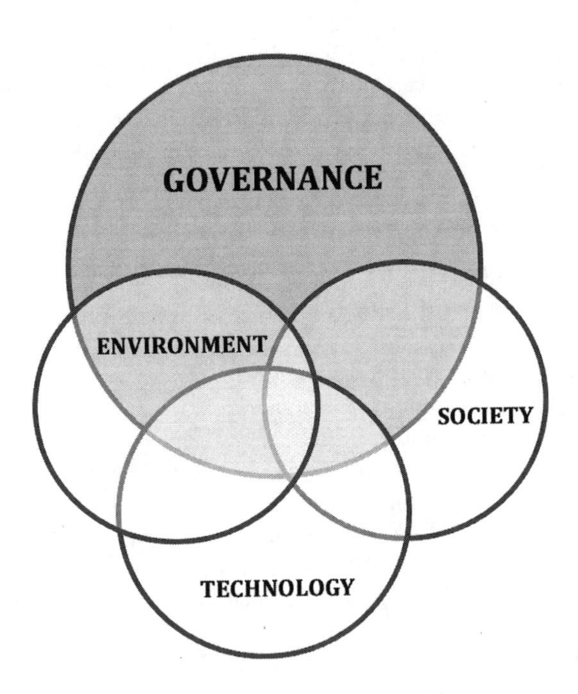

Governance: risk and opportunity

Governance is all about how we take care of each other at every level of society – or don't. And therein lies the rub. As Albert Einstein, one of the greatest scientists who ever lived, once said: "The world is a dangerous place. Not because of the people who are evil; but because of the people who don't do anything about it".[1] Reinhold Niebuhr, the great thinker, adds the following flourish: "Man's capacity to justice makes democracy possible but man's inclination to injustice makes democracy necessary".[2]

Speaking of large-scale, global and national governance concerns, Angela Merkel, one of the longest-serving post-WWII German leaders, has warned that: "When the generation that survived the war is no longer with us, we'll find out whether we have learned from history".[3] Which is something that Walter Shaub, former director of the US Office of Government Ethics, has also become alarmed about, as evidenced in his statement: "It's hard for the US to pursue international anticorruption and ethics initiatives when we're not even keeping our own side of the street clean. It affects our credibility … I think we are pretty close to a laughingstock at this point".[4]

But not all hope is lost – we can certainly cling to a famous statement from US Justice Louis D. Brandeis that helps us in our troubled times: "Publicity is justly commended as a remedy for social and industrial diseases. Sunlight is said to be the best of disinfectants".[5] Or as this statement – first ascribed to 19th-century Pastor Theodore Parker, then attributed to 20th-century civil-rights leader Martin Luther King Jr. and often used by 21st-century US President Barack Obama: "The arc of the universe is long but it bends towards justice".[6]

Chapter 5: Governance – summary overview

- 5.1 Introduction and overview
- 5.2 What is "governance"? Definitions and typologies
  - 5.2.a Definitions
  - 5.2.b Typologies, classifications and categories
    - 5.2.b.i Council on Foreign Relations Global Governance Monitor
    - 5.2.b.ii World Bank Governance Indicators
    - 5.2.b.iii Freedom House
    - 5.2.b.iv UN Sustainable Development Goals
    - 5.2.b.v G20/OECD Principles of Corporate Governance
    - 5.2.b.vi OECD Due Diligence Guidance for Responsible Business Conduct
    - 5.2.b.vii Allianz Risk Barometer
    - 5.2.b.viii MSCI
    - 5.2.b.ix RepRisk
    - 5.2.b.x UN Global Compact
    - 5.2.b.xi Transparency International
  - 5.2.c The way forward
- 5.3 Leaders navigating governance risks, crises and opportunities
  - 5.3.a International governance: navigating through the global chaos of geopolitical risk
  - 5.3.b National governance: transforming authoritarian risk into democratic opportunity
  - 5.3.c Corporate governance: transforming "the bad and the ugly" into "the good and the better"
    - 5.3.c.i Waves of scandal and reform, risk and opportunity from the 1970s to the 2020s

5.1 Introduction and overview

This chapter tackles a huge topic – in many ways the biggest in this book: governance. Why do I say this? Because it is my deeply held contention that within any kind of organization (from the smallest to the biggest) all things – policy, culture, behaviors, action, strategy, etc. – flow from the top (as some of the sayings listed at the beginning of this chapter underscore as well as those at the top of Chapter 2 on leadership). Let me explain a very personal journey that has made me such a strong believer in the criticality of good governance at every level.

As a child, I grew up in an authoritarian country – Generalissimo Francisco Franco's Spain. And, like most young children, as a pre-teen I was completely oblivious to any notion of politics, ideology and its effects on society. As I grew up into my teens, I started noticing a variety of things about the society and the culture I was living in. There was no freedom of expression. Newspapers were censored – the *International Herald Tribune* which my parents read would sometimes "not be published" (at least in Spain). There was only one official television channel. The media was state controlled. There were no elections. There were no political parties (except of course Franco's official state Falangist party). There were no (overt) demonstrations. Beyond the local police, there was a special paramilitary police force – the much-feared civil guard. There were few dissenting voices. Those who dared speak about politics did so in hushed tones and fearfully. With my eyes widening, I started to notice a few other things about this society – police brutality, the mistreatment of people, the ghettoization of certain "classes" of people like the poor and of ethnic minorities (gypsies).

I not only learned from my own life in Spain but I also learned from my parents – a German mother who had grown up in East Prussia and Nazi

Berlin and had survived two years of Allied bombings, hiding in bunkers nightly. She told me of her work as a young telephone operator for the Nazi press agency right before the war ended and as a junior journalist for ten years after that working for the East German Communist press agency run by the Russians in East Berlin. And I had an American father who volunteered to fight in WWII, was rumored to be an officer with the OSS (Office of Strategic Services, the precursor to the US Central Intelligence Agency formed during WWII) and fought in many battles, including the Battle of the Bulge in which he was wounded. Post-WWII, my dad dedicated the rest of his career as an intelligence officer for the US government in Japan and Germany. My parents met in postwar Berlin. The rest is family history, as they say. I share all this because thanks in so many ways to my parents and their background, I occupied a privileged front-row seat to history and to the macro-picture of leadership and culture that it provides. I got insights early on into what it was like to live in and under authoritarian, fascist and even totalitarian political leadership.

I have also learned a lot at the micro-level – organizational leadership. After having worked at two law firms and four companies over a period of 25 years, been on a number of boards and reported to a wide variety of bosses, many of whom were at the highest positions of power (CEOs, chief operating officers, board members, managing partners, etc.), it also became clear to me (especially in hindsight) that the culture and the mood at each of these workplaces were pretty much determined by the personality, behavior and policies actively emitted or tolerated by the leader *and permitted or tolerated by the leaders' oversight body* (whether a committee, a board or some other "chief of chiefs"). In other words, governance, leadership and culture are intimately intertwined by the acts and omissions of leaders and their bosses (overseers).

Finally, I also lived through a few exceptionally positive governance, leadership and culture experiences as well – one as a general counsel reporting to a great CEO and a highly responsible and responsive board of directors (made up of luminary CEOs of publicly traded companies). The other was as a senior global executive trying to implement a super-challenging global change management program (the company's first global ethics and compliance program) for 100,000 employees worldwide at a very large, highly decentralized and diversified Fortune 250 corporation.

In each of these cases I learned different positive lessons – how leaders and boards create the safe spaces for others to speak up, innovate and change, and the accountability and responsibility to get things done in the right way and for the right reasons – even when it costs financially. Many of the lessons I have learned are encapsulated in this book and especially in Chapters 7 and 8, "Metamorphosis" and "Boom", respectively, but their importance needs to be underscored here in this chapter, which is all about the various aspects of governance.

Governance matters whether it is global, national, corporate, governmental, societal, local or even personal. Leaders represent and drive governance for good, for better, for bad and for evil. Governance comes from those who govern. Governance is as good as the people who occupy leadership positions. And that is why those responsible at the highest levels of governance of any organization have the ultimate responsibility to proactively ensure that proper culture, integrity and stakeholder care and management are taking place.

In this chapter we will review some definitions and typologies (as we did in the previous two chapters on environment and society) and then we will delve into a variety of examples and cases illustrating governance at different levels and in different incarnations – at the international, national and local levels and through governmental, corporate and nonprofit experiences – all with an eye to understanding how leaders are navigating some of the most important governance issues and risks confronting their organizations and hopefully transforming those risks into opportunity and value for their stakeholders.

5.2 What is governance? Definitions and typologies

5.2.a Definitions

"Governance:"

> Governance is all of the processes of governing, whether undertaken by a government, a market or a network, over social system (family, tribe, formal or informal organization, a territory or across territories) and whether through the laws, norms, power or language of an organized society. It relates to "the processes of interaction and decision-making among the actors involved in a collective problem that lead to the creation, reinforcement, or reproduction of social norms and institutions".[7]
>
> The action or manner of governing a state, organization, etc.[8]
>
> Establishment of policies, and continuous monitoring of their proper implementation, by the members of the governing body of an organization. It includes the mechanisms required to balance the powers of the members (with the associated accountability), and their primary duty of enhancing the prosperity and viability of the organization. See also corporate governance.[9]

"Country Governance" from the World Bank:[10]

> Governance consists of the traditions and institutions by which authority in a country is exercised. This includes the process by which governments are selected, monitored and replaced; the capacity of the government to effectively formulate and implement sound policies; and the respect of citizens and the state for the institutions that govern economic and social interactions among them.

"Corporate Governance:"[11]

> The framework of rules and practices by which a board of directors ensures accountability, fairness, and transparency in a company's relationship with its stakeholders (financiers, customers, management, employees, government, and the community). The corporate governance framework consists of (1) explicit and implicit contracts between the company and the stakeholders for distribution of responsibilities, rights, and rewards, (2) procedures for reconciling the sometimes-conflicting interests of stakeholders in accordance with their duties, privileges, and roles, and (3) procedures for proper supervision, control, and information-flows to serve as a system of checks-and-balances.

"Governance" from the investor community:[12]

> Governance refers to a set of rules or principles defining rights, responsibilities and expectations between different stakeholders in the governance of corporations. A well-defined corporate governance system can be used to balance or align interests between stakeholders and can work as a tool to support a company's long-term strategy.

Governance happens at many different levels, from the smallest family unit all the way up to the largest global organizations – from a tiny local village to the largest transnational corporation, from the smallest family-owned business to the largest global nonprofit. Figure 5.1 depicts the big picture examples of where governance needs to take place.

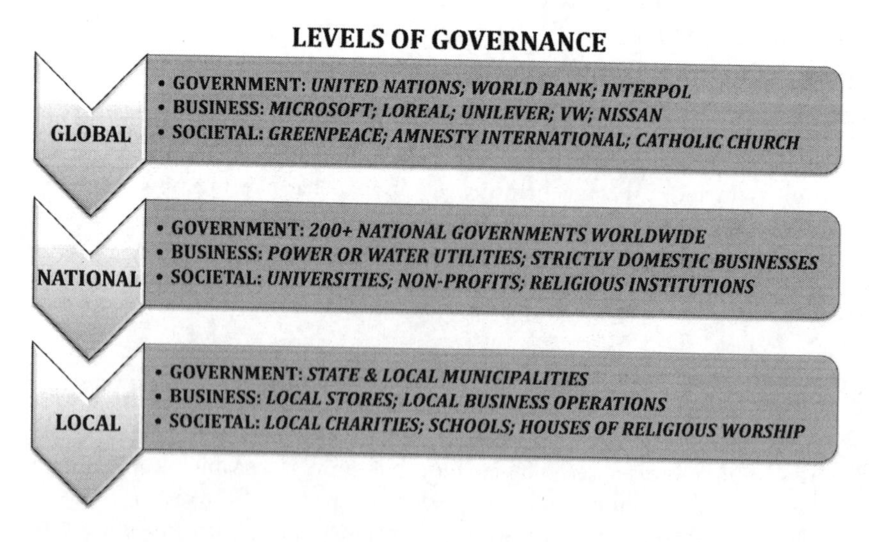

Figure 5.1 Governance: levels, types and examples.

Source: GEC Risk Advisory.

Everyone has and needs governance. But the real question is: How good is your governance? How effective is it in representing the core interests of the key stakeholders? Do you even know who your universe of stakeholders is beyond the most obvious? How good is your governance in holding the most powerful in your organization accountable and responsible? This chapter explores these and many related questions, but first let's turn to some typologies and classifications of governance issues.

5.2.b Typologies, classifications and categories

With these definitions and preliminary thoughts in mind, Table 5.1 provides an overview of the sources and resources that we believe are useful for leaders of any type of organization to keep in mind (a) as you think about your own organizational governance issues, and (b) as you tackle the identification of key governance issues relevant to your organization, the risks that may be associated with these issues, the crisis management that would need to be done in the event an issue or risk goes wrong for your organization and the opportunities embedded in those potential issues, risks and crises.

Table 5.1 On governance: sources and resources

Level of detail	*Resource*
FROM BIG PICTURE ...	i. Council on Foreign Relations Global Governance Monitor[13]
	ii. World Bank Governance Indicators[14]
	iii. Freedom House Freedom in The World Study[15]
	iv. UN Sustainable Development Goals[16]
	v. G20/OECD Principles of Corporate Governance[17]
... TO MORE FOCUSED ...	vi. OECD Due Diligence Guidance for Responsible Business Conduct[18]
	vii. Allianz Risk Barometer[19]
	viii. MSCI[20]
	ix. RepRisk[21]
	x. UN Global Compact[22]
... TO MORE GRANULAR	xi. Transparency International[23]

Source: Author.

5.2.b.i CFR Global Governance Monitor[24]

Major global categories of governance are generally captured in the CFR Global Governance Monitor, which the Council on Foreign Relations, a leading geopolitical think tank, has deployed and maintains. In its launch of this interactive research tool, CFR stated:

> International cooperation is crucial for coping with today's most pressing challenges. The Global Governance Monitor tracks global cooperation and recommends policy options to improve the world's capacity to tackle ten global challenges.[25]

The site goes on to show the ten global challenges that are the subject of their global governance monitoring, as presented in Table 5.2.

Table 5.2 Council on Foreign Relations Global Governance Monitor[26]

Category	*Description*
Armed Conflict	Preventing armed conflict, keeping peace, and rebuilding war-torn states remain the most difficult challenges for policymakers and government officials throughout the world.
Crime	Over the past two decades, the global impact of transnational crime has risen to unprecedented levels as criminal groups use new technologies and diversify their activities.
Nuclear Proliferation	The current nuclear nonproliferation regime must be reinforced to effectively address today's proliferation threats and pave the way for a world without nuclear weapons.
Global Finance	Regulating market volatility and economic risk has become fraught with difficulty following the 2008 financial crisis that plunged developed economies into recession.
Oceans	Nations around the world need to embrace multilateral governance to protect the world's oceans, which play a critical role in global climate, provide an avenue for commerce, and sustain life on earth.
Climate Change	Avoiding the consequences of climate change will require large cuts in global greenhouse emissions and significant efforts to mitigate and adapt to changing weather patterns.
Public Health	Despite medical advances and improvements in water and sanitation, nutrition, housing, and education, poor health still plagues hundreds of millions around the world.
Terrorism	The unprecedented reach and threat of terrorist networks constitutes a new danger to states and requires innovative counterterrorism efforts.
Human Rights	In shaping a human rights policy for the twenty-first century, states must carefully craft tactics consistent with their interests and values to protect victims of abuse.
The Internet	Collaborative Internet governance structures are emerging, but they are being outpaced by policy challenges arising from the Internet's rapid expansion and development.

Source: Council on Foreign Relations.

The main lesson to be taken from this list is that, as per the CFR, these are the ten big categories of global governance challenge that the nations of the world need to work on together to maintain a well governed and safe world. While not all of these categories are immediately relevant to the work of a given leadership team or organization, I would suggest that all of them can and will have a direct or indirect effect on the well-being of any given organization and its stakeholders at any given time.

5.2.b.ii World Bank Governance Indicators

Since 1996, the World Bank has been tracking six governance indicators applicable to the governance of more than 200 countries. Table 5.3 has a summary of the six dimensions of governance that they have studied and tracked for purposes of doing the work that they do focusing on national governance, deploying 40 data sources, produced by 30 different organizations and updated annually since 2002.

Table 5.3 World Bank Governance Indicators[27]

Voice and Accountability	Captures perceptions of the extent to which a country's citizens are able to participate in selecting their government, as well as freedom of expression, freedom of association, and a free media.
Political Stability and Lack of Violence	Measures perceptions of the likelihood of political instability and/or politically motivated violence, including terrorism.
Government Effectiveness	Captures perceptions of the quality of public services, the quality of the civil service and the degree of its independence from political pressures, the quality of policy formulation and implementation, and the credibility of the government's commitment to such policies.
Regulatory Quality	Captures perceptions of the ability of the government to formulate and implement sound policies and regulations that permit and promote private sector development.
Rule of Law	Captures perceptions of the extent to which agents have confidence in and abide by the rules of society, and in particular the quality of contract enforcement, property rights, the police, and the courts, as well as the likelihood of crime and violence.
Control of Corruption	Captures perceptions of the extent to which public power is exercised for private gain, including both petty and grand forms of corruption, as well as "capture" of the state by elites and private interests.

Source: The World Bank.

5.2.b.iii Freedom House "Freedom in the World 2019" study[28]

Freedom House is a think tank that specializes in measuring the degree of social and political freedom around the world along a series of key governance measures. They produce an annual "Freedom in the World Report" measuring democracy or its decline or absence, for which they apply a series of metrics each year.

The report for 2019 was alarming, as it registered for the 13th year running a decline in democracy around the world. Figure 5.2 shows a world map of where democracy exists in 2019 and later in this chapter Figure 5.3 is a chart showing the decline in democracy over the past 13 years.

Figure 5.2 Freedom House: freedom in the world 2019.

Source: Freedom House.

To gain a closer understanding of what democratic governance means at the national governmental level, it is worth looking at what Freedom House measures to assess the level of democratic governance in each country. Table 5.4 provides a synopsis of the questions asked to develop these metrics.

Table 5.4 Freedom in the world: methodology questions[29]

Political Rights (0–40 Points) **A. Electoral Process**	A1.	Was the current head of government or other chief national authority elected through free and fair elections?
	A2.	Were the current national legislative representatives elected through free and fair elections?
	A3.	Are the electoral laws and framework fair, and are they implemented impartially by the relevant election management bodies?
B. Political Pluralism and Participation	B1.	Do the people have the right to organize in different political parties or other competitive political groupings of their choice, and is the system free of undue obstacles to the rise and fall of these competing parties or groupings?
	B2.	Is there a realistic opportunity for the opposition to increase its support or gain power through elections?
	B3.	Are the people's political choices free from domination by the military, foreign powers, religious hierarchies, economic oligarchies, or any other powerful group that is not democratically accountable?
	B4.	Do various segments of the population (including ethnic, religious, gender, LGBT, and other relevant groups) have full political rights and electoral opportunities?
C. Functioning of Government	C1.	Do the freely elected head of government and national legislative representatives determine the policies of the government?
	C2.	Are safeguards against official corruption strong and effective?
	C3.	Does the government operate with openness and transparency?
Civil Liberties (0–60 Points) **D. Freedom of Expression and Belief**	D1.	Are there free and independent media?
	D2.	Are individuals free to practice and express their religious faith or nonbelief in public and private?
	D3.	Is there academic freedom, and is the educational system free from extensive political indoctrination?
	D4.	Are individuals free to express their personal views on political or other sensitive topics without fear of surveillance or retribution?
E. Associational and Organizational Rights	E1.	Is there freedom of assembly?
	E2.	Is there freedom for nongovernmental organizations, particularly those that are engaged in human rights– and governance-related work?
	E3.	Is there freedom for trade unions and similar professional or labor organizations?

(*Continued*)

Table 5.4 Continued

F. Rule of Law	F1. Is there an independent judiciary?
	F2. Does due process prevail in civil and criminal matters?
	F3. Is there protection from the illegitimate use of physical force and freedom from war and insurgencies?
	F4. Do laws, policies, and practices guarantee equal treatment of various segments of the population?
G. Personal Autonomy and Individual Rights	G1. Do individuals enjoy freedom of movement, including the ability to change their place of residence, employment, or education?
	G2. Are individuals able to exercise the right to own property and establish private businesses without undue interference from state or nonstate actors?
	G3. Do individuals enjoy personal social freedoms, including choice of marriage partner and size of family, protection from domestic violence, and control over appearance?
	G4. Do individuals enjoy equality of opportunity and freedom from economic exploitation?

Source: Freedom House.

5.2.b.iv UN Sustainable Development Goals

As we have commented in the previous two chapters, the backbone of the Sustainable Development Goals (SDGs) is that in order for any of them to happen there needs to be good and effective governance at many different levels. Looking at the summary of the SDGs in Table 5.5, it is possible to unequivocally state that governance is required for all 17.

Table 5.5 United Nations Sustainable Development Goals: *governance issues*

SDG – Governance Focused	*E*	*S*	*G*
1. No poverty		X	X
2. Zero hunger	X	X	X
3. Good health and well-being		X	X
4. Quality education		X	X
5. Gender equality		X	X
6. Clean water and sanitation	X	X	X
7. Affordable and clean energy	X	X	X
8. Decent work and economic growth		X	X
9. Industry, innovation and infrastructure	X	X	X
10. Reduced inequalities		X	X
11. Sustainable cities and communities	X	X	X
12. Responsible consumption and production	X	X	X
13. Climate action	X	X	X
14. Life below water	X		X
15. Life on land	X		X
16. Peace and justice – strong institutions		X	X
17. Partnerships for the goals		X	X

Source: United Nations.

5.2.b.v G20/OECD Principles of Corporate Governance

Diving now into more specific detail on corporate governance, let's look at the updated G20/OECD Principles of Corporate Governance, which were adopted in late 2015 to "create market confidence and business integrity … essential for companies that need access to capital for long-term investment." Additionally, these principles are aimed at "millions of households around the world" that have their savings in the stock market one way or another, thus creating a more expansive view of the beneficiaries or the potentially detrimentally affected in the marketplace beyond shareholders to stakeholders generally. Table 5.6 is a summary of these principles.

Table 5.6 G20/OECD Principles of Corporate Governance[30]

Principles	*Key Topics of Focus*
I. Ensuring Basis for Effective Corporate Governance Framework	• Transparency and fair markets • Efficient allocation of resources • Regulatory framework • Quality of supervision and enforcement • Role of stock markets in good corporate governance
II. The Rights and Equitable Treatment of Shareholders and Key Ownership Functions	• Basic shareholder rights • Right to information and key decision-making participation • Disclosure of control structures (including voting rights) • Use of information technology • Approval of related party transactions • Shareholder participation in executive remuneration decisions
III. Institutional Investors, Stock Markets and Other Intermediaries	• Sound economic incentives through investment chain • Disclose and minimize conflicts of interest of third parties • Cross border listing principles • Fair and effective price discovery in stock markets
IV. The Role of Stakeholders in Corporate Governance	• Active cooperation with stakeholders • Recognizing rights of stakeholders established by law or mutual agreement • Stakeholder access to information • Stakeholder right to redress for violations of rights

V. Disclosure and Transparency	• Key areas of disclosure to include:
	  • Financial and operating results
	  • Company objectives
	  • Major share ownership
	  • Remuneration
	  • Related party transactions
	  • Risk factors
	  • Board members
	  • Non-financial information

VI. Responsibilities of the Board	• Key functions • Review of strategy • Compensation management

(*Continued*)

Table 5.6 Continued

Principles	*Key Topics of Focus*
	• M&A and divestitures • Accounting integrity • Financial reporting integrity • Specialized committees for remuneration, audit & risk management

Source: Organization for Economic Cooperation and Development and G20.

5.2.b.vi OECD Due Diligence Guidance for Responsible Business Conduct

In 2018, the Organization of Economic Cooperation and Development (OECD) produced the "OECD Due Diligence Guidance for Responsible Business Conduct".[31] A publication in a long line of research and work that the OECD has done to support and encourage member nations and their corporations to engage in responsible business conduct, this piece provides practical guidance around a variety of topics derived from their original OECD Guidelines for Multinational Enterprises of 2008.[32]

The OECD Due Diligence Guidance may be summarized as follows:

> The Due Diligence Guidance is built around six core process expectations with respect to how companies will manage the impacts of their activities. Namely, companies should:
>
> 1. Embed responsible business conduct into policies and management systems
> 2. Identify and assess actual and potential adverse impacts associated with the enterprise's operations, products or services"
> 3. Cease, prevent and mitigate adverse impacts
> 4. Track implementation and results
> 5. Communicate how impacts are addressed
> 6. Provide for or cooperate in remediation when appropriate[33]

5.2.b.vii Allianz Risk Barometer

As we have in previous chapters in this Part II, let's take a look at what a leading global commercial insurance company has done in terms of identifying the top-ten global business risks for 2018 and see which ones have a strong governance component. See Table 5.7.

Table 5.7 Allianz Risk Barometer Top-Ten Business Risks 2017–2019:[34] focus on governance issues

Top-Ten Risks (with governance issues)	*2019 Ranking*	*2018 Ranking*	*2017 Ranking*	*ESGT Issue*
Business Interruption (Including Supply Chain Disruption)	#1	#1	#1	E, S, G, T

(*Continued*)

Table 5.7 Continued

Top-Ten Risks (with governance issues)	*2019 Ranking*	*2018 Ranking*	*2017 Ranking*	*ESGT Issue*
Cyber Incidents (e.g., Cybercrime, IT Failure, Data Breaches)	#2	#2	#3	G, T
Changes in Legislation and Regulation (e.g., Government Change, Economic Sanctions, Protectionism, Brexit, Eurozone Disintegration)	#4	#5	#5	G
New Technologies (e.g., Impact of Increasing Interconnectivity, Nanotechnology, Artificial Intelligence, 3D Printing, Drones)	#7	#7	#10	G, T
Loss of Reputation or Brand Value	#9	#8	#9	G
Shortage of Skilled Workforce	#10 (NEW RISK)	(*POLITICAL RISK #9*)	(*POLITICAL RISK #8*)	S, G

Source: Allianz.

5.2.b.viii MSCI

Pivoting once again to the view of the institutional investor community – what does corporate governance mean to them? In Table 5.8, we provide the governance typology used by MSCI.

Table 5.8 MSCI categories of ESG Issues:[35] *governance*

Corporate Governance

- Board diversity
- Executive pay
- Ownership & control
- Accounting

Corporate Behavior

- Business ethics
- Anti-competitive practices
- Tax transparency
- Corruption & instability
- Financial system instability

Source: MSCI.

5.2.b.ix RepRisk

Getting down to an even more granular level, let's turn to what RepRisk has done this time vis-à-vis research scope, including ESG Issues and Topic Tags linked to "Corporate Governance" factors. Table 5.9 provides an overview.

Table 5.9 RepRisk ESG Issues: *corporate governance issues*[36]

Corporate Governance
• Corruption, bribery, extortion, money laundering
• Executive compensation issues
• Misleading communications, e.g., "green-washing"
• Fraud
• Tax evasion
• Tax optimization
• Anti-competitive practices
Cross-Cutting Items
• Controversial products and services
• Products (health and environmental issues)
• Violation of international standards
• Violation of national legislation
• Supply chain issues

Source: RepRisk.

Drilling down a bit deeper, Table 5.10 provides the Topic Tags RepRisk tracks as an extension of its core research scope of 28 ESG issues. RepRisk screens for these issues in over 80,000 media and stakeholder sources every day in 20 languages to identify ESG and business conduct risks related to more than 115,000 public and private companies worldwide.

Table 5.10 RepRisk Topic Tags for ESG issues with a primary focus on governance issues[37]

Alcohol
Asbestos
Automatic and semi-automatic weapons
Biological weapons
Chemical weapons
Cluster munitions
Cyberattack
Depleted uranium munitions
Drones
Gambling
Genetically modified organisms (GMO)
Land mines
Marijuana/cannabis
Negligence
Nuclear power
Nuclear weapons
Palm oil
Pornography

(*Continued*)

Table 5.10 Continued

Predatory lending
Security devices
Soy
Tax havens
Tobacco

Source: RepRisk.

Lastly, to illustrate some of the results of these analyses, Table 5.11 presents some examples of governance-related serious reputation risk incidents as compiled by RepRisk in their Most Controversial Companies 2017, and Table 5.12 presents their summary of Most Controversial Projects 2017. An interesting observation: every one of the top most controversial companies was on that list for governance issue failures (in additional to other ESG issues).

Table 5.11 RepRisk Most Controversial Companies 2017:[38] governance related

MCC 2017 Rank	*Company Name*	*Peak RepRisk Index 2017 (1–100)*[39]	*Sector*	*Country of Head-Quarters*	*Primary ESGT Issue(s)*
#1	**The Weinstein Company**	92	Media	USA	S & G
#2	**Kobe Steel**	87	Industrial metals	Japan	E & G
#3	**J&F Investmentos SA**	83	Food and beverage, personal and household goods	Brazil	S & G
#4	**Appleby Global Group Services Ltd (Paradise Papers)**	82	Support services (industrial goods and services)	Bermuda Islands (UK)	G
#5	**Stalreiniging Barneveld (Chickfriend)**	82	Support services (industrial goods and services)	Netherlands	E S & G
#6	**Equifax Inc.**	79	Financial support services (industrial goods and services)	USA	S G & T
#7	**Rolls Royce Holdings Plc**	75	Aerospace and defense industrial engineering	UK	G
#8	**Odebrecht SA**	74	Construction and materials	Brazil	E S & G
#9	**Petroleos de Venezuela SA**	73	Oil and gas	Venezuela	E S & G
#10	**Transnet Soc Ltd**	73	Industrial transportation	South Africa	G

Source: RepRisk.

Table 5.12 RepRisk Most Controversial Projects 2017:[40] governance related

MCC 2017 Rank	*Project Name*	*Peak RepRisk Index 2017 (1–100)*	*Sector*	*Country of Head-Quarters*	*Primary ESGT Issue(s)*
#4	**Brook House Immigration Removal Centre**	68	Support services (industrial goods & services)	UK	S & G
#6	**Jalabiya Cement Works**	59	Financial support services (industrial goods & services)	Syria	S & G
#7	**OPL 245 Oil Block**	59	Aerospace and defense industrial engineering	Nigeria	S & G

Source: RepRisk.

5.2.b.x UN Global Compact

In the case of a governance issue per se, the UN Global Compact only contains one principle that relates directly to governance, and it is their last one (#10), which was only added after the first nine were originally launched – focusing on anti-corruption – and which reads as quoted in Table 5.13.[41]

Table 5.13 United Nations Global Compact 10 principles:[42] *governance issues*

Anti-Corruption	**Principle 10**	Businesses should work against corruption in all its forms, including extortion and bribery.

Source: United Nations Global Compact.

As noted in prior chapters in this Part II, the UN Global Compact has useful resources for organizations that want to align their internal policies with either or both the UN Global Compact Principles and the SDGs helping to tie each SDG to actual, measurable goals and objectives.[43]

5.2.b.xi Transparency International[44]

Before we turn to a variety of governance issues, risks and opportunities in the global marketplace, I would be remiss not to mention the extraordinary work that another organization – Transparency International (TI) – has done in the field of global anti-corruption. Best known for their annual Corruption Perceptions Index (CPI) survey, mapping and ranking the world's nations along a most-to-least-corrupt perceptions index, TI is also known for the development – often in collaboration with a variety of public and private partners – of an array of tools, research and materials.

These are extraordinarily useful to leaders who are intent on identifying and understanding the challenge of corruption to their organizations and the opportunity to transform this serious and destructive risk into an opportunity for change, improvement and maybe even value creation, as we will discuss later. I share some of TI's useful materials on anti-corruption here, starting with a very helpful definition of "corruption" presented in Table 5.14.

Table 5.14 Transparency International: *How do you define corruption?*[45]

Generally speaking as "**the abuse of entrusted power for private gain**". Corruption can be **classified as grand, petty and political**, depending on the amounts of money lost and the sector where it occurs.

Grand corruption consists of acts committed at a high level of government that distort policies or the central functioning of the state, enabling leaders to benefit at the expense of the public good.

Petty corruption refers to everyday abuse of entrusted power by low- and mid-level public officials in their interactions with ordinary citizens, who often are trying to access basic goods or services in places like hospitals, schools, police departments and other agencies.

Political corruption is a manipulation of policies, institutions and rules of procedure in the allocation of resources and financing by political decision makers, who abuse their position to sustain their power, status and wealth.

Source: Transparency International.

TI goes on to provide a vast wealth of resources to combat corruption. And why? Table 5.15 encapsulates their compelling business case as to why it is essential to combat corruption in all of its forms and shapes around the world:

Table 5.15 Transparency International: *What are the costs of corruption?*[46]

The cost of corruption can be divided into four main categories: political, economic, social and environmental.

On the political front, corruption is a major obstacle to democracy and the rule of law. In a democratic system, offices and institutions lose their legitimacy when they're misused for private advantage. This is harmful in established democracies, but even more so in newly emerging ones. It is extremely challenging to develop accountable political leadership in a corrupt climate.

Economically, corruption depletes national wealth. Corrupt politicians invest scarce public resources in projects that will line their pockets rather than benefit communities, and prioritize high-profile projects such as dams, power plants, pipelines and refineries over less spectacular but more urgent infrastructure projects such as schools, hospitals and roads. Corruption also hinders the development of fair market structures and distorts competition, which in turn deters investment.

(*Continued*)

Table 5.15 Continued

Corruption corrodes the social fabric of society. It undermines people's trust in the political system, in its institutions and its leadership. A distrustful or apathetic public can then become yet another hurdle to challenging corruption.
Environmental degradation is another consequence of corrupt systems. The lack of, or non-enforcement of, environmental regulations and legislation means that precious natural resources are carelessly exploited, and entire ecological systems are ravaged. From mining, to logging, to carbon offsets, companies across the globe continue to pay bribes in return for unrestricted destruction.

Source: Transparency International.

5.2.c The way forward

Now that we have reviewed a variety of governance definitions, characteristics and classifications, let's turn to some of the big governance issues of our day through three distinct lenses – the lens of international governance, national governance and corporate governance.

In the corporate governance section (which is broadly applicable in its content to governance of other kinds of entities like nonprofits, agencies, universities and others), we examine in some detail organizational governance issues and examples as follows:

- Waves of scandal and reform, risk and opportunity from the 1970s to the 2020s.
- Financial sector governance since 2008.
- The global auto industry: a race to the bottom, reputation risk contagion or both? With a look at Volkswagen and the latest scandal at the time of this writing involving the resignation of the CEO of Nissan-Renault-Mitsubishi.
- Silicon Valley – transforming immature company governance into stakeholder value – with a look at "bad" governance (Theranos), "unattractive" governance (Tesla), and "improving" governance (Uber), and providing lessons learned for other companies.
- We conclude with a somewhat humorous but quintessentially accurate review of individual cases of "bad governance" – i.e., directors that don't rate – as expertly assembled by *Directorship Magazine*.

5.3 Leaders navigating governance risks, crises and opportunities

The following section is divided into three levels of governance – international, national and corporate or organizational.

5.3.a International governance: navigating through the global chaos of geopolitical risk

While beyond the scope of this book and a much larger subject for other experts to tackle, it bears mentioning that trends – as described more fully in

Chapter 1 – "Gloom" – are moving in the direction of the erosion of global governance and the world order as we have known it since WWII and the Cold War. A mere look at the headlines in which long-standing trade, economic and military institutions (the EU, NATO, NAFTA) are under assault not only from within but also from without, demonstrates this point.

In other words, some of the foundational institutions, norms and practices established after WWII are under siege partly because of the unusual and unexpected US attack on many of these institutions (via President Trump and his administration's policies) and partly because of other international and national pressures exemplified by the international role Putin has assumed, the rise of Xi's China and the growth of populistic, illiberal democracy in the West (not to mention other tectonic changes taking place, as partly explored in Chapter 1).

The work of international thinkers like Richard Haass in *A World in Disarray*, Ian Bremmer in *US versus Them: The End of Globalism*, and Timothy Snyder in both *Tyranny* and *The Road to Unfreedom* along with the collective works of leading authors in the April/May 2018 Issue of *Foreign Affairs* titled "Letting Go: Trump, America and the World" and in their June/July 2018 issue titled "Is Democracy Dying? A Global Report"[47] all lead to a similar place: we are living in a time of great disruption both geopolitically and technologically, and the outcome of these times is far from clear.

Suffice it to say that the international governance that has been in place for a very long time is under attack, eroding and/or under pressure to transform into something else, with odds fairly high that this something else might be less stable, less broadly acceptable and potentially more volatile for economic, financial, geopolitical and technological governance globally.[48]

The job of world leaders now? To help stabilize some of the more chaotic tendencies, to listen closely to various key stakeholders around the world on what ails them (inequality, institutional mistrust, war, migration, climate change, corporate greed, loss of jobs to technology, etc.), and begin to build a new world order that addresses the maximum number of stakeholders and the most critical and important risks facing them.

Take a look at Table 5.2 in this chapter for a succinct and useful summary by the Council on Foreign Relations of the key international governance cross-cutting themes of our time.

5.3.b National governance: transforming authoritarian risk into democratic opportunity

I will take another moment to address some of my own personal and professional experience as it relates to what I am writing about. As I mentioned earlier in this chapter, I grew up in an authoritarian, fascist country (Spain) and was literally a Cold War baby, born into a family that had experienced both WWII and the Cold War up close and personally from both sides (the American and the German). If we want to go back even further, growing up I heard stories from parents and grandparents about their family's respective experiences living in WWI and Weimar Germany and enduring the US Depression of the 1930s.

As a young woman, I was lucky to be accepted into the Columbia University Graduate School of Arts and Sciences PhD program where I ended up working for and studying with two luminaries of the world of political science – Dr. Zbigniew Brzezinski and Dr. Seweryn Bialer (also East Europeans who themselves or whose families had fled to the West).

My family background and the great people I studied with and worked for all fed into my life-long passion – how do societies build lasting democracy? – and led me to write my PhD dissertation on a then-rare occurrence: a peaceful transition from authoritarianism to democracy, something that had only occurred a few times before in history – and it happened to be taking place in the country I loved most: Spain.

Spain turned out to be the tip of the spear of a large wave of worldwide democratizations beginning in southern Europe (Greece, Portugal and Spain) in the 1970s, continuing in Latin America and Eastern Europe in the 1980s and 1990s as well as in a number of countries in Asia Africa through the end of the century and the beginning of the 21st. We even saw a glimmer of democratic hope in the Middle East in the early 2000s with Tunisia leading the way (and, sadly, it turns out, it was the only long-term beneficiary of that particular wave of democratization and a problematic one at that). As I finish writing this book in mid-2019, some additional and fascinating pro-democracy developments are brewing in Algeria and Sudan, of all places, and in the Ukraine, which just elected a novice politician/TV comedian for its President. Spain, led by Social Democrats, seems to have countered the movement toward populistic, illiberal or more chaotic democracy we have been witnessing in Europe of late by soundly voting for a coalition of center-left parties. And surprising pro-democracy demonstrations (that are dangerous to demonstrators' life and limb) are taking place in Moscow and Hong Kong in the summer of 2019.

For all these reasons – the positive and the more negative – it's worth looking at some key governance lessons that we learned from the prior waves of democratization which Mark Brzezinski and I summarized in our article "Mideast Shift: Lessons from Europe" for Politico in 2011. In that article, we laid out the most important criteria for a successful transition to democracy that are applicable to the Middle East based on our respective experience, research and books on Southern and Eastern European democratizations, which are summarized in Table 5.16.

Table 5.16 The governance of political democratization: key elements[49]

1. Lasting reform is not possible without the rule of law. Democracy is not just holding elections. Rather, sustainable democracies and complex market economies depend on legal certainty and the assurance that law is not subordinate to politics.
2. Non-democratic leaders are ill-suited to building democracies.
3. Corruption can't be stamped out unless the state's administration is modernized and its bureaucrats retrained.

(*Continued*)

Table 5.16 Continued

4. Financial assistance alone does not determine success of democratic and market transition.
5. Policy implications:
• First, democratization involves the creation of juridical structures and guarantees, especially constitutional changes that institutionalize respect for human rights and the rule of law.
• Second, reform has to build on the global anti-corruption effort now underway elsewhere.

Source: Mark Brzezinski and Andrea Bonime-Blanc. Politico.

The waves of democratization the world witnessed from the late 1970s through the beginning of the 21st century represented nothing short of a revolution in governance at the highest national levels – with movements initiated by various portions of civil society (the government's "stakeholders") and led by more or less enlightened leaders to create more open, less repressive political systems where the rule of law, elections, freedoms of expression, separation of powers were among the objectives. Surely this represented a transformation of authoritarian risk into democratic value. A look at Table 5.16 and Table 5.4 (earlier in this chapter), in which we list some of the key indicia of democratic governance as expertly assembled by Freedom House, demonstrates the fundamental nature of governance at the governmental level.

Where are we on national governance as I write this book? We are in a very mixed and concerning place, if you look at what the experts are saying. The advent of Trumpism in America, Brexit in the UK, a variety of populist and anti-democratic right-wing parties throughout the European Union and a variety of other authoritarian or authoritarian wannabe leaders throughout the world in the past few years is cause for alarm for those who cherish democracy and its rights, responsibilities, norms and rule of law.

A look at another chart from Freedom House from 2019 (see Figure 5.3) paints a troubling longer-term picture. Over the past 13 years, the world has experienced a slow but steady decline in democratic governance, where the number of countries improving such governance has declined steadily and those declining in democratic governance have increased dramatically. The data in Figure 5.3 for 2019 however does offer a glimmer of positivity.

One can hope the overall trend is a temporary blip on the longer-term screen, but there is larger cause for concern that actually goes to the even bigger governance picture – that of the geopolitical balance of power around the world: the post-WWII, post-Cold War international order (i.e., governance) appears heavily challenged (maybe even crumbling), and we are clearly on the road to a much more uncertain time where the institutions we have relied on heavily for peaceful economic, financial, political and geographical governance may become weakened.[50]

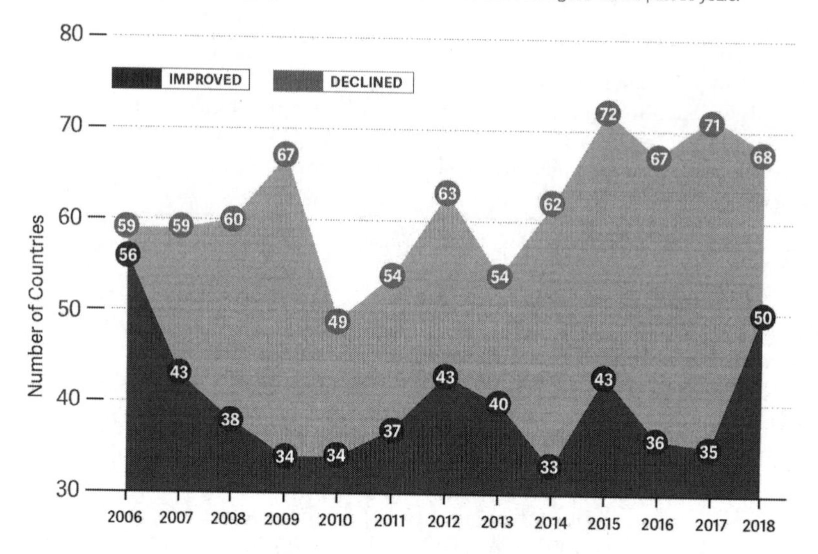

Figure 5.3 Freedom House: 13 years of democratic decline.

Source: Freedom House.

To conclude this segment on a somewhat more positive note, I would like to share the results of the *World Happiness Report* produced by the United Nations pursuant to its SDG initiatives, which in essence correlates a number of criteria (including perceptions of corruption or the absence thereof) with higher levels of happiness.

The top-ten countries for 2018 and the six criteria determining their top rankings are summarized in Table 5.17.[51]

Table 5.17 World Happiness Report 2018: top countries and criteria

The six criteria	*Top-ten countries*
• GDP per capita/income	1. Finland
• Healthy life expectancy	2. Norway
• Social support	3. Denmark
• Freedom to make life choices	4. Iceland
• Perceptions of corruption	5. Switzerland
• Generosity	6. Netherlands
	7. Canada
	8. New Zealand
	9. Sweden
	10. Australia

Source: *The World Happiness Report.*

How shocking to discover that the happiest countries in the world are all democracies! Now there is a powerful governance lesson to be applied by governments, business and society.

5.3.c *Corporate governance: transforming "the bad and the ugly" into "the good and the better"*

Now let's narrow our view somewhat to the next level of governance – corporate governance. When I reference "corporate governance" I generally mean the governance of an organization – it could be a corporation, but it could also be a nonprofit, an academic institution, a religious entity, a nongovernmental organization, and even a government agency, as every type of organization must have some form of self-governance, has leaders, has some form of oversight and has stakeholders.

5.3.c.i *Waves of scandal and reform, risk and opportunity from the 1970s to the 2020s*

Management and the board are responsible for understanding and managing risk, having a deep knowledge and oversight of strategic risk and opportunity. What does that mean? In essence, it means having a robust and well-resourced system of risk identification, management, mitigation, on the one hand, and strategic risk governance, on the other.

Let's review some of the key big-picture strategic risks gone wrong that we have seen in the marketplace over the past few decades in what appear to be waves of scandal and scandal response. See Table 5.18 for an overview. Sadly, there is a lot to draw on when we talk about corporate scandals, certainly in the more than three decades I have been a practitioner in various areas of corporate legal, ethical and social responsibility.

Our overview starts with the wave of corruption and bribery scandals that enveloped the US defense industry in the 1970s that led to the adoption of the US Foreign Corrupt Practices Act (FCPA) in 1977. Others might remember the savings and loan scandals of the 1980s. One might also recall in the 1990s the health care scandals that led to greater compliance and board oversight requirements in the healthcare industry in the US.

Table 5.18 summarizes a series of waves of industry-wide scandals many in the US that is not intended to be exhaustive but certainly illustrative of my main point: that strategic governance risk and crises occurred in these industries and companies for more than one reason, but one of the core reasons is the failure of strategic risk governance at the highest levels of management in terms of managing risk and the board in terms of oversight.

Table 5.18 Waves of scandal and reform: how the pendulum swings since the 1970s

Point/Scandal	*Counterpoint/Solution*
1970s **The US Defense Industry foreign bribery and corruption scandals**	• US Congress adopts US Foreign Corrupt Practices Act of 1977
1980s **The US Defense Industry fraud scandals**	• Congressional investigations and Packard Commission Report • Defense Industry led by Jack Welch of GE establishes "Defense Industry Initiative" first voluntary internal ethics program to create internal governance of ethics
1980s **The US Savings and Loans Bank scandals**	• Greater banking and regulatory oversight of savings and loan industry
1990s–2000s **Global corporate corruption**	• World Bank economists founded transparency international in 1993 to combat corruption • The OECD passes the 1997 Anti-Bribery and Anti-Corruption Convention (40+ signatories to date) • Countries increase investigations and prosecutions for violations and collaborate in global fight against corruption and kleptocracy • The US Congress creates non-partisan commission – the us Federal Sentencing Commission – part of whose job is to establish guidelines for corporate ethics and compliance programs
1990s **The US Healthcare Industry Government fraud scandals of the 1990s**	• Healthcare fraud leads to more regulations and more severe compliance oversight requirements under Delaware courts applicable to the industry broadly • Creation of first association of ethics and compliance professionals - the Ethics Officer Association (EOA)
2001 **"Enron et al." – wave of accounting fraud scandals including WorldCom, Adelphi, Parmalat, Tyco and others in the early 2000s**	• Sarbanes-Oxley is adopted in 2002 • Establishment of the Public Company Accounting Oversight Board (PCAOB) • Voluntary ethics and compliance programs continue to evolve into stronger & more effective in the US and globally
2007–2008 **The financial sector scandals erupting in 2007/8 but dating back to "inventions" and practices starting in the 1980s (mortgage backed securities, securitization, collateralized mortgage obligations, collateralized debt obligations, derivatives, etc.)**	• The US Congress adopts Dodd-Frank in 2010 • Many consumer and other protection regulations and rules are enacted • Basel II regulations are enacted • Global, National and Banking Bodies Develop Financial Conduct Programs Chief among them the UK Financial Conduct Authority, the Netherlands Banking Authority and The New York Federal Reserve Bank

(*Continued*)

Table 5.18 Continued

Point/Scandal	*Counterpoint/Solution*
2015 to date **Auto industry emissions testing scandals first VW, then most german auto companies and several other in Japan, US and EU**	• Fines and continuing investigations but no concerted response yet though some of the auto companies have appointed new chief ethics and compliance officers (CECOs)
2017 to date **Panama Papers 1 and 2 and Paradise Papers lay bare opaque and criminal corporate vehicles**	• International initiatives from Transparency International and others to create accountability and transparency • Some initiatives at national and local levels to push for ownership transparency – in Delaware and the UK
Second decade of 2000s **Various technology sector privacy and other perceived excesses**	• EU GDPR 2018 • California and Vermont state privacy regulations • Attempts in Washington, DC, to deal with various tech-related threats, including cyber, fake news, fake accounts, social engineering, etc.

Source: Author.

Let's ask another key question: why is it that when one scandal is uncovered in an industry it so often happens that it is an industry-wide practice? Let's take a look at two industries with industry-wide scandals in recent years – the financial and the auto sectors.

5.3.c.ii *Financial sector governance: better 10 years after the great recession?*

> The important lesson of the (2008) crisis is not that markets are fallible ... It is that essential regulations ... are stymied by fractured government machinery and rapacious lobbies. Even today, the financial system has multiple overseers answerable to multiple congressional committees, because all this multiplying produces extra opportunities for lawmakers to extract campaign contributions. Vast government subsidies still encourage Americans to take big mortgages; Fannie Mae and Freddie Mac still operate, despite endless talk of breaking them up. And although post-2008 regulations have ensured that banks are better capitalized, the lobbyists are pushing back. Merely a decade after the Lehman bankruptcy brought the world economy to its knees, the Trump administration is listening to them.[52]

Who can forget the financial industry mega-scandal of 2008 led by the US financial sector and closely followed by a variety of UK and European banks? In an excellent interactive graphic on the Council on Foreign Relations' website, summarized in Table 5.19, CFR provides a historical timeline and details dating back to 1992 and showing some of the milestones that got us there (2008) and got us here, 10 years later (2018).

Table 5.19 Council on Foreign Relations: US financial crisis timeline 1992–2018 – summary[53]

1992	New rules for Fanny & Freddie
1995–1999	Subprime market grows
1999	Glass-Steagall weakened
2000–2001	Federal Reserve cuts interest rates
2004	Wall Street places riskier bets
2007 – FEBRUARY	US housing bubble bursts
2007 – APRIL	Subprime bankruptcies proliferate
2007 – AUGUST	Subprime woes go global
2007 – SEPTEMBER	Fed slashes rates and market peaks
2008 – MARCH	Fire sale of Bear Stearns
2008 – SEPTEMBER 14	Government nationalizes Fannie and Freddie
2008 – SEPTEMBER 15	"Moral hazard" and Lehman's collapse
2008 – SEPTEMBER 16	Federal Reserve bails out AIG
2008 – SEPTEMBER 19	Paulson unveils TARP rescue plan
2008 – SEPTEMBER 25	Bank failures signal end of an era
2008 – OCTOBER 10	Dow finishes worst week as Fed intervenes
2008 – NOVEMBER 25	Fed announces quantitative easing
2008 – DECEMBER	Bush launches auto bailouts
2009 – FEBRUARY	Obama signs $787 billion stimulus package
2009 – APRIL	G20 Summit pushes financial regulation
2009 – MAY	US Banks stress tested
2010 – JULY	Dodd-Frank financial overhaul signed
2011 – SEPTEMBER	Occupy Wall Street taps into discontent
2013 – DECEMBER	Regulators approve Volcker Rule
2014 – DECEMBER	TARP ends
2017 – JUNE	Dodd-Frank repeal passes house
2017 – NOVEMBER	Mulvaney takes over Consumer Financial Protection Bureau
2018 – MAY	Trump signs Dodd-Frank reform bill

Source: Council on Foreign Relations.

In the category of a "picture speaks louder than words" Table 5.20 shows the aftermath of the financial industry scandal which is now just 10 years old. The Table shows a tally of bank fines levied so far of upward of US$321 billion since the financial scandal, with Bank of America leading the pack by far.[54]

Table 5.20 Partial list of largest banking sector fines since 2008[55]

Bank	*Total fines in $ billions to date*
Bank of America	76.1
JP Morgan Chase	43.7
Citigroup	19.0
Deutsche Bank	14.0
Wells Fargo	11.8

(*Continued*)

Table 1.5 Continued

Bank	*Total fines in $ billions to date*
RBS	10.1
BNP Paribas	9.3
Credit Suisse	9.1
Morgan Stanley	8.6
Goldman Sachs	7.7
UBS	6.5

Source: MarketWatch.

The reverberations of those days continue to this day – as some have pointed out (and I would agree with them) the impact of the financial scandal of 2008 was much larger and longer term than the previous industry scandals we had experienced.

Indeed, some have argued that the downturn occasioned by the crisis had a tectonic generational impact in two ways: first, on the generation of young millennials who were just graduating from college to a precipitously declining job market with fast growing unemployment, and second, on a vast swath of older workers who permanently lost well-paying or at least decent jobs with benefits that they never got back. In fact, recent research has gone a step further to posit that the US financial crisis also damaged the US economy in another deep and hard-to-reverse way: separate and apart from income inequality trends, it has also drastically increased wealth inequality.[56]

Some of the longer-term implications of this economic hit (in the US at least) may also be seen in other social impacts, such as the increase in health problems, opioid addiction, suicide rates, lowering of white male life expectancy and other such negative social implications that also feed into the rise of populism and Trumpism in America. Some studies show a deep interconnection between the financial crisis and the deepening of income inequality in the US. Similar commentary can also be made about other mature democracies with advanced industrial economies.

Getting back to the financial sector before the crisis. Until scandal hit, these banks were not thinking about proactive governance or a culture of integrity and ethics as part of how they measured success, and if they did it was truly meaningless and pro forma. Clearly these were not coveted principles or practices in this sector. Indeed, while the financial sector has always been heavily regulated and as a result compliance heavy, it has always been ethics and integrity light.

But I digress a little. The principal point I make in sharing the reality and implications of all of these waves of scandal and response, including the financial industry, is that governance or the failure thereof is at the heart of each and every one of these scandals. And the failure of governance can be pinpointed on the failure of boards and oversight bodies to hold executive management accountable for anything but the numbers. Point in fact: until it was dictated by the regulator, financial institutions in the US were not required to have a board-level risk committee.

To this day, the biggest banks in the US and globally, despite the deep-seated cultural and ethical issues that have beset them, do not have robust internal ethics and integrity programs – yes, they have expensive and highly populated compliance functions, but compliance alone does not get at what *The Economist*, as shown in the following quotes, has called a "crisis of culture" in which the metrics of financial institutional culture did not look good:[57]

- Most firms have attempted to improve adherence to ethical standards … Global institutions, from Barclays to Goldman Sachs, have launched high-profile programs that emphasize client care and ethical behavior.
- Industry executives champion the importance of ethical conduct … Despite a spate of post-crisis scandals that suggest continued profit-chasing behavior, large majorities agree that ethical conduct is just as important as financial success at their firm.
- Executives struggle to see the benefits of greater adherence to ethical standards … Fifty-three percent think that career progression at their firm would be difficult without being flexible on ethical standards.
- To become more resilient, financial services firms need to address knowledge gaps … The increasingly complex risk environment has made advancing and updating knowledge of the industry crucial for those working in or serving the financial services industry.
- A lack of understanding and communication between departments continues to be the norm. Many argue that ignorance was a key contributor to the global financial crisis.

I have often wondered whether the reason the abuses in the financial sector – in volume, size, recidivism and extent (not only recently but going back to the 1980s when I first started working on Wall Street as a lawyer and the mantra from the movie *Wall Street* was "greed is good") doesn't have something to do with:

- The actual, physical proximity to money.
- The pipeline of talent coming almost exclusively from the most competitive MBA schools that cater to a certain, mostly male, competitive ethos and cadre of self-selecting students.
- The fact that the singular be-all and end-all of the financial sector is making money – it's not building a tower, it's not producing a widget, it's not inventing a new cure – it's managing and making money – as much of it as possible.

There's some interesting academic research that has been conducted on this topic, which in essence says the closer people are to actually seeing and touching money, the more willing they are to cut corners to get it. In other words, as the sight of money increases, the willingness to compromise ethical standards also increases.

In Table 5.21 the reader will find a summary of the measures Credit Suisse has promised to undertake in response to what they called in 2018 their "legacy cases" of non-compliance from 2006 to 2014. While all the measures announced reflect what should be considered a high-quality compliance program (see discussion of high-quality ethics and compliance programs in Chapter 2) – two questions should be raised:

- First, we've been to this rodeo before where a bank that has suffered the consequences of poor or inconsistent compliance is caught and fined and in response throws money and people at the problem (but doesn't necessarily address the elephant in the room – culture at the very top). What will make this stick this time?
- Second, is this program sustainable or will it fall apart or flounder as new leadership takes over the bank over time? This last issue is one of the most important for leaders everywhere – maintaining good practices over time – institutionalizing them – above and beyond the current leadership – making them part of the long-term, sustainable institutional ethos.

Table 5.21 Transforming bank culture risk into stakeholder value? *Credit Suisse compliance reforms announced in September 2018 in response to regulatory review of ongoing review of "legacy cases" of non-compliance 2006–2014*[58]

- **Separation of legal and compliance functions**, "creating a dedicated group compliance and regulatory affairs function that reports directly to the CEO, with a new head sitting on the executive board".
- **Increase in global compliance headcount by** "42% in less than three years, hiring over 800 additional compliance specialists".
- **More than 10,000 bank-wide control enhancements** to correct legacy weaknesses, while pivoting investments into next-generation technology and data analytics.
- **Deployment of "state-of-the-art systems and industry-leading capabilities to mitigate unwanted risks,** including three labs that continually enhance our prevention and detection capabilities. We now have 40 data scientists in compliance alone who manage and enhance our ability to detect risks on a real time basis".
- **Creation of "a single client view program,** giving our compliance teams an integrated view of clients for the first time in the history of our bank".
- **Implementation of "new compensation framework that enables performance to be assessed against specific risk and conduct metrics** across the entire organization".
- **Creation of "conduct and ethics board chaired by two of our executive board members** to ensure that we carry out our business according to the highest standards".
- **Plan for "board of directors ... to establish a dedicated compliance, conduct and culture board committee** that will consolidate the board's ongoing intensive efforts to address these important topics and ensure they continue to receive the highest strategic focus and priority at all times".

Source: Credit Suisse.

On the better-angels-of-our-nature side of the ledger, it is also important to emphasize that leaders at many different levels – governmental, academic, private – have also been working hard to find solutions to protect the most important and vulnerable stakeholders in the financial marketplace. Since the financial crisis hit in 2008, there have been a number of important efforts undertaken mainly by the public sector in several countries to get at the root cause of financial misconduct – human behavior. The following is a partial list of several such efforts in both the public and private sectors that are worth mentioning and watching:[59]

- UK Financial Conduct Authority.
- UK Revised Corporate Governance Code of 2018.
- Netherlands Financial Conduct.
- Basel II.
- New York Federal Reserve Bank.
- The US Consumer Financial Protection Bureau.

5.3.c.iii The global auto industry: a race to the bottom, reputation risk contagion or both?

And who can forget the automotive industry emissions scandals of the past few years erupting with Volkswagen in late 2015, then appearing to be an isolated case until a broad cross section of German, Japanese and other car companies started to fess up to being guilty of the same or similar emissions test cheating?

Even at the time of this writing, the scandal was widening in Germany to include other auto companies beyond Volkswagen.[60] Indeed, the beat goes on as I finish writing this book in mid-2019 and the US Department of Justice has just announced a criminal investigation into US automaker Ford for possible emissions cheating.

Meanwhile, in Japan, several companies have admitted to emissions cheating dating as far back as the 1990s in the case of Mitsubishi Motors. Here's just a short list of relevant headlines as well as a probably non-exhaustive list of auto companies that have been implicated in emissions testing cheating allegations over the past several years (Table 5.22):

- The September 26, 2015, edition of *The Economist* cover story: "A Scandal in the Motor Industry: Dirty Secrets".[61]
- "Mitsubishi: We've Been Cheating on Fuel Tests for 25 Years".[62]
- "Nissan Admits Falsifying Emissions Tests in Japan".[63]
- "Why Much of the Car Industry is Under Scrutiny for Cheating".[64]
- "Dieselgate Leaves UK's Car Industry in Crisis".[65]

Table 5.22 Auto companies implicated/questioned on emissions testing cheating since late 2015[66]

- Volkswagen
- Daimler
- Audi
- BMW
- Porsche
- Opel (GM in Germany)
- Chevrolet/GMC/Buick (in Germany)
- RSA and Renault
- Mitsubishi
- Nissan
- Subaru
- Fiat Chrysler

Source: *The Telegraph*, *Road and Track*, *Motor 1*.

Why do we so often encounter not an isolated case, but a bunch of similar cases clustered around a particular industry? Is it coincidence or is something more alarming and nefarious happening? Does it go back to human nature? To improper market incentives? To the wrong performance management system within an organization? Is there some sort of contagion that is taking place in these industries prior to the discovery of the scandal that then appears to be something everybody was doing before they were caught – some kind of a well-kept secret within the industry, a race to the bottom to see what you can get away with in pursuit of the almighty dollar or euro or yen?

5.3.c.iii.A SPOTLIGHT ON VOLKSWAGEN: THE PERFECT GOVERNANCE STORM

It is not the purpose of this book to detail everything that seemed to have gone wrong at Volkswagen with the revelation in 2015 that it had engaged in emissions test cheating in the US, because definitive works and reports have been written on this topic.[67] However, there are two sets of conclusions that are important to note as lessons to be learned beyond Volkswagen for the larger corporate governance community – one has to do with actual governance lessons and the other with infusing boards with the stakeholder view – in other words, getting boards and executive management to see their companies through the lens of their main stakeholders, not just their most important stakeholder (i.e., shareholders for corporations, beneficiaries for nonprofits and citizens for governments). There are other stakeholders who are critically important to the proper and sustainable governance of an organization.

Staying with the Volkswagen example, it presents a fascinating case from a governance standpoint for a variety of reasons. It pretty much reads like a textbook case of governance done on remote, in other words, the separation of the governing body from the reality on the ground, which is a very common and alarming problem with corporate boards. This case illustrates in gory detail the financially hyper-focused management team led by a super-aggressive CEO who pretty much had carte blanche to run the company as he saw fit with single-minded pursuit of one goal: become the largest auto company in the world. This ex-CEO, Martin Winterkorn, has been, at the time of this writing, charged by German authorities for aggravated fraud – the first such criminal indictment of an individual in connection with the VW scandal.

From a governance standpoint, it is possible to single out a variety of concerns preceding the outbreak of the emissions scandal, as follows (which continue to be present to varying degrees despite a raft of reforms):

- The two-tier German governance system allows the highest or supervisory board of directors to remain fairly removed from more proactive oversight (than one might perhaps find in a one-tier system, as in the UK or US).
- Strategic and enterprise risk governance and governance of intangible topics like ethics, compliance, corporate responsibility and reputation are not or rarely on the agenda.
- The absence of truly independent directors at the time of the scandal and a continuing lack of independence and diversity on the board to this day (though this has improved somewhat). Even by 2017 (after the major shockwaves of this scandal had been felt), the supervisory board was still characterized as follows:
  - 17 members of either German or Austrian descent.
  - 2 non-German members (Qatari and Swedish).
  - 2 members of the German State of Lower Saxony.
  - 5 seats from the Porsche and Piëch families.
- The presence, as required by German law, of the workers' council representative on the Supervisory Board.

Additionally, according to entrepreneur and governance and risk expert Michael Marquardt, there were three cultural and strategic things present at Volkswagen:

- The existence of a supervisory board comprised of non-independent directors who were either unwilling or unable to effectively challenge management.
- VW's corporate culture has been described as one that discourages dissent and is characterized by an outsized respect for company hierarchy.
- Management made the business decision to "bet the company" on diesel engines, going against the prevailing trend toward hybrid electric vehicles, and putting tremendous pressure on engineers to meet stricter-than-expected emissions standards.[68]

As Marquardt states in one of his articles about VW:

> I would respectfully submit that any board with such a composition may be unwilling or unable to effectively challenge management when necessary.

It is or should be, at the end of the day, a principal remit, even mandate, of any board of directors, to proactively oversee management, operations, risk and strategy. Not challenging management defeats the main purpose of governance just as not challenging a government defeats the purpose of citizenship. But, of course, when the board is controlled directly or indirectly by management or a powerful owner, then all bets for good governance are off.

5.3.c.iv *Silicon Valley: transforming immature company governance into stakeholder value*

Theranos, Uber, Tesla. Just a few names of technology companies in the headlines in recent months and years. Just a few governance headaches for the boards of those companies over the past few months and years. Or maybe not, until and unless a scandal or crisis erupts. And in each of these cases we have seen varying degrees of crisis and scandal. Let's take a look and see what we can learn from a governance standpoint – starting with "bad" governance at Theranos, "unattractive" governance at Tesla and "improving" governance at Uber.

5.3.c.iv.A THERANOS: "BAD" GOVERNANCE GONE WRONG

> [S]he has probably one of the most mature and well-honed senses of ethics, personal ethics, managerial ethics, business ethics, medical ethics that I've ever heard articulated" the retired general gushed.

This is a quote from none other than retired General James Mattis, former US Defense Secretary, about Elizabeth Holmes, the now disgraced founder and CEO of Theranos, a blood testing start-up and once "unicorn" (i.e., multi-billion-dollar private company).

Theranos had a meteoric rise and precipitous crash-and-burn over a period of years, chronicled meticulously by *Wall Street Journal* reporter, John Carreyrou, first in his newspaper and more recently in his brilliant book *Bad Blood: Secrets and Lies in a Silicon Valley Start-Up*.

Mattis was one of a roster of "luminary" board members CEO Holmes was able to rustle up and convince to serve on the board of her "revolutionary" Silicon Valley start-up.[69] Table 5.23 shows the "who's who" who served on the Theranos board at the zenith of the company's "success", when it was valued (some would say wildly) at its peak of US$9 billion.

Table 5.23 Theranos board of directors circa 2015: "before the fall"

Who	*Prior Career*
George Shultz	Former US Secretary of State
Henry Kissinger	Former US Secretary of State
William Perry	Former US Secretary of Defense
Richard Kovacevich	Former Wells Fargo CEO
Sam Nunn	Former US Senator
David Boies	Founder of Boies Schiller Law Firm
General James Mattis	Retired General and formerly Trump's Defense Secretary

Source: John Carreyrou, *Bad Blood*.

In his book, Carreyrou tells the full story of how Holmes, a 19-year-old college drop-out who fancied herself to be the female version of Steve Jobs (sans his brilliance, it turns out), peddled what seemed to be a good idea (tiny blood-testing kits that automatically, almost magically, performed a variety of tests) but wasn't, all the way up to "Unicorn" status, building a company valued at its peak at US$9 billion and in the process convincing or bamboozling a number of heavy-hitting investors, board members, customers and others into believing the hype about what turned out to be a total fraud.

Without intending to insult any of the members of the Theranos board and at the risk of stating the obvious, I would point out the following about the Theranos board (all of which are clearly not great governance practices):

- All of the members of the board were of the same gender.
- All of the members of the board were of a similar older age group.
- All of the members of the board were of the same race/ethnicity.
- All of the members of the board were of the same nationality.
- None appear to have extensive or specialized technology or biology background or expertise.
- Most are former highest-level military or executive branch cabinet secretaries (a.k.a. "luminary directors").
- One of the members of the board also acted as the company and the CEO's personal lawyer (a potentially damaging conflict of interest).

What this discussion underscores is a variety of serious governance issues and/or red flags, as follows:

- No personal diversity whatsoever – not gender, not age, not ethnicity, not geography (international).
- Limited professional diversity.
- No truly independent directors.
- No deep expertise in the industry of the company or in any of its aspects.

- The appearance of a rubber stamp board without teeth or independence.
- No typical early stage investor, venture capital type board members.
- A number of latent, potential or actual conflicts of interest.

And what should probably remain unsaid but won't in this book is that, based on my reading of the Carreyrou book, Elizabeth Holmes, as a manipulator and a grifter clearly and specifically, deliberately, calculatedly and literally "hit on" this cadre of luminary homogenous males to dress her board up in spectacular clothes (so to speak) intended to dazzle her audience: customers, potential customers, regulators, suppliers, never intending to create real, effective board governance.

Other telltale signs of the troubles at the company involved her sidekick and then live-in lover, obscure Silicon Valley denizen Sunny Balwany. Sunny, according to *Bad Blood*, acted as a bully and enforcer on Holmes's behalf, scaring existing and former employees (and others) into silence and confidentiality agreements through thuggish behaviors and legal tactics. The enormous and rapid employee turnover was another telltale sign of trouble – something the board should have been on top of but clearly was completely oblivious to.

These and other red flags of a problematic culture to say the least were captured in the investigative work that Carreyrou did that led to the increasing spotlight on Theranos, which eventually led to where it is today at the time of this writing summarized in the following headlines:

- "How a Wannabe Steve Jobs Duped Investors of Millions of Dollars".
- "Theranos Founder Elizabeth Holmes Indicted on Fraud Charges".
- "She Absolutely Has Sociopathic Tendencies": Elizabeth Holmes, Somehow, is Trying to Start a New Company![70]

Finally, maybe there were red flags about Holmes early on – the following is an interesting anecdote from Carreyrou's book:

> When Holmes was just 9 or 10, one of her relatives asked her, "What do you want to do when you grow up?". Holmes replied, "I want to be a billionaire." "Wouldn't you rather be president?" Holmes was asked. She replied, "No, the president will marry me because I'll have a billion dollars.[71]

5.3.c.iv.B TESLA: "UNATTRACTIVE" GOVERNANCE IN NEED OF REFORM

Who doesn't love Elon Musk, the genius, inventor, investor, wannabe colonizer of Mars and rocket man who not only started Tesla, the groundbreaking (but somewhat troubled) electric car company but also SpaceX and the Boring company? He is the second coming of Christ to some people, a great investment bet to others, a bit crazy to yet others and frustrating to quite a few people as well (some of his shareholders and stakeholders). But say what you want about him, he is different and he is revolutionary.

If you were on the board of directors of any one of his companies but especially Tesla (since it is a publicly traded company that is very much under the microscope of not only the investor community but the regulatory community as well), you might have a bit of a challenge on your hands.

Table 5.24 provides a chronology of events at Tesla from early 2017 to August 2018, which Musk himself has acknowledged as being "the most difficult and painful year of my career … It was excruciating". Read this chronology and think about being on the board of directors of Tesla during this time.

Table 5.24 A Tesla chronology 2017–2018: assembled from *The New York Times*[72]

Date	*Developments*	*Comments*
April 10, 2017	**Tesla becomes the most valuable American carmaker**	Over the course of one week, Tesla leapfrogged Ford and General Motors in terms of market capitalization—a measure of a public company's value. As Mr. Musk articulated it, Tesla "is going to change the world".
July 28, 2017	**Tesla unveils Model 3, but Musk warns of coming "manufacturing hell"**	"The whole point of this company was to make a really great, affordable electric car, and we finally have it". – Elon Musk
July 30, 2017	**Musk says his mental health is suffering from the burdens of running his business**	"Bad feelings correlate to bad events, so maybe real problem is getting carried away in what I sign up for". – Elon Musk
November 15, 2017	**Identifying the enemy: short sellers**	"They're jerks who want us to die". – Elon Musk
January 3, 2018	**Tesla says Model 3 production is far behind expectations**	From 20,000 projected Models 3s to 2,425
March 23, 2018	**A fatal crash raises safety concerns**	A Tesla Model X crashed in Northern California when the car's semiautonomous driving mode called autopilot was engaged, killing the driver.
April 16, 2018	**Model 3 production is temporarily halted**	"Excessive automation at Tesla was a mistake" and "humans are underrated". – Elon Musk
May 2, 2018	**Flustered, Musk publicly berates financial analysts**	"Excuse me. Next. Boring bonehead questions are not cool. Next?" – Elon Musk

(*Continued*)

Table 5.24 Continued

Date	*Developments*	*Comments*
June 17, 2018	**Musk turns his attention back to shorts**	"they're not dumb guys, but they're not super-smart. They're o.k. they're smartish". – Elon Musk
June 28, 2018	**Musk spends his birthday in a factory**	"All night—no friends, nothing." – Elon Musk
July 7, 2018	**Shifting his attention, Musk sends help to trapped children in a Thai cave**	A diver involved in the operation called Mr. Musk's plan a "P.R. stunt" and told him to "stick his submarine where it hurts". Mr. Musk lashed back at the diver, implying in one tweet that he was a pedophile. Mr. Musk later deleted the tweets and apologized.
July 31, 2018	**Musk meets with Saudi fund to discuss privatization**	Mr. Musk came away with impression that the Saudi fund would provide financing for a tesla privatization. But the Saudi fund had not committed to provide any cash, two people briefed on the discussions have since said.
August 7, 2018	**A tweet starts a storm**	"Mr. Musk sent his fateful tweet suggesting that he would take Tesla private at $420 a share. The stock soared, trading on the Nasdaq was halted and the Tesla board, blindsided by Mr. Musk's announcement, went into damage control mode".
August 12, 2018	**Azealia Banks wades in**	"I waited around all weekend while Grimes coddled her boyfriend for being too stupid to know not to go on Twitter while on acid". Tesla board members have grown concerned about Mr. Musk's use of twitter, but Mr. Musk said in the interview that he had no plans to stop.
August 16, 2018	**Musk opens up (interview with the *NYT*)**	"The worst is over from a Tesla operational standpoint," he said, continuing: "but from a personal pain standpoint, the worst is yet to come".

Source: Author and *The New York Times*.

The crux of the matter with Tesla, as it stands at the time of this writing and as governance experts have stated, has to do with the fact that "Tesla, which began selling stock to the public in 2010, has a governance structure that more resembles a startup than a company with $11.76 billion in revenue and sales of 101,000 vehicles last year".[73]

Table 5.25 provides a summary as chronicled by Bloomberg in late September 2018.

Table 5.25 Elon Musk "antics" in the summer of 2018[74]

- Elon Musk, the CEO, accused a British spelunker of being a pedophile amid an argument over … the utility of a metal tube (or "mini-submarine" depending on your perspective) in rescuing some stranded children in a cave in Thailand;
- Elon Musk, the CEO, apologized for that, then doubled down on the accusation in an exchange with Buzzfeed and went on to speculate that it was "strange" that the spelunker hadn't sued him—a few weeks before the spelunker then sued him;
- Elon Musk, the CEO, gave an interview to *The New York Times* in which he talked of his exhaustion and at one point was "seemingly overcome by emotion," and then later appeared on a podcast during which he took a drag on a spliff;
- Elon Musk, the CEO, touched off a spike in Tesla's stock with tweets saying he might take the company private at $420 a share, having secured funding and investor support;
- Elon Musk, the CEO, then indicated that funding wasn't that secured and investors weren't wholly on board, before ditching the plan in a *late-Friday news dump* 17 days later;
- Elon Musk, the CEO, by sending out those tweets, exposed Tesla to a class-action lawsuit, SEC scrutiny and, it emerged Tuesday, an inquiry from the Department of Justice;
- Elon Musk, the CEO, said goodbye to his latest chief accounting officer after just 29 days on the job, as well as his HR chief and VP for worldwide finance and operations;
- And Tesla made and sold some cars and things.

Source: *Bloomberg Businessweek*.

Some of the key governance measures that the board of Tesla should consider going forward are the following:

- Bring in more independent, expert board members who understand the worlds of industry and technology.
- Appoint a chairman of the board who is independent.
- Consider appointing a right-hand operational executive who can take over the day-to-day as well as the strategic planning initiatives from Musk and relieve him to continue being the luminary thinker and dreamer that he is.
- Understand the risk universe that applies to Tesla, including the leadership and culture risk that a luminary and unpredictable CEO like Musk presents.
- Is there a chief risk or compliance office in place? If not, consider appointing one and have him/her report to the board.
- Undertake a reputation risk analysis of the primary risks and opportunities Tesla is facing and make sure the company has a crisis management plan, team and exercises in place.
- Understand the stakeholder expectations – all important stakeholders, not just shareholders – customers, regulators, suppliers, employees, prospective employees.

With Tesla as with other rapidly growing companies – whether they are private or become public – it's not just about the luminary CEO. It's about everyone who has invested money, blood, sweat and tears into the development of the luminary's incredible idea – the stakeholders. And thus, the board – the governance – has to adapt to the growth of the company to be able to cater to the interests of these stakeholders. Granted that shareholders are king in the corporate for-profit world, but others are critical, too, as they can and will bring down a luminary and his or her company if enough damage takes place.

5.3.c.iv.C UBER: ACHIEVING NEW GOVERNANCE ALTITUDE

In the case of Uber, we saw a long litany of improperly managed risks gone wrong around the world, mediocre to illegal business practices taking place, unethical and illegal conduct both at the company level and among some of its executives, a pervasive workplace culture of sexual harassment and other bad behaviors and missteps for several years. All were presided over by Travis Kalanick, the founder and visionary behind Uber, whose frat-boy antics and overall demeanor (his mantra for Uber – "be hustling") led to a widespread culture of cutting corners and a general lawless attitude towards winning and growing business at all costs.

All this came to a head in the spring of 2017 when a couple of incidents proved the tipping point – a blogpost by a female ex-employee about the pervasive culture of sexual harassment inside Uber and a video gone viral of Kalanick berating an Uber driver who secretly recorded the berating and released it online.[75]

What has happened since is a combination of boardroom drama, investigations, actions, boardroom reform (with the appointment of new, independent directors) and a major management shake-up with the resignation of Kalanick as CEO (he remains on the board) and the appointment of a more professional CEO, Dara Khosrowshahi. The new CEO has done a lot of positive things to rebuild the reputation of Uber and rebuild the trust of the company's many unhappy stakeholders.[76] While the jury is still out on where all this goes, the work of the new CEO has certainly stemmed the bleeding and is showing signs of healing the patient.

Among some of the constructive lessons from this experience are the following:

- If a founder/CEO is unable to run a business, perhaps the board should consider a replacement for day-to-day executive management.
- A board of "friends and family" (i.e., sycophants) does not work in the best interests of all stakeholders, including shareholders.
- A "dependent" board together with aggressive, untested management can preside over a series of mounting ethical, cultural and legal problems and even nightmares that can take place over time and significantly erode stakeholder trust and financial value.

- Reaching out and listening to the primary stakeholders to repair the relationship and rebuild the trust is the job of both management and the board.
- At the time of this writing, Uber has recently gone public, and despite the fact that it continues to operate within a high-risk environment (as they themselves describe in an unusually lengthy and open risk-disclosure section of their first US securities public filing), trust by stakeholders seems to be improving and much of the credit goes to the new CEO, the executive team he has assembled as well as the heavy-culture improvement work they have been focused on internally.

5.3.c.iv.D SILICON GOVERNANCE STRANDS: LESSONS FOR THE FUTURE

What are the common threads and/or lessons learned from these three admittedly very different cases? One of the key challenges for start-ups that are no longer start-ups but actually very valuable companies with real products and services, and employees and other stakeholders that are very dependent on the company's success, is that their governance is often quite poor or even nonexistent.

Companies and other types of organizations that are reaching certain levels or stages of development really need to pause and think about whether their governance is on a par with their stage of development. This is the job of both management and the board, and if one or the other are uninformed, derelict or worse, bad things can happen. But when at least one side of this equation is paying attention and is savvy about needing to reform and improve, good things can happen.

Here are some of the common themes of these companies:

- The board of directors of most up-and-coming small and midcap companies are not properly constituted, do not have professional governance and have little or no independence from the founder.
- If there is a board of directors, the board is pretty much made up of friends and family – literally or figuratively.
- Founders are often visionaries who don't believe in having parameters put around their style of leadership, business development or communication.
- Founders don't always have management skills – they may be brilliant at creating and inventing but not necessarily at executing and operating.
- While visionaries frequently attract talented employees and partners, they may suffer brain drain and/or exoduses if the proper employment and benefits parameters aren't in place.
- Early-stage investors don't necessarily bring a different and value-added perspective to the governance of the entity and thus as the company grows, it grows without some essential governance and culture frameworks in place.
- It is incumbent on the "grown-ups" in the room – whether in management or the board – to make the business case for improved governance. I suggest they take a look at Chapters 7 and 8 of this book to put that business case together.

5.3.c.v Transforming corruption risk into reputation opportunity

We have seen many waves of scandal and reform, as illustrated in Table 5.18 earlier in this chapter. We have witnessed a continuing parade of large and shocking corruption scandals over the years as well. In some cases, companies have learned from their mistakes and transformed those mistakes into better practices, tighter controls, better ethics and compliance programs, board oversight, etc.

Siemens is a stellar case in point of a company that learned from its mistakes and transformed corruption and reputation risk into opportunity – both reputationally and in the overall fight against corruption. At the time of their big corruption scandal (2006) they were the largest and most dramatic example of a company caught for systemic global corporate corruption. Since then, sadly there have been many additional (and ongoing) cases of extensive business corruption as Table 5.26 illustrates.

Table 5.26 Largest US Foreign Corrupt Practices Act enforcement actions (through 2018)[77]

Year	*Company*	*Country*	*Fine in $US*
2018	**1. Petrobras**	Brazil	1.78 Billion
2017	**2. Telia Company AB**	Sweden	965 Million
2008	**3. Siemens**	Germany	800 Million
2016	**4. Vimpelcom**	Holland	795 Million
2014	**5. Alstom**	France	772 Million
2018	**6. Societe Generale SA**	France	585 Million
2009	**7. KBR/Halliburton**	United States	579 Million
2016	**8. Teva Pharmaceutical**	Israel	519 Million
2017	**9. Keppel Offshore & Marine Ltd.**	Singapore	422 Million
2016	**10. Och-Ziff**	United States	412 Million

Source: FCPA Blog.

But what did Siemens do that put them back on track as a respected company after their scandal? Besides going through a massive intercontinental criminal investigation with all the lessons learned that come from that experience and paying a grand total cost (between fines and related expenses) of $2.5 billion, they did a whole bunch of things to restore the trust of their stakeholders.

The following is a summary of the things Siemens did to restore trust that is based on an excellent piece published by the Institute for Business Ethics in which they profile several cases of restoration of trust. Here's what they did:[78]

- They appointed one of the cofounders of Transparency International to serve as an adviser.
- They hired an executive with a stellar background to become their global CEO, Dr. Peter Loscher.

- From a management standpoint, they streamlined and de-complexified their global structure.
- From a culture standpoint, they recognized that they had a systemic global culture of corruption that went deep into their past and present where staff was made to feel that bribes were "not only acceptable but implicitly encouraged".
- Loscher declared a month-long amnesty with 40 whistleblowers coming forward at that time.
- They created and implemented a massive, systemic anti-corruption compliance initiative.
- They hired 500 full-time compliance officers (up from 86 in 2006).
- They hired a former Interpol official to head up their internal investigation unit.
- They established compliance hotlines.
- They established a global external ombudsman office.
- They created a web portal to evaluate client and supplier risk.
- They launched a comprehensive employee anti-corruption training and education program.
- They announced they would avoid competing in certain high corruption risk locations.
- They voluntarily suspended applying for World Bank funds for two years.
- They agreed to pay $100 million to anti-corruption nonprofits for 15 years.
- They undertook 900 internal disciplinary actions, including terminations.

As I write these lines, there are several pending multinational corruption scandals, some of which may end up making it into the annals of most extensive, most expensive and most dangerous corruption cases ever:

- Odebrecht, involving corruption and bribery in over a dozen countries (and most recently apparent murder of a whistleblower and his son in Colombia) and, as chronicled elsewhere in this book, shockingly, some alleged murders and suicides as well.
- The Petrobras Lava Jato case emanating in Brazil and affecting other countries as well but most notably the entire political establishment in Brazil (it would seem).
- 1MDB in Malaysia, involving the former prime minister (now jailed) and over $1 billion in corruption, it appears, also now ensnaring Goldman Sachs and a number of their current or past executives in continuing investigations.

But true to form, I would like to signal that there is hope in the despair of such corruption and that, just like Siemens was able to get its act together after its historic scandal, others can too. It requires leadership and tone from the top first and foremost – it is almost always the case that major corruption is correlated with absentee, negligent or willful leadership that either isn't interested, doesn't care or intentionally seeks to gain or retain business through any means possible, including bribery and corruption.

For leaders who care, Table 5.27 provides a synopsis of some of the better practices that an organization can undertake to transform corruption risk into reputation value.

Table 5.27 Anti-corruption reputation opportunity and resilience: a few risk management tools

Anti-corruption risk tool	*Associated reputation risk management activities*
Tone from the Top	Boards and C-Suites need to be committed to anti-corruption not only from a marketing and branding standpoint but also from a resource and budgetary standpoint and a "doing the right thing" approach when and if a potential corruption situation is discovered. This would include actions up to and including voluntary disclosure to appropriate government and/or enforcement agencies.
Global Anti-Corruption Policy and Plan	Coordination with all relevant internal functions owning a piece of this policy: legal, compliance, risk, finance, PR, external relations, HR, and communications.
Anti-Corruption Due Diligence Plan for all Forms of M&A	Coordination with legal and finance to ensure incorporation of ethics and compliance measures (including anti-corruption) into due diligence activities.
Gifts and Entertainment Policy	Ensure monitoring and testing of this policy is ongoing and periodic – enlist support of finance and internal audit.
Third Party Management	Ensure this tool is properly embedded, ownership of aspects of it are well understood, training and testing takes place and, preferably, that a shared technology management tool is used.
Audits and Continuous Self-Improvement	Robust and periodic audits – planned and spontaneous – that lead to root cause analysis of events, integration with compliance and risk management, and continuous process and culture improvements as necessary

Source: Author.

5.3.c.vi Individual cases of "bad governance": directors that don't rate

The cover story of the Summer 2018 edition of the *NACD Directorship Magazine* was about the various forms of dysfunctional board members that populate many boards. While somewhat humorous in its treatment, the categories are worth mentioning here, as boards are only as good as the people that constitute them, and if any one or more of the characters depicted in the *Directorship* article are on your board, that might be a signal for a call to action for your organization (Table 5.28).

Table 5.28 NACD Directorship Magazine's "A field guide to bad directors"[79]

The Technocrat	These bad directors are exceptional narrow-band experts, true geniuses in their fields, therefore contemptuous of anyone who cannot match their clear intellectual dominance and expecting deference because of their technological capacity and knowledge.
The Representative	Directors who represent only one point of view, one specific group of shareholders, or one narrow perspective.
The Hanger-On	These bad directors have stayed too long and are too emotionally invested in their corporate directorship to be effective any longer. Age usually plays a dominant role here, but not universally.
The Authority	Frequently found on the boards of technology companies, these bad directors possess a fairly robust technical knowledge in the field, but they express their authority with resonance far beyond actual competence.
The Sidekick	These bad directors are especially frustrating. They are effectively the "second vote" for one director with a particularly strong personality,. more often than not the board's chair.
The Financial Stumbler	These bad directors are less than wholly financially literate, and it shows.
The Unprepared	These bad directors are "fifteen-percenters." They pre-read and understand about 15 percent of the board materials, are about 15 percent attentive in committee or board conversations, are 15 percent knowledgeable.
The Questioner	These are the only bad directors with their own often-articulated universal motto: "I ask the tough questions." Board meetings and committee meetings become captured, mired down by their inquiries, which mostly demonstrate the extent to which they are unprepared.
The Partisan	In a nation that remains deeply divided following the presidential election of 2016, these bad directors are intent upon interjecting partisan politics into the boardroom.
The Consensus Denier	The most often-employed weapon in their arsenal is their denial of consensus until accommodated. They are passive-aggressive negative people, and they effortlessly make corporate board life miserable for their colleagues.
The Bully	The worst type of bad director, they are straight-up mean, conflating aggression with toughness, hiding their deep-seated insecurities with volatility, and denying effective oversight by means of threats and intimidation.
The Storyteller	These bad directors constantly tell war stories, unrelentingly and to distraction … about their past experiences in business and on corporate boards. Every issue or action undertaken by the board will trigger a past experience.

(*Continued*)

Table 5.28 Continued

The Lawyer-Reliant	Either "defers to legal counsel and mistrusts their own judgment and the board's collective wisdom" or demands "that the lawyers find a way to accomplish whatever dubious action they may desire, rather than engaging in a reasoned board-level discourse about the law and where its fences may stand".
The Celebrity	Marquee directors—big names with major cachet, high Q-score, and seven-figure numbers of social media followers—are not always bad directors, but there are enough of them to confirm the type.

Source: *NACD Directorship Magazine.*

5.4 Endnote: exercising personal governance – on being a citizen

No discussion of governance would be complete without a discussion of the most important role each of us plays within the governance of our own countries, nations, political systems. So, after everything else is said and done in this chapter, I'd like to conclude on a more personal note, once again.

The ultimate form of governance comes from our own personal self-governance – what each of us does as a person, in our family milieu, at school, at work, in society and as a citizen of a community, a city, a nation and the globe. One of the most critically important roles anyone can play anywhere is that of a responsible stakeholder in our own society and in our own polity – i.e., being a citizen. At these times of tectonic change, it behooves us to behave like an engaged and responsible stakeholder in our own communities and governments, which are hopefully democracies. But even democracies can be flawed.

I conclude this chapter with a synopsis in Table 5.29 of the very wise, historically based admonishments by the respected historian and thinker Tim Snyder, who published *On Tyranny*, a short but powerful book about what ordinary citizens should be doing in the face of possible or actual tyranny. In my humble opinion, this synopsis can certainly serve as a citizen's governance stakeholder manifesto.

Table 5.29 Timothy Snyder's 20 lessons from the 20th century[80]

1. Do not obey in advance.
2. Defend institutions.
3. Beware the one-party state.
4. Take responsibility for the face of the world.
5. Remember professional ethics.
6. Be wary of paramilitaries.
7. Be reflective if you must be armed.
8. Stand out.
9. Be kind to our language.

(*Continued*)

Table 5.29 Continued

10. Believe in truth.
11. Investigate.
12. Make eye contact and small talk.
13. Practice corporeal politics.
14. Establish a private life.
15. Contribute to good causes.
16. Learn from peers in other countries.
17. Listen for dangerous words.
18. Be calm when the unthinkable arrives.
19. Be a patriot.
20. Be as courageous as you can.

Source: Timothy Snyder, *On Tyranny*.

Notes

1 Albert Einstein. https://inspirational-quotes2.weebly.com/albert-einstein.html. Accessed on August 17, 2019.

2 Jon Meacham. *Soul of America: The Battle for the Better Angeles of our Nature*. Random House 2018. https://www.amazon.com/gp/product/0399589813/ref=dbs_a_def_rwt_bibl_vppi_i0https://www.amazon.com/gp/product/0399589813/ref=dbs_a_def_rwt_bibl_vppi_i0. Accessed on July 8, 2018.

3 Quote by Angela Merkel at a press conference held on July 20, 2018, as reported on Twitter on July 21, 2018.

4 Eric Lipton and Nicholas Fandos. "Outgoing US Ethics Chief: US Is 'Close to a Laughingstock'". *The New York Times*. July 17, 2017. https://www.nytimes.com/2017/07/17/us/politics/walter-shaub-ethics.html?smprod=nytcore-iphone&smid=nytcore-iphone-share. Accessed on August 23, 2018.

5 Justice Louis Brandeis. https://en.wikiquote.org/wiki/Louis_Brandeis. Accessed on August 17, 2019.

6 Theodore Parker and Martin Luther King Jr. https://quoteinvestigator.com/2012/11/15/arc-of-universe/. Accessed on August 17, 2019.

7 Wikipedia. Governance. https://en.wikipedia.org/wiki/Governance. Accessed on August 20, 2018.

8 *Oxford Dictionaries*. https://en.oxforddictionaries.com/definition/governance. Accessed on August 18, 2018.

9 *The Business Dictionary*. http://www.businessdictionary.com/definition/governance.html. Accessed on July 20, 2018.

10 The World Bank. World Governance Indicators. http://info.worldbank.org/governance/WGI/#home. Accessed on July 30, 2018.

11 *The Business Dictionary*. http://www.businessdictionary.com/definition/corporate-governance.html. Accessed on July 20, 2018.

12 RobecoSAM. https://www.robeco.com/en/key-strengths/sustainability-investing/glossary/esg-definition.html. Accessed on June 20, 2018.

13 Council on Foreign Relations. Global Governance Monitor. https://www.cfr.org/interactives/global-governance-monitor#!/global-governance-monitor. Accessed on August 18, 2018.

14 World Bank. Worldwide Governance Indicators. Worldbank.org. http://info.worldbank.org/governance/wgi/#doc. Accessed on August 20, 2018.

15 Freedom House. *Freedom in the World 2018: Democracy in Crisis*. January 2018. https://freedomhouse.org/report/freedom-world/freedom-world-2018. Accessed on August 20, 2018.
16 UN Sustainable Development Goals. https://sustainabledevelopment.un.org/?menu=1300. Accessed on July 29, 2018.
17 G20/OECD Principles of Good Corporate Governance. September 2015. http://www.oecd.org/corporate/principles-corporate-governance.htm. Accessed on August 19, 2018.
18 OECD. *OECD Due Diligence Guidance for Responsible Business Conduct*. http://www.oecd.org/investment/due-diligence-guidance-for-responsible-business-conduct.htm. Accessed on August 18, 2018.
19 Allianz. The Allianz Risk Barometer 2019. https://www.agcs.allianz.com/content/dam/onemarketing/agcs/agcs/reports/Allianz-Risk-Barometer-2019.pdf. Accessed on April 12, 2019.
20 MSCI. https://www.msci.com/esg-investing. Accessed on July 18, 2018.
21 RepRisk. https://www.reprisk.com/our-approach. Accessed on July 17, 2018.
22 UN Global Compact. https://www.unglobalcompact.org/. Accessed on July 29, 2018.
23 Transparency International. The Corruption Perceptions Index for 2017. https://www.transparency.org/news/feature/corruption_perceptions_index_2017. Accessed on August 19, 2018.
24 Council on Foreign Relations. Global Governance Monitor. https://www.cfr.org/interactives/global-governance-monitor#!/global-governance-monitor. Accessed on August 18, 2018.
25 Council on Foreign Relations. Global Governance Monitor. https://www.cfr.org/interactives/global-governance-monitor#!/global-governance-monitor. Accessed on August 20, 2018.
26 Council on Foreign Relations. Global Governance Monitor. https://www.cfr.org/interactives/global-governance-monitor#!/global-governance-monitor. Accessed on August 20, 2018.
27 World Bank. Worldwide Governance Indicators. Worldbank.org. http://info.worldbank.org/governance/wgi/#doc. Accessed on August 20, 2018.
28 Freedom House. *Freedom in the World 2018: Democracy in Crisis*. January 2018. https://freedomhouse.org/report/freedom-world/freedom-world-2018. Accessed on August 20, 2018.
29 Freedom House. *Freedom in the World 2018: Democracy in Crisis*. January 2018. https://freedomhouse.org/report/freedom-world/freedom-world-2018. Accessed on August 20, 2018.
30 G20/OECD Principles of Good Corporate Governance. September 2015. http://www.oecd.org/corporate/principles-corporate-governance.htm. Accessed on August 19, 2018. The OECD describes itself as follows: "The OECD is a unique forum where the governments of 30 democracies work together to address the economic, social and environmental challenges of globalization. The OECD is also at the forefront of efforts to understand and to help governments respond to new developments and concerns, such as corporate governance, the information economy and the challenges of an ageing population. The Organization provides a setting where governments can compare policy experiences, seek answers to common problems, identify good practice and work to co-ordinate domestic and international policies".
31 As summarized by Foley Hoag LLP in *CSR and the Law*. http://www.csrandthelaw.com/2018/06/26/oecd-releases-due-diligence-guidance-for-responsible-business-conduct/. Accessed on August 18, 2018.
32 OECD. *OECD Guidelines for Multinational Enterprises*. 2008. http://www.oecd.org/investment/mne/1922428.pdf. Accessed on August 19, 2018.
33 As summarized by Foley Hoag LLP in *CSR and the Law*. http://www.csrandthelaw.com/2018/06/26/oecd-releases-due-diligence-guidance-for-responsible-business-conduct/. Accessed on August 18, 2018.

34 Allianz. *The Allianz Risk Barometer 2019*. https://www.agcs.allianz.com/content/dam/onemarketing/agcs/agcs/reports/Allianz-Risk-Barometer-2019.pdf. Accessed on April 12, 2019.

35 MSCI. https://www.msci.com/esg-investing. Accessed on July 18, 2018.

36 RepRisk. https://www.reprisk.com/our-approach. Accessed July 17, 2018.

37 RepRisk . https://www.reprisk.com/our-approach. Accessed July 17, 2018.

38 RepRisk. *Most Controversial Companies 2017*. https://www.reprisk.com/content/5-publications/1-special-reports/53-most-controversial-companies-of-2017/mcc-2017.pdf. Accessed on August 5, 2018.

39 According to RepRisk, "Peak RRI: equal to the highest level of the RRI over the last two years – a proxy for overall reputational exposure related to ESG and business conduct risk." https://www.reprisk.com/our-approach.

40 RepRisk. *Most Controversial Projects 2017*. https://www.reprisk.com/content/5-publications/1-special-reports/1-most-controversial-projects-of-2017/reprisk-most-controversial-projects-of-2017-report.pdf. Accessed on August 5, 2018.

41 UN Global Compact Principles. https://www.unglobalcompact.org/what-is-gc/mission/principles. Accessed on July 17, 2018.

42 United Nations Global Compact. https://www.unglobalcompact.org/what-is-gc/mission/principles. Accessed on August 7, 2018.

43 UN Global Compact. https://www.unglobalcompact.org/sdgs/17-global-goals. Accessed on July 17, 2018.

44 Transparency International. The Corruption Perceptions Index for 2017. https://www.transparency.org/news/feature/corruption_perceptions_index_2017. Accessed on August 19, 2018.

45 Transparency International. What Is Corruption? https://www.transparency.org/what-is-corruption?gclid=CjwKCAjwzenbBRB3EiwAItS-u9q6AtK2FrIPvrqIDIclY5rGMl-nfJiFJNq6JPSfwJqvBiV9A-t660xoCsbMQAvD_BwE#define. Accessed on August 20, 2018.

46 Transparency International. What Are the Costs of Corruption? https://www.transparency.org/what-is-corruption?gclid=CjwKCAjwzenbBRB3EiwAItS-u9q6AtK2FrIPvrqIDIclY5rGMlnfJiFJNq6JPSfwJqvBiV9A-t660xoCsbMQAvD_BwE#define. Accessed on August 20, 2018.

47 Foreign Affairs. 2018. https://www.foreignaffairs.com/issues/2018/97/2. Accessed on August 22, 2018.

48 The resources mentioned in this paragraph are the following: Richard Haass. *A World in Disarray: American Foreign Policy and the Crisis of the Old Order*. Penguin Books 2017; Ian Bremmer. *Us Versus Them: The Failure of Globalism*. Portfolio 2018; Timothy Snyder. *On Tyranny: Twenty Lessons from the Twentieth Century*. Tim Duggan Books 2017; Timothy Snyder. *The Road to Unfreedom: Russia, Europe, America*. Tim Duggan Books 2018; and Foreign Affairs. 2018. https://www.foreignaffairs.com/issues/2018/97/2. Accessed on August 22, 2018.

49 Mark Brzezinski and I thought that the lessons of eastern and southern European democratizations would be applicable to the then apparent seedlings of democracy in the Middle East but we were proved ultimately wrong. Here's what we said: Mark Brzezinski and Andrea Bonime-Blanc. "Mideast Shift: Lessons from Europe". Politico. February 6, 2011. https://www.politico.com/story/2011/02/mideast-shift-lessons-from-europe-048921. Accessed on August 22, 2018.

50 There are several notable books that have come out in the past couple of years which make these points much better than I can here and which I highly recommend. Among them: Richard Haass. *A World in Disarray: American Foreign Policy and the Crisis of the Old Order*. Penguin Books 2017; Ian Bremmer. *Us Versus Them: The Failure of Globalism*. Portfolio 2018; Timothy Snyder. *On Tyranny: Twenty Lessons from the Twentieth Century*. Tim Duggan Books 2017; Timothy Snyder. *The Road to Unfreedom: Russia, Europe, America*. Tim Duggan Books 2018.

51 World Happiness Index 2018. United Nations. http://worldhappiness.report/. Accessed on August 22, 2018

52 Sebastian Mallaby. "The Dangerous Myth We still Believe of the Lehman Brothers Bust". *The Washington Post*. September 9, 2018. https://www.washingtonpost.com/opinions/the-dangerous-myth-we-still-believe-about-the-lehman-brothers-bust/2018/09/09/5a2f8a9c-b2bc-11e8-9a6a-565d92a3585d_story.html?utm_term=.e1a76b153f3d&wpisrc=nl_opinions&wpmm=1. Accessed on September 10, 2018.

53 Council on Foreign Relations. "Timeline: US Financial Crisis". https://www.cfr.org/timeline/us-financial-crisis?utm_medium=email&utm_source=twtw&utm_content=091418&sp_mid=57366168&sp_rid=YWJvbmltZWJsYW5jQGdlY3Jpc2suY29tS0. Accessed on September 14, 2018.

54 DW.com (Deutsche Welle). "Financial Crisis Bank Fines Hot Record 10 Years After Market Collapse". https://www.dw.com/en/financial-crisis-bank-fines-hit-record-10-years-after-market-collapse/a-40044540. Accessed on August 21, 2018; graphic from: BCG. Staying the Course in Banking. March 2017. http://image-src.bcg.com/BCG_COM/BCG-Staying-the-Course-in-Banking-Mar-2017_tcm9-146794.pdf. Accessed on August 21, 2018.

55 Steve Goldstein. "Here's the Staggering Amount Banks Have Been Fined since the Financial Crisis". MarketWatch. February 24, 2018. https://www.marketwatch.com/story/banks-have-been-fined-a-staggering-243-billion-since-the-financial-crisis-2018-02-20. Accessed on November 25, 2018.

56 Moritz Kuhn, Moritz Schularick, and Ulrike Steins. "Research: How the Financial Crisis Drastically Increased Wealth Inequality in the U.S.". *Harvard Business Review*. September 13, 2018. https://hbr.org/2018/09/research-how-the-financial-crisis-drastically-increased-wealth-inequality-in-the-u-s. Accessed on September 14, 2018.

57 The CFA Institute and The Economist Intelligence Unit. "A Crisis of Culture: Valuing Ethics and Knowledge in Financial Services". 2013. file:///Users/andreabonime-blanc/Desktop/crisis-of-culture-report%20(1).pdf. Accessed on August 24, 2018.

58 Credit Suisse. "Press Release Credit Suisse Response to Closure of Legacy Case Review by FINMA and Update on Progress in Compliance". September 17, 2018. https://www.credit-suisse.com/corporate/en/articles/media-releases/press-release-global-201809.html. Accessed on September 18, 2018.

59 Greg Ip. "10 Years After: The Crisis Made Us Afraid of Risk – For A While". *The Wall Street Journal*. September 8, 2018. https://www.wsj.com/articles/the-financial-crisis-made-us-afraid-of-riskfor-a-while-1536379306?mod=hp_lead_pos5. Accessed on September 8, 2018.

60 Jack Ewing. "Diesel Scandal Deepens as German Authorities Target Audi Chief and Daimler." June 11, 2018. https://www.nytimes.com/2018/06/11/business/audi-rupert-stadler-diesel.html. Accessed on August 21, 2018.

61 *The Economist*. "Dirty Secrets". September 26, 2015. https://www.economist.com/leaders/2015/09/26/dirty-secrets. Accessed on August 22, 2018.

62 Charles Riley. "Mitsubishi: We've Been Cheating on Fuel Tests for 25 Years". CNN Money. April 26, 2016. https://money.cnn.com/2016/04/26/news/companies/mitsubishi-cheating-fuel-tests-25-years/index.html. Accessed August 21, 2018.

63 "Nissan Admits Falsifying Emissions Tests in Japan". BBC. July 9, 2018. https://www.bbc.com/news/business-44763905. Accessed on August 21, 2018

64 Kartikay Mehrotra and David Welch. "Why Much of the Car Industry Is Under Scrutiny for Cheating". *Bloomberg Businessweek*. January 10, 2018. https://www.bloomberg.com/news/articles/2017-08-02/why-it-seems-like-open-season-on-car-companies-quick-take-q-a. Accessed on August 22, 2018.

65 Karl West. "Dieselgate Leaves UK's Car Industry in Crisis". *The Guardian*. April 21, 2018. https://www.theguardian.com/business/2018/apr/21/dieselgate-uk-car-industry-sales-slump. Accessed on August 22, 2018.

66 *The Telegraph.* "German Car Giants May Have Colluded on Emissions". July 21, 2017. https://www.telegraph.co.uk/business/2017/07/21/german-car-giants-may-have-colluded-emissions/. Accessed on August 22, 2018; Bob Sorokanich. "The Facts Behind Every Major Auto Maker Emissions Scandal since VW". *Road & Track.* https://www.roadandtrack.com/new-cars/car-technology/a29293/vehicle-emissions-testing-scandal-cheating/. Accessed on August 22, 2018; Anthony Alaniz. "Subaru Boss to Retire Amid Emissions Scandal". *Motor 1.* June 6, 2018. https://www.motor1.com/news/244206/subaru-ceo-retire-emission-scandal/. Accessed on August 22, 2019.

67 Particularly good is the work of Jack Ewing, a *New York Times* reporter who followed this story carefully and then wrote a detailed and revealing history of the car company in *Faster, Higher, Farther: The Volkswagen Scandal*. WW Norton & Company 2017. https://www.amazon.com/Faster-Higher-Farther-Volkswagen-Scandal/dp/039325450X/ref=tmm_hrd_swatch_0?_encoding=UTF8&qid=&sr. Accessed on August 23, 2018.

68 Michael Marquardt. "Why the VW Board Appears Stuck in Low Gear". *Directorship Magazine.* March/April 2016. https://www.nacdonline.org/insights/magazine/article.cfm?itemnumber=26111. Accessed on August 24, 2018; and Michael Marquardt. "A Year and a Half Later, VW Board Still Stuck in Low Gear". *Directorship Magazine.* October 2017. https://read.nxtbook.com/nacd/directorship/september_october_2017/index.html#viewpoint. Accessed on August 24, 2018.

69 John Carreyrou. *Bad Blood: Secrets and Lies of a Silicon Valley Start-up.* Knopf 2018. https://www.amazon.com/Bad-Blood-Secrets-Silicon-Startup/dp/152473165X. Accessed July 8, 2018.

70 Peter Griffin. "How a Steve Jobs Wannabe Duped Investors out of Millions of Dollars". Noted. August 1, 2018. https://www.noted.co.nz/tech/how-elizabeth-holmes-duped-investors-millions-of-dollars/. Accessed on August 24, 2018; Ephrat Livni. "Elizabeth Holmes and other Famous Grifters Expose the Myth of Quick and Easy Success". Quartz. August 2, 2018. https://qz.com/1345502/elizabeth-holmes-and-other-famous-grifters-expose-the-myth-of-quick-and-easy-success/. Accessed on August 24, 2018; Reed Abelson. "Theranos Founder, Elizabeth Holmes, Indicted on Fraud Charges". *The New York Times.* June 15, 2018. https://www.nytimes.com/2018/06/15/health/theranos-elizabeth-holmes-fraud.html. Accessed on August 24, 2018.

71 As quoted in Yashar Ali. "The Reporter Who Took Down a Unicorn". *New York Magazine.* May 2018. http://nymag.com/daily/intelligencer/2018/05/john-carreyrous-new-book-on-silicon-valley-bad-blood.html. Accessed on August 24, 2018.

72 David Gelles, Kate Kelly and Jessica Silver-Greenberg. "Elon Musk's No Good, Very Bad Year: A Tesla Timeline". *The New York Times.* August 18, 2018. https://www.nytimes.com/2018/08/18/business/elon-musk-tesla-timeline.html. Accessed on August 24, 2018.

73 Associated Press. "Experts Say Tesla Board May Have too many Ties to CEO Musk". *The New York Times.* August 21, 2018. https://www.nytimes.com/aponline/2018/08/21/us/ap-us-tesla-board-of-directors-.html.

74 Liam Denning. "Are You All Caught Up on Tesla: There's Always More". *Bloomberg Businessweek.* September 18, 2018. https://www.bloomberg.com/view/articles/2018-09-18/tesla-tsla-justice-department-probe-adds-to-drama. Accessed on September 19, 2018.

75 Eric Newcomer and Brad Stone. "The Fall, and Fall, and Fall of Uber's Travis Kalanick". *Bloomberg.* January 18, 2018. https://www.bloomberg.com/news/features/2018-01-18/the-fall-of-travis-kalanick-was-a-lot-weirder-and-darker-than-you-thought. Accessed on August 24, 2018; and Susan Fowler. "Reflecting on One Very, Very Strange Year at Uber". Susan Fowler Blog. February 19, 2017. https://www.susanjfowler.com/blog/2017/2/19/reflecting-on-one-very-strange-year-at-uber. Accessed on August 24, 2018.

76 Sheelah Kolhatkar. "At Uber, A New CEO Shifts Gears". *New Yorker Magazine*. April 9, 2018. https://www.newyorker.com/magazine/2018/04/09/at-uber-a-new-ceo-shifts-gears. Accessed on August 24, 2018.
77 Richard Cassin. "Petrobras Smashes the Top Ten List (and We Explain Why)". FCPA Blog. September 28, 2018. http://www.fcpablog.com/blog/2018/9/28/petrobras-smashes-the-top-ten-list-and-we-explain-why.html. Accessed on November 30, 2018.
78 Graham Dietz and Nicole Gillespie. Institute for Business Ethics. February 2012. "The Recovery of Trust: Case Studies of Organizational Failure and Trust Repair". https://www.ibe.org.uk/userfiles/op_trustcasestudies.pdf. Accessed on November 30, 2018.
79 Michael Pocalyko. "A Field Guide to Bad Directors". *NACD Directorship Magazine*. August 2018. https://read.nxtbook.com/nacd/directorship/july_august_2018/a_field_guide_to_bad_director.html. Accessed on October 21, 2018.
80 Timothy Snyder. *On Tyranny: Twenty Lessons from the Twentieth Century*. Tim Duggan Books, 2017.

6 Technology

Transforming technology risk into opportunity from cyber-fear to trusted tech

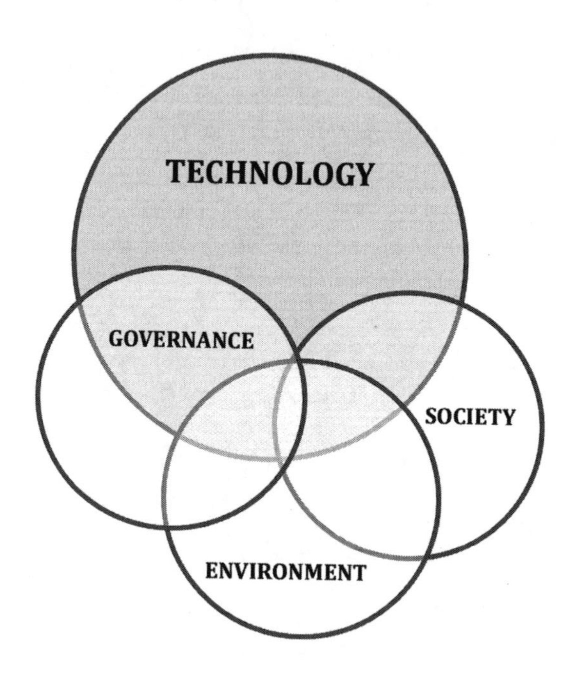

Technology – risk and opportunity

Probably no other topic in this book brings more emotional and personal concern to the fore than what will happen to humanity with all this new technology that is bursting on the scene apparently unbridled. Mark Zuckerberg, Facebook founder and CEO, said early in his career as a technologist: "Move fast and break things. Unless you are breaking stuff, you are not moving fast enough."[1]

And that's exactly what engenders reactions of fear, concern and even loathing. And this includes warnings from those who know best, such as this one from the late great scientist Stephen Hawking: "Unless we learn how to prepare for, and avoid, the potential risks, AI could be the worst event in the history of our civilization. It brings dangers, like powerful autonomous weapons, or new ways for the few to oppress the many."[2]

Inventor and author Ray Kurzweil has put it in somewhat more neutral and "factual" terms: "Artificial intelligence will reach human levels by around 2029. Follow that out further to, say, 2045, we will have multiplied the intelligence, the human biological machine intelligence of our civilization a billion-fold."[3]

In the midst of the doom and gloom scenarios, there is more than a sliver of light, hope, opportunity and value for mankind in all this, as summarized by Klaus Schwab, founder and executive chairman of the World Economic Forum: "We must address, individually and collectively, moral and ethical issues raised by cutting-edge research in artificial intelligence and biotechnology, which will enable significant life extension, designer babies, and memory extraction."[4]

Or as stated by Dr. Bonnie Wintle of the Centre for the Study of Existential Risk, University of Cambridge: "The growth of the bio-based economy offers the promise of addressing global environmental and societal challenges, but ... it can also present new kinds of challenges and risks. The sector needs to proceed with caution to ensure we can reap the benefits safely and securely."[5] Dr. Jenny Molloy, of the Synthetic Biology Strategic Research Initiative, University of Cambridge: "One theme that emerged repeatedly was that of inequality of access to the technology and its benefits. The rise of open source, off-patent tools could enable widespread sharing of knowledge within the biological engineering field and increase access to benefits for those in developing countries."[6]

Chapter 6: Technology – summary overview

6.1 Introduction and overview

Why do we have a stand-alone chapter on the topic of technology situated after three chapters on ESG (environmental, social and governance) topics? Quite simply: it belongs here and deserves to have a chapter of its own. The utterly disruptive and revolutionary impact that technology (of all kinds) is having and will continue to have on everything in the world today invites the question of why technology hasn't been properly included in ESG discussions before.

Looking at the definitions and typologies of the three other categories we have examined, it becomes abundantly clear that there are overlaps and interrelationships between these categories. Technology – especially the life-altering nature of the technologies produced under the Fourth Industrial Revolution – has become a pervasive dimension of life on earth and with the "singularity" coming around 2045 according to Kurzweil - when human intelligence and machine will become one - we have no choice but to add technology as a fourth dimension to ESG or ESGT.[7]

Tech, without a doubt, requires its own analysis and consideration whichever way one looks at it – whether you are a business getting disrupted, a citizen whose data is being used (and abused), a regulator who is overwhelmed, under-resourced and unprepared to manage the tech barrage, a young person trying to figure it all out – what to study (or not), what to do professionally (or not) or whether to trust that universal basic income will come along to help in the not too distant future.

In this chapter we address technology in a fashion similar to the other three categories of ESG. However, since we are dealing with a lot of novelty here (and, truth be told, your author is no technologist or scientist), we come at this subject from a broader liberal arts perspective. Because the "T" in ESGT is still not a thing in the ESG world, we will be digressing, extrapolating and/or projecting a bit more about this subject in this chapter than we did in the preceding three with the more established ESG subjects. For one thing, some of the ESG-focused classifications and typologies we used previously do not apply here because they simply don't exist. Either way, we hope to inform and spark further thinking and doing on this topic.

One other caveat: because I am no technologist, I am certain that this chapter will disappoint anyone expecting to see a thorough, comprehensive and/or technologically sophisticated treatment of the subject. You won't find that here. What you will find is the following: an attempt, first, to place technology on the "all things" ESGT map and second, to provide the perspective of the non-technologist (i.e., of someone from the world of governance, geopolitics, risk, ethics and policy). Indeed, one of the underlying themes of this chapter is that we – the non-technologists – are as needed at the proverbial table as the technical guys are. Period. But don't just take my word for it – author and venture capitalist Scott Hartley not only makes this case with facts and figures in his excellent book *The Fuzzy and the Techie* but he actually claims that liberal arts types will "rule the digital world."[8] While I won't necessarily be as bold as Scott in my prediction, I wholeheartedly concur that we all deserve a seat at the tech table.

The chapter begins with definitions and typologies that should help us get our arms around the vast treasure trove of issues, sub-issues, risks and opportunities that lie in plain sight or just beneath the surface. Next, the core of the chapter will involve an exploration of three powerful topics that are of greatest concern in social discourse today and which leaders at all levels must focus on – misinformation in the age of super-connectivity, cyber-insecurity and the need to create trustworthy tech.

6.2 What is "technology"? Definitions and typologies

6.2.a Definitions

"Technology" has its etymological roots in two ancient Greek words: "technos" (which means art or skill) and "ology" (which signifies branch or knowledge).

One of the first "modern" definitions of "technology" can be found in this definition from Jacob Bigelow in 1829 who described it as:

> principles, processes, and nomenclatures of the more conspicuous arts, particularly those which involve applications of science, and which may be considered useful, by promoting the benefit of society, together with the emolument [compensation] of those who pursue them.[9]

The *Business Dictionary* states that "technology" is:

> The purposeful application of information in the design, production, and utilization of goods and services, and in the organization of human activities.

Technology can be described in the following ways:

> Tangible: blueprints, models, operating manuals, prototypes.
>
> Intangible: consultancy, problem-solving, and training methods.
>
> High: entirely or almost entirely automated and intelligent technology that manipulates ever finer matter and ever powerful forces.
>
> Intermediate: semiautomated partially intelligent technology that manipulates refined matter and medium level forces.
>
> Low: labor-intensive technology that manipulates only coarse or gross matter and weaker forces.[10]

And from Lego Engineering we get the following definition based in part on the ancient Greek words we discussed above:

> Technology is the knowledge or expert skill of how to make things, from the Greek words technos (art or skill) and –ology (branch of knowledge), typically through the use of tools. Typical examples of technologies include such things as baking, weaving, soldering, injection molding, welding, glass making, or semiconductor manufacturing.
>
> Technology (process/tool skills) should be distinguished from the **products** of technology, which can be referred to as **technological artifacts**. Indeed, most allegations of technological theft refer not to theft of physical items, but rather to unauthorized copying of **processes**, from weaving technology in the early Industrial Revolution to Samsung's creation of smartphones highly similar to those of Apple.[11]

6.2.b Typologies, classifications and categories

As we did in the prior three chapters, in Table 6.1 we share an overview of the sources and resources that we utilize in our discussion of types, classes of technology issue, risk and opportunity in the following sections of this chapter.

Table 6.1 On technology: sources and resources

Level of detail	*Resource*
FROM BIG PICTURE ...	i. Imperial College of London Disruptive Technology Periodic Table[12]
	ii. Centre for the Study Of Existential Risk[13]
	iii. World Economic Forum[14]
	iv. UN Sustainable Development Goals and UN Global Pulse[15]
... TO MORE FOCUSED ...	v. Gartner 2018 Top Five Strategic Technology Trends[16]
	vi. Plug and Play Start-Up Ecosystem Accelerator Investment Categories[17]
	vii. CSER Synthetic Biological And Bioengineering Risks and Opportunities
	viii. Allianz Risk Barometer[18]
... TO MORE GRANULAR	ix. RepRisk[19]
	x. UN Global Compact[20]

Source: Author.

6.2.b.i Imperial College of London Disruptive Technology Periodic Table[21]

In the research I conducted for this book, I came across a wonderful chart created by the Imperial College of London in which, somewhat tongue in cheek, a group of academics created a "periodic table of mind-blowing tech" organized to look like the classic scientific Periodic Table and presented in such a way that we see the shorthand description of 100 technologies from the least to the most disruptive socioeconomically and the sooner to the later in terms of time frame adoption.

I highly recommend this resource and have recreated two smaller tables showing excerpts from this larger "periodic table" – Table 6.2 shows the 16 most current and least disruptive technologies, and Table 6.3 shows the 16 most disruptive and futuristic technologies.

A quick look at the first table will show how socioeconomically disruptive the current least disruptive technologies are, and they are all already underway – robotic cars, distributed ledgers (blockchain), deep ocean wind farms, cultured meats, wireless energy transfer, delivery robots. While not all of us have seen these in action, we are very likely to have experienced direct or indirect effects – benefits and detriments – of some of these technologies already.

Table 6.2 Current least disruptive technologies: assembled from the Imperial College of London, table of 100 disruptive technologies[22]

HIGHER ↑ ***SOCIO-ECONOMIC DISRUPTION*** ↓ ***LOWER***				
	Dl DISTRIBUTED LEDGERS	Pa PRECISION AGRICULTURE	Av AUTONOMOUS VEHICLES	Id INTENTION DECODING ALGORITHMS
	Rc ROBOTIC CAR COMPANIONS	Sc SMART CONTROLS & APPLIANCES	Cm CULTURED MEAT	Ro DELIVERY ROBOTS AND PASSENGER DRONES
	Cr CRYPTO-CURRENCIES	So CONCENTRATED SOLAR POWER	Pp PREDICTIVE POLICING	Eh MICRO-SCALE AMBIENT ENERGY HARVESTING
	Sn SMART NAPPIES	Dw DEEP OCEAN WIND FARMS	Va VERTICAL AGRICULTURE	We WIRELESS ENERGY TRANSFER
	SOONER ←	***TIME***	→	***LATER***

Source: Imperial College of London.

If, however, we turn our attention to the upper right quadrant of the original 100 disruptive technologies periodic table Imperial College put together, we are sure to have our minds blown. Half-jokingly, the team designated *the top most disruptive* and *most futuristic* technology as "We can't talk about this one". The bottom line is that the more secretive and powerful defense industrial complexes of the most powerful countries and regions of the world today – the US, China, Russia and the EU – are developing technologies like or similar to these and more which we cannot even imagine – yet.

Take a look at Table 6.3 and try to figure out what the socioeconomic, political and personal disruption technologies like the "Internet of DNA", "Dream Reading and Decoding", "Whole Earth Virtualization", and "Telepathy" might have on our daily lives. Now imagine what the lesser angels of our nature – whether authoritarian or dictatorial leaders or trans-border terrorist and criminals – might do with such power if and when they get their hands on it. Then, imagine what the world might look like when these forces of human evil unleash their misdeeds. The evil characters in James Bond movies will have nothing on them.

Table 6.3 Future most disruptive technologies: assembled from the Imperial College of London, table of 100 disruptive technologies[23]

HIGHER	EI SPACE ELEVATORS	Vr FULLY IMMERSIVE VIRTUAL REALITY	Co ARTIFICIAL CONSCIOUSNESS	Qt WE CAN'T TALK ABOUT THIS ONE
	Is INVISIBILITY SHIELDS	Ph FACTORY PHOTOSYNTHESIS	Th TRANS-HUMAN TECHNOLOGIES	Te TELEPATHY
SOCIO-ECONOMIC DISRUPTION	Qs QUANTUM SAFE CRYPTO-GRAPHY	Cp COGNITIVE PROSTHETICS	Ud DATA UPLOADING TO THE BRAIN	Rd REACTIONLESS DRIVE
	Me INTERNET OF DNA	Tc THOUGHT CONTROL MACHINE INTERFACES	Dr DREAM READING AND RECORDING	Wh WHOLE EARTH VIRTUALIZATION
LOWER	***SOONER*** ⟸	***TIME***	⟹	***LATER***

Source: Imperial College of London.

So we begin with this somewhat provocative but not totally farfetched consideration of technology. Now let's turn to some additional categories and typologies to put a little more immediate and realistic meat on the bone of this topic.

6.2.b.ii Centre for the Study of Existential Risk

Our friends at the Centre for the Study of Existential Risk (CSER), who do important work on existential risks in general, devote a good chunk of their work to future disruptive technologies and the existential risk that they may present. Table 6.4 summarizes some of their work focused on extreme technological risk, AI risks and catastrophic biological risks.

Table 6.4 Centre for the Study of Existential Risks: University of Cambridge[24]: technology issues

Category of risk	*The challenge*	*Risks and opportunities*	*Examples*
Managing Extreme Technological Risks	Existential risks share common methodological challenges: how to horizon-scan for and	Risks associated with emerging and future technological advances, and	Risks associated with human technology and activity, such as

(*Continued*)

Table 6.4 Continued

Category of risk	*The challenge*	*Risks and opportunities*	*Examples*
	evaluate low probability/high impact events; how to encourage responsible innovation among technologists and a safety culture among scientists.	impacts of human activity, threaten human extinction or civilizational collapse. Managing these extreme technological risks is an urgent task – but one that poses particular difficulties and has been comparatively neglected in academia.	nuclear war, engineered pandemic, climate change, ecological collapse, or advanced artificial intelligence.

Risks from Artificial Intelligence	Recent years have seen dramatic improvements in artificial intelligence, with even more dramatic improvements possible in the coming decades. In both the short-term and the long-term, AI should be developed in a safe and beneficial direction.	• Super-intelligent AI:
		  • Safety
		  • Security
		• Privacy
		• Bias
		• Inequality
		• Safety
		• Security
		• Scientific discoveries
		• Cheaper & better goods & services
		• Medical advances
			• Open AI
			• DeepMind
			• The Centre for Human-Compatible AI

Global Catastrophic Biological Risks	Pandemics are as old as humanity, but in today's interconnected world we are more vulnerable than ever. The increase in the capability and spread of biotechnology poses new risks, from accidental release to intentional misuse.	• Natural pandemic (bird flu, swine flu) • Manufactured, more infectious pathogens • Genome editing CRISPR-Cas9 • New & better drugs • Improvements in agricultural productivity • Need better surveillance mechanisms • Need more national & international health systems • Need better stockpiles of vaccines and medical countermeasures	• Biotechnology, analyzing gene drives • Debating gain-of-function research • Biosafety and responsible research & innovation • Biosecurity – next steps for Biological Weapons Convention

Source: Centre for the Study of Existential Risk.

6.2.b.iii World Economic Forum

An interesting trend embedded in the WEF GRR for 2018 and 2019 is that technological risks appear in the top-five most serious threats for the first time – in 2018 "Cyber-attacks" as the #3 and "Data Fraud or Theft" as the #4 most likely risks. In 2019, the trend continued with both of these risks appearing in the top-five again – cyber-attacks as #5 and data fraud or theft as #4. And in 2019 both of these risks also appeared as #7 and #8 respectively in the most impactful top-ten risks of the year.

Indeed, historically, technology risks are the least frequent to show up of the five categories of global risk that the WEF tracks each year (economic, environmental, social, geopolitical, societal and technological). In the past, technology risks have only showed up a few times, as follows:

- In 2012, for the first time, one technology risk "cyber-attacks" comes in as the #4 most likely global risk
- In 2013, there is no technology risk listed in the top five most likely or highest impact
- In 2014, two technology risks appear on the list: "cyber-attacks" at #5 most likely risk, and a new one: "Critical Information Infrastructure Breakdown" as the #5 most impactful risk
- In neither 2015 nor 2016 are there any top-five technology risks listed
- In 2017, a new one shows up as the #5 most likely: "Massive Incident of Data Fraud/Theft"

I'm not a betting woman, but I would venture to guess that in future years – 2020 and beyond – we will continue to see two or more global technology risks making it to the WEF Global Risks Report (GRR) top ten most likely or impactful risks.

6.2.b.iv UN Sustainable Development Goals and UN Global Pulse

At the risk of appearing to oversell the importance of technological change and the issues, risks and opportunities relating to technology, I would argue that each and every one of the Sustainable Development Goals (SDGs) has a technology component associated with it. Why? Simply because there are technology aspects, solutions, problems and opportunities associated with every one of the 17 SDGs. Think about it as you read through each of them in Table 6.5.

Indeed, if we go back to the Imperial College "periodic" table of technological disruption, we could conduct an interesting exercise of mixing and matching which current and future technologies might support, further or indeed turbocharge some of the SDGs, such as the following:

- In support of SDG #1 – *No Poverty* – perhaps "Vertical Agriculture", "Precision Agriculture" and "Cultured Meat"?
- In support of SDG #2 – *Zero Hunger* – perhaps "Delivery Robots" and "Autonomous Vehicles" can help deliver food to isolated places?
- In support of SDG #7 – *Affordable and Clean Energy* – perhaps "Deep Ocean Wind Farms" and "Concentrated Solar Power" can help?

- In support of SDG #10 – *Reduced Inequalities* – perhaps "Distributed Ledgers" can help bring transparency in banking and trade to less advanced locations?
- In support of SDG #16 – *Peace and Justice/Strong Institutions* – perhaps "Predictive Policing" and "Distributed Ledgers" can help?

I wouldn't even try to match the more futuristic and disruptive technologies to any of the SDGs, partly because they are still far off, but also because I have no idea how some of these crazy-sounding but eminently possible inventions ("Space Elevators", "Trans-human Technologies", "Invisibility Shields"?) will affect our daily lives. But I'm sure of one thing – there will be technologies we haven't yet imagined that will help to further the SDG goals and whatever replaces them when they come due in 2030.

Table 6.5 United Nations Sustainable Development Goals: technology-related issues

SDG – Technology related	*E*	*S*	*G*	*T*
1. No poverty		X	X	X
2. Zero hunger	X	X	X	X
3. Good health and well-being		X	X	X
4. Quality education		X	X	X
5. Gender equality		X	X	X
6. Clean water and sanitation	X	X	X	X
7. Affordable and clean energy	X	X	X	X
8. Decent work and economic growth		X	X	X
9. Industry, innovation and infrastructure	X	X	X	X
10. Reduced inequalities		X	X	X
11. Sustainable cities and communities	X	X	X	X
12. Responsible consumption and production	X	X	X	X
13. Climate action	X	X	X	X
14. Life below water	X		X	X
15. Life on land	X		X	X
16. Peace and justice – strong institutions		X	X	X
17. Partnerships for the goals		X	X	X

Source: United Nations.

Focused specifically on the role of big data in the SDGs, another important UN initiative is UN Global Pulse. It actually looks at harnessing big data in furtherance of the SDGs and humanitarian objectives.[25] Indeed, they provide great examples of how big data can achieve these goals, and we will explore these in detail later in this chapter.

6.2.b.v Gartner 2018 Top Technology Trends

In 2018, Gartner identified 17 technologies as important areas under development and organized them into "five major trends in the technology hype cycle".

Each of these five trends are summarized in Table 6.6 and include "democratized artificial intelligence (AI)", "digitalized ecosystems", "do-it-yourself biohacking", "transparently immersive experiences" and "ubiquitous infrastructure".

Table 6.6 Gartner's Five Trends in the Technology Hype Cycle 2018[26]

Trend #1 Democratized AI	AI … will become more widely available due to cloud computing, open source and the "maker" community. While early adopters will benefit from continued evolution of the technology, the notable change will be its availability to the masses. These technologies also foster a maker community of developers, data scientists and AI architects, and inspire them to create new and compelling solutions based on AI.
Trend #2 Digitalized Ecosystems	Emerging technologies in general will require support from new technical foundations and more dynamic ecosystems. These ecosystems will need new business strategies and a move to platform-based business models", and there will be a "shift from compartmentalized technical infrastructure to ecosystem-enabling platforms … laying the foundation for entirely new business models that are forming the bridge between humans and technology … For example, blockchain could be a game changer for data security leaders, as it has the potential to increase resilience, reliability, transparency, and trust in centralized systems".
Trend #3 Do-It-Yourself BioHacking	[The year] 2018 is just the beginning of a 'trans-human' age where hacking biology and "extending" humans will increase in popularity and availability. This will range from simple diagnostics to neural implants and be subject to legal and societal questions about ethics and humanity. These biohacks will fall into four categories: technology augmentation, nutrigenomics, experimental biology and grinder biohacking.
Trend #4 Transparently Immersive Experiences	Technology, such as that seen in smart workspaces, is increasingly human-centric, blurring the lines between people, businesses and things, and extending and enabling a smarter living, work and life experience. In a smart workspace, electronic whiteboards can better capture meeting notes, sensors will help deliver personalized information depending on employee location, and office supplies can interact directly with IT platforms.
Trend #5 Ubiquitous Infrastructure	In general, infrastructure is no longer the key to strategic business goals. The appearance and growing popularity of cloud computing and the always-on, always-available, limitless infrastructure environment have changed the infrastructure landscape. These technologies will enable a new future of business. For example, quantum computing, with its complicated systems of qubits and algorithms, can operate exponentially faster than conventional computers. In the future, this technology will have a huge impact on optimization, machine learning, encryption, analytics and image analysis … A second new technology in this trend is neuromorphic hardware. These are semiconductor devices inspired by neurobiological architecture, which can deliver extreme performance for things like deep neural networks, using less power and offering faster performance than conventional options.

Source: Gartner.

Clearly, these trends are here to stay and are only moving forward. However, significant ethical and social implications of such currently mind-blowing technologies as biohacking and nano biological implants present huge challenges. These serious ethical and social concerns are why it is so important that all sectors – whether government research, corporate R&D, university or nonprofit work in technology development – incorporate people from other walks of life (e.g., liberal arts types like ethicists, lawyers, sustainability experts) into the earliest stages of technology development.

6.2.b.vi Plug and Play start-up ecosystem accelerator investment categories

A different angle that may also be useful for organizational leaders to think about while trying to figure out their specific technology issues, risks and opportunities, is to look at what the start-up investor and venture capital community is looking at in terms of categories of new products and services. Table 6.7 shows the current list of investment categories of global start-up accelerator Plug and Play Tech Center (for whom I serve as a governance, risk and ethics mentor).

Table 6.7 Plug and Play Tech Center ecosystem: categories of start-ups[27]

Fintech
Insurtech
Brand and Retail
Internet of Things
Mobility
Health
Supply Chain
Real Estate Tech
Food and Beverage
Travel and Hospitality
Energy and Sustainability
New Materials and Packaging
Cyber-Security
Enterprise 2.0
Start-Up Autobahn
Retailtech Hub
Fashion For Good – Plug And Play

Source: Plug and Play Tech Center.

6.2.b.vii Synthetic biology and bioengineering risks and opportunities

University of Cambridge's CSER has also been involved in specific biological and bioengineering research and study, and Table 6.8 provides a summary of some of the top issues, risks and opportunities that they have identified for further study and coverage. It is clear from these five categories of biological inquiry that there are both grave threats and risks as well as opportunities to turbocharge the betterment of the world. Therein lies one of the great paradoxes of our time as discussed in Chapter 1 – the paradox of extreme good and extreme bad lying side by side right in front of us.

Table 6.8 Synthetic biology and bioengineering: risks and opportunities from the Cambridge Centre for the Study of Existential Risk[28]

Artificial photosynthesis and carbon capture for producing biofuels if technical hurdles can be overcome, such developments might contribute to the future adoption of carbon capture systems and provide sustainable sources of commodity chemicals and fuel.

Enhanced photosynthesis for agricultural productivity – synthetic biology may hold the key to increasing yields on currently farmed land – and hence helping address food security.

Synthetic gene drives – gene drives promote the inheritance of preferred genetic traits throughout a species, for example, to prevent malaria-transmitting mosquitoes from breeding. However, this technology raises questions about whether it may alter ecosystems, potentially even creating niches where a new disease-carrying species or new disease organism may take hold.

Human genome editing – genome engineering technologies such as crispr/ CAS9 offer the possibility to improve human lifespans and health. However, their implementation poses major ethical dilemmas. It is feasible that individuals or states with the financial and technological means may elect to provide strategic advantages to future generations.

Defense agency research in biological engineering -the areas of synthetic biology in which some defense agencies invest raise the risk of 'dual-use'. For example, one program intends to use insects to disseminate engineered plant viruses that confer traits to the target plants they feed on, with the aim of protecting crops from potential plant pathogens – but such technologies could plausibly also be used by others to harm targets.

Source: Centre for the Study of Existential Risk.

6.2.b.viii Allianz Risk Barometer

The 2019 Allianz Barometer distinguishes several categories of issues as top risks for businesses, and the ones that contain substantial technology components are identified in Table 6.9.

Table 6.9 Allianz Risk Barometer top 10 business risks 2017–2019:[29] focus on technology issues

Top 10 Risks (w/Technology Issues)	*2019 Ranking*	*2018 Ranking*	*2017 Ranking*	*ESGT Issue*
Business Interruption (Including Supply Chain Disruption)	#1	#1	#1	E, S, G, T
Cyber Incidents (e.g., Cyber-Crime, IT Failure, Data Breaches)	#3	#3	#4	T, G

(*Continued*)

Table 6.9 Continued

Top-Ten Risks (w/Technology Issues)	*2019 Ranking*	*2018 Ranking*	*2017 Ranking*	*ESGT Issue*
Changes in Legislation And Regulation (e.g., Government Change, Economic Sanctions, Protectionism, Brexit, Eurozone Disintegration)	#4	#5	#5	G, T
New Technologies (e.g., Impact of Increasing Interconnectivity, Nanotechnology, Artificial Intelligence, 3D Printing, Drones)	#7	#7	#10	T, G
Loss of Reputation or Brand Value	#9	#8	#9	G, T

Source: Allianz.

6.2.b.ix RepRisk

Although RepRisk does not distinguish a separate "T" category in its ESG typology and nomenclature, there are a few ESG Topic Tags and ESG issues they use to reference technological issues directly. There are two specific "technology" Topic Tags identified in the RepRisk research scope – namely, "Drones" and "Privacy Violations" – although as with some of the other classifications we deal with in this book, there is a lot of overlap and interconnection between a variety of issues regardless of whether they are ESG or T.[30]

For example, using RepRisk's "Drone" Topic Tag, we find an excellent example in the WEF GRR for 2018 in the following mini-case study.

Mini-case study from the World Economic Forum Global Risks Report 2018*: precision extinction – AI-piloted drone ships wipe out a large proportion of global fish stocks*[31]

A third of the fish consumed in the world are already caught illegally. AI and drone technologies are increasingly commonplace. Add to these facts the automation of illegal fishing, and the impact on fish stocks could be devastating – particularly in international waters where oversight is weaker.

A rapid collapse of fish stocks could engender cascading failures across marine ecosystems. Communities reliant on fishing for their incomes might struggle to survive, leading to fiscal pressures and/or displacement.

A sufficiently large surge in the supply of illegal fish might distort global food markets, leading to disruption in the agriculture and food-production sectors.

If illegal drone fishing crossed national maritime boundaries and was perceived to be state-sanctioned, retaliatory measures might lead to diplomatic or military tensions.

Targeted schemes such as genetic markers to track fish throughout the supply chain might limit demand for illegally caught fish. So might better vessel observation.

But key to progress in this and similar areas of hybrid technological disruption will be new global governance norms and institutions, particularly those designed to protect the global commons and prevent the destructive deployment of emerging technologies.

Source: World Economic Forum.

In the above WEF example, a combination of AI, drone technology and age-old fishing industry practices, have all come together to create what could become a monster global governance challenge: something that, on the one hand, may be useful to humanity (more food) but that, on the other, may be destructive of life on earth (depleting fish stocks, destroying fragile communities, disrupting or depleting our already delicately balanced food supply in the long run). This is exactly the kind of complex ESGT and ethical challenge facing humanity today that will require smart people of all walks of life working together to solve the puzzle – in a socially beneficial way.

Another example of a technology risk issue covered by RepRisk is that of the massive cyber-breach at Equifax, a company that experienced a very significant cyber-incident in 2017 resulting in the theft of the personal and credit information of almost 150 million people. This risk incident moved Equifax to RepRisk's top-ten "Most Controversial Companies of 2017", coming in at #6 with a Peak RepRisk Index of 79 (out of a maximum of 100, denoting extremely high-risk exposure).

6.2.b.x UN Global Compact

Finally, the UN Global Compact does not have any Principles that are primarily or specifically related to technology issues. However, and as discussed in the SDG section earlier, there are technology aspects one could associate with many of the Principles. As set forth in Table 6.10, they do incorporate a technology element into their Principle 9, which relates to the environment, and specifically note the following:

> In order to fulfil obligations under the three environmental principles, companies and other organizations may want or need to deploy environmental technology to assist in achieving their strategic and tactical goals and objectives.

Table 6.10 United Nations Global Compact 10 Principles:[32] technology issues

Principle 9	Businesses should encourage the development and diffusion of environmentally friendly technologies.	E, T

Source: United Nations Global Compact.

6.2.c The way forward

As we did in the previous three ESG chapters, we will now turn to an exploration of three critical technology themes. These are pervasive and important topics that affect all of us in our individual capacities as well as our capacity as stakeholders in society, business and civic duty. And these are large topics that all leaders of any kind of organization – whether governmental, business or nonprofit – must pay attention to because critical and in some cases existential sociopolitical, economic, environmental, health, safety and security issues all hang in the balance, literally. The three overarching topics I have selected for this discussion are:

- (Dis)information warfare: the weaponization of facts and fakes
- Cyber-insecurity: from cyber-threat to cyber-resilience
- Building sustainable tech value: transforming tech fear into tech trust

6.3 Leaders navigating technology issues, risks and opportunities

Technology is not only here to stay – it is here to dominate, infiltrate and permeate every aspect of life on earth for better *and for worse*. It is precisely for this reason that leaders at all levels – public, private, nonprofit, global, national and local – need to organize themselves into working, diverse, collaborative groups to address, analyze, explore and solve the many and multiplying technology and technology-related issues, risks, crises and opportunities that can, will and are already happening at every level.

Humankind has rarely – if ever – faced such a tsunami of change all at once and actually known that it was coming, known that much of it (and its consequences) are unknown and unknowable and also known that now more than ever before humanity must consciously, proactively and conscientiously address the overwhelming disruption that is upon us. And the good news is that never before have we been as connected, informed, interdependent and able to help each other to address the potentially grave existential threats *and opportunities* ahead.

One of the key benefits of recent technology is the unprecedented democratization of information because of what I have called in previous writings the age of hyper-transparency (we can see everything) and super-connectivity (we can get it all, right here, right now). But this age also brings with it serious negative intended and unintended consequences. Many developments are for the greater good (see discussion later in this chapter of the work of UN Global Pulse) with, e.g., greater information availability and exponential data analytics allowing for enormous progress in some of the most remote and difficult places to reach on such desperately needed things as deploying health care through drones or understanding how best to serve the needs of populations suffering from water or food shortages.

However, this democratization of information is also fueling the creation of dark uses by a host of nasty actors – from rogue nation states to criminal gangs and from drug dealers to the 400-pound troublemaker sitting in his mother's basement. Thus, the democratization of information is creating asymmetric power for good and for bad.

A thread that runs through this chapter is the following: we are at a time where no single genius or group of geniuses are capable of even getting their collective brains around everything that is coming at us like a fast and furious video game. And yet we must collaborate, because the bad and the ugly may very well outpace the good and the beautiful.

There are so many technology topics that are critically important, terribly complex and upon us, as the preceding brief survey shows, and it's just the tip of the iceberg. For the remainder of this chapter we deal with topics within this author's reach. There are so many other topics that are as important (or more) but we simply don't have the capacity to address these in this book and hope others will in their work, including issues of biohacking, nanotechnology, genetic engineering, quantum computing, etc.

6.3.a (Dis)Information warfare: the weaponization of facts and fakes

> Falsehood flies, and the Truth comes limping after it.
>
> Jonathan Swift

Reality, recently, seems to have become stranger than fiction. The events surrounding Brexit and the 2016 US elections (and their aftermath), the most powerful man in the world wielding power via Twitter, ubiquitous cyber-attacks, the unleashing of fake news, fake voice, fake video, deep fakes, facial recognition, biometrics, nano-implants – you get it. It all makes you wonder if we haven't already descended into Alice in Wonderland's rabbit hole or maybe crash-landed on George Orwell's *1984* (which should maybe be renamed "2019"?).

We are living in an age of "(dis)information warfare". Here is a simple and useful definition:

> Information warfare is the "conflict or struggle between two or more groups in the information environment".[33]

But information warfare is also disinformation warfare – in other words, we are not only subject to warfare at different levels concerning the facts – we are also immersed in a new and dangerous tunnel of disinformation warfare where the fight isn't between fact and fact but between fact and fiction or, better said, fact and deliberate, intentional fakes, all turbocharged by social media and supercharged connectivity.

(Dis)information warfare has several different layers to it – the global geopolitical, the national, the social, the corporate and the personal. Let's take a look at each of these levels.

6.3.a.i Global geopolitical (dis)information warfare

After the publication of the Mueller Report in the spring of 2019, it is now incontrovertible that Russia had unleashed a (dis)information and social engineering campaign against the US for many years that was most starkly revealed through the 2016 US election campaign. Some of this is no longer conjecture but fact. As part of his team's work, US Special Counsel Robert S. Mueller III meticulously outlined detailed events in chapter and verse in two indictments filed as part of the Russia/2016 US election investigation: one in the winter of 2018 against the Russian intelligence front and social media troll factory known as the "Internet Research Agency" and others, and the other in the summer of 2018 against 12 Russian intelligence operatives from the GRU (the Russian government intelligence agency) in absentia.[34]

The social, cyber and fake news interference by Putin's Russia in America for years leading up to the 2016 elections is excellently documented in four recent, fact-intensive and expertly reported books written by respected veteran journalists and former US intelligence officers:

- *Russian Roulette: The Inside Story of Putin's War on America and the Election of Donald Trump*
- *The Plot to Destroy Democracy: How Putin and his Spies are Undermining America and Dismantling the West*
- *The Plot to Hack America: How Putin's Cyberspies and WikiLeaks Tried to Steal the 2016 Election*
- *Messing with the Enemy: Surviving in a Social Media World of Hackers, Terrorists, Russians and Fake News*[35]

Such geopolitical information warfare by the Russians and others – most notably Iran, China and North Korea – is something that is increasingly apparent and part and parcel of our daily lives. Technology companies and governments in democracies are finally beginning to work together to address this alarming and growing development.

The following are a couple of excerpts from recent news reports about these incursions one from Buzzfeed regarding Russian operations and the other from Reuters regarding Iranian operations: [36]

> The Russian government discreetly funded a group of seemingly independent news websites in Eastern Europe to pump out stories dictated to them by the Kremlin ... Russian state media created secret companies in order to bankroll websites in the Baltic states—a key battleground between Russia and the West—and elsewhere in Eastern Europe and Central Asia.

***

> An apparent Iranian influence operation targeting internet users worldwide is significantly bigger than previously identified ... encompassing a sprawling network of anonymous websites and social media accounts in 11 different languages. Facebook and other companies said ... that multiple social media accounts and websites were part of an Iranian project to covertly influence public opinion in other countries. A Reuters analysis

has identified 10 more sites and dozens of social media accounts across Facebook, Instagram, Twitter and YouTube.

This relatively new and concerning set of global geopolitical developments can be explained partly from an international relations standpoint, partly because of the explosion of social media technology and partly from the rise of the politics of populism and even authoritarianism in recent times. Political actors both within and outside of national borders can now virtually attack various national and international targets indiscriminately and without much ado, deploying information or disinformation warfare in a wide multiplicity of new ways. Indeed, democracies have been more vulnerable to the information wars than controlled societies for obvious reasons – authoritarian regimes can more tightly control information within their borders and can thus interfere more boldly inside the more open, freedom-of-speech-loving democracies. The corollary is true as well – because of the tighter control internally within authoritarian regimes, it is more difficult for outside influences to mess with controlled, censored and more fearful political environments.

Thus, stakeholders – meaning citizens and others present in a country – must demand that the highest levels of leadership – the heads of government, their cabinets, their department heads all the way from the top national level to the local municipal levels – act responsibly vis-à-vis these new tools of information gathering and disinformation distribution.

In democracies, we supposedly have the checks and balances of a constitution and the rule of law which should counter any despotic or authoritarian tendencies (but we know that this isn't always the case). And in more authoritarian or despotic regimes we know this isn't the case, certainly not domestically, but can occur via outside stakeholder pressures through organizations like Amnesty International, Human Rights Watch and others who champion the causes of democracy, rule of law, protection of rights, etc. and sometimes the moral and economic pressure that democracies can exert.

The following is an example about Twitter and the promise of the democratization of discourse in Saudi Arabia which, given events taking place there since Crown Prince Mohammad Bin Salman ruthlessly took over in late 2017, has proven to be wrong:

> Many Saudis had hoped that Twitter would democratize discourse by giving everyday citizens a voice, but Saudi Arabia has instead become an illustration of how authoritarian governments can manipulate social media to silence or drown out critical voices while spreading their own version of reality.[37]

6.3.a.ii National political (dis)information warfare

The next level of (dis)information warfare is what takes place within countries. Information distortion, fake news, and other devices can also be used strictly domestically, within national borders by political parties and their allies in what can only be called domestic information wars. An example is the site of Alex Jones, the creator of InfoWars, the until now wildly successful peddler of fake news to the far right in the US in which among other things he has promoted conspiracies about 911 and other events.

Another source of distorted facts are shows on US television cable networks that are peddled as news but are right-wing entertainment mainly for the benefit of President Trump and his base supporters. Some of these information wars can also sadly spill into the "real" world as the dangerous example of the threat of gun violence that occurred at "Pizzagate" at the heels of fake news about Hillary Clinton being a pedophile and the pedophile ring being run out of a pizza store sadly demonstrates.[38]

One particularly stark example of deliberate national disinformation occurred in Myanmar with the local society and government's use of Facebook against the minority Rohingya people as part of the terrible persecution and crimes against humanity that are taking place there. Here's part of the problem:

> Reuters reported that Facebook has outsourced most of its monitoring in Myanmar to a firm called Accenture in an operation dubbed "Project Honey Badger." Former monitors told Reuters they did not search for offending material themselves but reviewed a giant backlog of flagged posts. Facebook does not have a single employee in Myanmar, where its active user base has grown to 18 million, according to Reuters.[39]

Finally, another example of domestic information action is that deployed in China under its highly controlled information regime where the government itself is gathering information – the good, the bad and the in-between – on all of its population. China is moving swiftly to gather all possible data on their populace, including biometrically, to rate every one of their almost 1.5 billion people by 2020. As a result, it is becoming impossible to jaywalk across the street in a Chinese city with impunity because the authorities – with a click of their smartphone camera – can quickly detect who you are, where you live and your social rating. Yes, in China, there is/will be a social score for every single person. Is this information warfare against your own people? Is it disinformation if the information gathered is incorrect, distorted, manipulated? Yes, we are reaching beyond a brave new world.

6.3.a.iii Social (dis)information warfare

Social (dis)information happens when one or more social groups use social, ethnic, religious, economic or political facts or fake information and take to social media to disparage or injure the reputation (or worse) of others. This has led in some remote locations to horrible consequences for specific people who have been targeted for injury or killing or for persecution and sometimes death, often based on incorrect information or intentional lies.

A terrible example of this phenomenon occurred via the WhatsApp texting service in India, described in the following excerpt:

> Suspected child lifters are carrying sedatives, injections, spray, cotton and small towels. They speak Hindi, Bangla and Malyali. If you happen to see any stranger near your house immediately inform the local police as he could be a member of the child lifting gang," said a message circulated in India on WhatsApp, the Facebook-owned messaging app. The message led to mob violence that killed seven people in Jharkhand state in eastern India last year.[40]

6.3.a.iv Corporate (dis)information warfare

This can occur by organized groups or a single person against a corporation to point out a product or service flaw (tweeting against an airline that has mistreated you or others, for example), or can be used for more nefarious purposes such as spreading lies about someone, something, a product or service as part of a negative campaign or as part of an outright crime to extort money, perhaps via ransomware or the threat of reputational damage via negative views on social media.

This new reality has led to a whole new practice area within the general category of public relations and social media management that most well-prepared organizations are ready to deal with at a moment's notice. Again, I'm thinking of airlines since there's so much to complain about and so many people actually do that via social media. Companies and other organizations in the possible crosshairs of competitors, social groups and individuals need to be prepared to react at a moment's notice both to factual information and fake information that goes viral and gets socialized exponentially.

6.3.a.v Personal (dis)information warfare

A better name for this may be bullying, and it can happen at all levels – from the most powerful man in the world (President Trump) using his twitchy Twitter finger to bully a wide assortment of individuals, organizations and countries to the 9-year-old kid in school taunting and bullying another child which, very sadly, has led to horrible and unacceptable consequences such as child suicide.

There are many other examples of this – countless indeed – but another one that bears mentioning is when a famous and influential person falsely accuses someone else. This can also have serious, even devastating consequences. Tesla founder Elon Musk gratuitously accused one of the rescuers who saved several children stuck in the Thai Cave saga of the summer of 2018 of being a pedophile without the facts to prove it. And he retweeted it and didn't back off when questioned about it. The *Fortune* magazine story headline says it all: "Elon Musk Triples Down on Pedophilia Claims Against British Diver in Thailand".[41]

6.3.a.vi The rise of all things "fake" and "post-truth"

In 2016, the new *Oxford Dictionary* added a new word to its collection – "post-truth" and defined it as follows:

> Relating to or denoting circumstances in which objective facts are less influential in shaping public opinion than appeals to emotion and personal belief.
>
> in this era of post-truth politics, it's easy to cherry-pick data and come to whatever conclusion you desire.
>
> some commentators have observed that we are living in a post-truth age.[42]

Add to all of the foregoing the fact that there is a whole industry worldwide of fake-making whether it is written, visual, audio or otherwise. The rise of all things fake that is meant to emulate the real deal and or confuse the consumer of the fake is an alarming and fraught development that is only going to get worse (or better in quality and thereby trickery).

And then there is the alarming phenomenon of "deep fakes". In their excellent research report, "Deep Fakes: A Looming Challenge for Privacy, Democracy and National Security", authors Bobby Chesney and Danielle Citron, define "deep fakes" as follows:

> "deep fake" technology … makes it possible to create audio and video of real people saying and doing things they never said or did. Machine learning techniques are escalating the technology's sophistication, making deep fakes ever more realistic and increasingly resistant to detection. Deep-fake technology has characteristics that enable rapid and widespread diffusion, putting it into the hands of both sophisticated and unsophisticated actors.[43]

While the authors definitely see beneficial aspects to deep fakes (entertainment, art and autonomy), they also make an impassioned call to arms, as it were, for awareness and action against the multitude of deep fake harmful potential, as follows:

- Harm to Individuals or Organizations
  - Exploitation
  - Sabotage
- Harm to Society
  - Distortion of democratic discourse
  - Manipulation of elections
  - Eroding trust in institutions
  - Exacerbating social divisions
  - Undermining public safety
  - Undermining diplomacy

- Jeopardizing national security
- Undermining journalism
- Deep-fake news

Table 6.11 provides a summary of some of the fake things that are happening in our world today.

Table 6.11 The Rise of all Things Fake: some examples[44]

What	*Description*
Fake Twitter Accounts	Fake accounts and Twitter bots created to overstate followership, likes, influence, etc. Twitter is now working hard to counter this phenomenon.
Fake Facebook Accounts	Russian, Iranian, North Korean and other fake accounts created in various perceived enemy countries and used for sociopolitical manipulation. Facebook is working to counter this since the Cambridge Analytica scandal erupted.
Fake News Sites	Creation of fake news websites to spread manipulated and fabricated news or other information that then gets spread and picked up on more mainstream "non-fake" sites (example: Moldovan youth employed to create fake news for profit).
Fake Audio Including "Deep Fakes"	Audiovisuals of Donald Trump and Barack Obama have already been circulating on the Internet. There is technology today that can allow you to manipulate your voice to sound like someone else's as well.
Fake Video Including "Deep Fakes"	The newest techniques allow for a variety of 3D head moves, rotation, eye gaze and blinking. The new system uses AI in the form of generative neural networks taking data from the signal models and calculating, or predicting, the photorealistic frames for the given target actor. Impressively, the animators don't have to alter the graphics for existing body hair, the target actor body, or the background.[45]
Fake Groups	Through the creation of fake Facebook and Twitter accounts or social media sites, fake groups are created to stir up social unrest (in 2016 Russia created two fake politically opposing groups in a Texas town that became reality when actual people signed up and confronted each other).
Fake Goods	These are a little better known as we have all lived with fake goods and products in our economies best exemplified perhaps by the knock-offs of well-known fashion brands that we are sometimes tempted to buy on the streets of our cities.
Fake Services	There are certainly many examples of these as well, perhaps a little subtler and sometimes more difficult to identify but also exemplified dramatically by the many fake phone calls for real estate, loans, health and other services that don't really exist but are nevertheless trying to get your money.

Source: Author.

6.3.a.vii *The fight against fakes is only just beginning*

As the above section makes amply clear, we are in a (dis)information warfare no man's land, where the aggressors have the upper hand. But not all hope is lost, as there are efforts underway to deal with some of this chaos. Here are a few such examples.

6.3.a.vii.A FACEBOOK

Though apparently mostly under severe duress resulting from the Cambridge Analytica scandal, Facebook appears to have finally started to shut down fake Facebook pages and deploy a variety of additional tools to combat fakes generally – though critics continue to berate them for doing much less than they could. This may indeed be true given all the bad press they continue to get about their lack of cooperation, indeed subterfuge, possible cover-up and opposite actions as revealed in an outstanding investigative report by the *New York Times* published in November 2018 aptly titled "Delay, Deny, Deflect: How Facebook's Leaders Fought Through Crisis".[46]

Among the proactive measures Facebook seems to be taking, however, whose effectivity and reach is yet to be understood, are the following:[47]

- Limiting the number of forwards available on WhatsApp.
- Fact checkers checking to see if information is false.
- Reducing the distribution of content that is flagged.
- Checking whether information will advocate for harm – real harm, physical harm.
- Adopting a hate-speech policy that is unclear for now.
- Partnering with local civil society groups (anonymously) to understand local issue dynamics.

The fate of Facebook may very well hang on some of these deep trust and ethical issues that continue to swarm around them. Repeated reputational hits from these issues may also have a longer term negative effect on Fracebook's financial well-being (and that of its stakeholders, of course) as this example shows:

> Many of the changes that are being put in place to clean up the Facebook platform will be expensive and could have an impact on growth, putting a brake on the ad-revenue machine that Ms. Sandberg built. In July, when Facebook reported that a surprise slowdown in revenue for the second quarter was likely to continue along with an unexpected increase in costs for security and privacy, investors shaved almost $120 billion in value from the company's valuation – the biggest one-day loss ever for a U.S.-listed company.[48]

6.3.a.vii.B TWITTER

Twitter has rolled out new rules to shut down bots and other fake tweeters as well:[49]

"We don't want to incentivize the purchase of followers and fake accounts to artificially inflate follower counts, because it's not an accurate measure of someone's influence on the platform or influence in the world," said Del Harvey, Twitter's vice president for trust and safety. "We think it's a really important and meaningful metric, and we want people to have confidence that these are engaged users that are following other accounts".

In addition, Twitter has taken a number of additional measures, including rooting out "fake and automated accounts", locking almost 10 million suspicious accounts a week and removing others for spam violations. By shutting down some of these accounts they are also shutting down the "follower" sellers. In a major investigative report, *The New York Times* revealed that:

> one company, Devumi, sold over 200 million Twitter followers, drawing on an estimated stock of at least 3.5 million automated accounts, each sold many times over.

By policing accounts more proactively, Twitter has been able to start curtailing some of the egregious violations that are taking place:

> One such victim, a teenager named Jessica Rychly, had her account information – including her profile photo, biographical information and location – copied and pasted onto a fake account that retweeted crypto-currency advertisements and graphic pornography.

6.3.a.vii.C MICROSOFT

Microsoft has not found itself at the receiving end of some of the bad news regarding information warfare – in fact, it has distinguished itself by being quite proactive in both the cyber-security space as well as with AI-related issues. Two cases come to mind. First, Microsoft President Brad Smith has been busy publicly representing what can only be called a proactive corporate responsibility position on a number of technology issues.

Broadly speaking and especially as they relate to AI, these issues can be encapsulated in the themes listed in Table 6.12.

Table 6.12 Five take-aways from Microsoft's Brad Smith speech on artificial intelligence[50]

1. Giving computers an ethical compass.
2. Preparing employees for a digital workplace.
3. Tapping AI to solve the world's big problems.
4. Improving accessibility for people with disabilities.
5. Governing AI: a hippocratic oath for coders?

Source: Microsoft.

In addition, Microsoft has been at the forefront of another serious AI-related technology issue – facial recognition software – leading the tech sector in discussing this issue publicly as a topic that needs to be engaged in by a cross-disciplinary, cross-sector collaboration (including regulators). Here is an excerpt from one of Microsoft's Brad Smith's writings:[51]

> We live in a nation of laws, and the government needs to play an important role in regulating facial recognition technology … A world with vigorous regulation of products that are useful but potentially troubling is better than a world devoid of legal standards.

What Microsoft is doing in this arena is not only doing the right thing by its stakeholders (consumers, regulators, investors), it is also making a savvy move for competitive advantage as this reporting would underscore:

> With many of its rivals under fire, Microsoft has aggressively tried to position itself as the moral compass of the industry. Company executives have been outspoken about safeguarding users' privacy as well as warning about the potential discriminatory effects of using automated algorithm to make important decisions like hiring.[52]

6.3.a.vii.D THE "NEW GUARD": NEWSGUARD ET AL.

Finally, there are a few start-ups and other ventures addressing the growing need to counter some of the worst abuses of technology that are so clearly starting to crop up. This is indeed a land of opportunity for creative minds and start-ups and a great opportunity for a wide variety of actors – for-profit, civil society, governmental, international agencies – to get busy, develop and address this serious global social need.

One of these efforts – NewsGuard, whose tagline is "Restoring Trust and Accountability" – is designed to provide news sites with the equivalent of a "Good Housekeeping" seal of approval to help internet surfers distinguish legitimate news sites that don't traffic in fake news from illegitimate ones that do.[53] This is their opening statement on their website:

> NewsGuard uses journalism to fight false news, misinformation, and disinformation. Our trained analysts, who are experienced journalists, research online news brands to help readers and viewers know which ones are trying to do legitimate journalism—and which are not.

This is what Axios reported about NewsGuard in August 2018:

> The product is launching at a time when trust in news is at an all-time low, due in large part to the fact that some of the most popular web platforms that refer most news—like Google Search, Facebook, Twitter and Google Search—are being flooded with misinformation.

> This is a real solution that is available now that charts a middle path between two bad alternatives: government regulation or technology companies censoring content themselves or non-transparently using secretive algorithms to suppress certain websites.
>
> —Steven Brill, CEO and Co-Founder of NewsGuard[54]

Those of us who cherish the need for clear and unbiased information in our societies and internationally certainly wish NewsGuard and others success in the quest to separate the proverbial wheat (unbiased information) from the chaff (fake news) that has invaded our means of communicating and sharing information.

Axios also reported recently on a wide variety of additional efforts which bear mentioning (see Table 6.13). Keep your eye on these efforts – perhaps one or more of them will become a household name in helping to restore trust in information and news.

Table 6.13 The fight against fakes is only just beginning: Axios roundup of initiatives as of October 2018[55]

The Trust Project	Made up of dozens of global news companies, announced that the number of journalism organizations using the global network's "trust indicators" now totals 120, making it one of the larger global initiatives to combat fake news. Some of these groups (like NewsGuard) work with trust project and are a part of it.
News Integrity Initiative	Facebook, Craig Newmark Philanthropic Fund, Ford Foundation, Democracy Fund, John S. and James L. Knight Foundation, Tow Foundation, Appnexus, Mozilla and Betaworks.
Newsguard	Longtime journalists and media entrepreneurs Steven Brill and Gordon Crovitz.
The Journalism Trust Initiative	Reporters without Borders, and Agence France Presse, the European Broadcasting Union and the Global Editors Network.
Internews	Longtime international nonprofit.
Accountability Journalism Program	American Press Institute.
Trusting News	Reynolds Journalism Institute.
Media Manipulation Initiative	Data and society.
Deepnews.AI	Frédéric Filloux.
Trust and News Initiative	Knight Foundation, Facebook and Craig Newmark In. affiliation with Duke University.
Our.News	Independently run.
Wikitribune	Wikipedia founder Jimmy Wales.

Source: Axios.

6.3.b *Cyber-insecurity: from cyber-threat to cyber-resilience*

> If you know the enemy and know yourself, you need not fear the result of a hundred battles. If you know yourself but not the enemy, for every victory gained, you will also suffer a defeat. If you know neither the enemy nor yourself, you will succumb in every battle.
>
> Sun Tzu, *The Art of War*

6.3.b.i *Critical cyber-concepts*

When it comes to cyber-insecurity there are a great variety of levels of insecurity that we can slice and dice – from the global geopolitical to the individual and personal. Let's first define what we mean by "cyber-risk", "cyber-risk governance", "cyber-security" and "cyber-resilience" as these are critical concepts all leaders need to know about.

What is cyber-risk?

> 'Cyber-risk' means any risk of financial loss, disruption or damage to the reputation of an organization from some sort of failure of its information technology systems.[56]

What is cyber-security?

> Cyber-security is the practice of protecting systems, networks and programs from digital attacks. These attacks are usually aimed at accessing, changing or destroying sensitive information; extorting money from users; or interrupting normal business processes. Implementing effective cyber-security measures is particularly challenging today because there are more devices than people, and attackers are becoming more innovative.[57]
>
> Or
>
> Cyber-security is the ongoing application of best practices intended to ensure and preserve confidentiality, integrity, and availability of digital information and the safety of people and environments.[58]

What is cyber-risk governance?

> Cyber-risk governance is a framework adopted within an organization to deal with the new and evolving risks relating to cyberspace both within the organization and as the organization interfaces with the outside world. In this framework, the critical actors are the board, the C-Suite or executive team, and frontline top management in charge of executing cyber-risk management. This cyber-risk governance triangle:
>
> - Adopts, oversees and promotes an appropriate, concerted and coordinated philosophy or approach to cyber-risk and cyber-security for the organization;

- Develops the necessary and appropriate strategy (and budget, resources and incentives) to execute on that philosophy or approach; and
- Implements that strategy in the most nimble and effective manner possible at an operational and tactical level.[59]

What is cyber-resilience?

Cyber-resilience is what happens when an organization has all of the elements in place from tone from the top (governance), effective cyber-risk management, cyber-hygiene culture and an internal attitude of continuous improvement and opportunity seeking (Figure 6.1).[60]

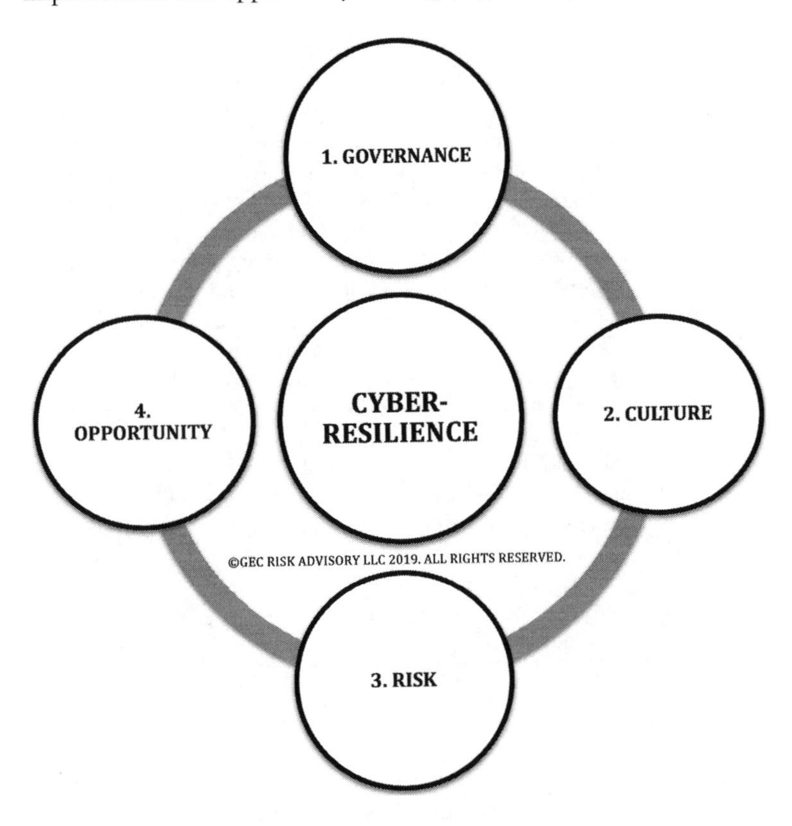

Figure 6.1 The Cyber-Resilience Virtuous Cycle.

Source: GEC Risk Advisory.

As Gideon Rose, editor in chief of *Foreign Affairs* magazine stated in the introduction of its September/October 2018 issue, which was appropriately named "World War Web":

> The last few decades have witnessed the growth of an American-sponsored Internet open to all, and that has helped tie the world together, bringing

wide- ranging benefits to billions. But that was then; conditions have changed. ...Whatever emerges from this melee, it will be different from, and in many ways worse than, what we have now.[62]

6.3.b.ii The tools of cyber-warfare

To get a general sense of the panoply of possible cyber-attacks, Table 6.14 shows a few of such categories prevalent today, according to Cisco.

Table 6.14 A few types of cyber-attack[62]

Ransomware	Ransomware is a type of malicious software. It is designed to extort money by blocking access to files or the computer system until the ransom is paid. Paying the ransom does not guarantee that the files will be recovered or the system restored.
Malware	Malware is a type of software designed to gain unauthorized access or to cause damage to a computer. Malware is intrusive software that is designed to damage and destroy computers and computer systems. Malware is a contraction for "malicious software". Examples of common malware include viruses, worms, trojan viruses, spyware, adware, and ransomware.
Social Engineering	Social engineering is a tactic that adversaries use to trick you into revealing sensitive information. They can solicit a monetary payment or gain access to your confidential data. Social engineering can be combined with any of the threats listed above to make you more likely to click on links, download malware, or trust a malicious source.
Phishing	Phishing is the practice of sending fraudulent emails that resemble emails from reputable sources. The aim is to steal sensitive data like credit card numbers and login information. It's the most common type of cyber-attack. You can help protect yourself through education or a technology solution that filters malicious emails.

Source: Cisco.

Now that we have a few concepts under control, let's take a look at cyber-insecurity in its various shapes, sizes and flavors from the largest macrocosm (global cyber-war) to the smallest microcosm (individual personally identifiable information).

6.3.b.iii Levels and examples of cyber-insecurity

6.3.b.iii.A INTERNATIONAL GEOPOLITICAL CYBER-WAR

There is the overarching geopolitical cyber-risk best exemplified (to date) by the conflict between Russia and the Ukraine and summarized in the following excerpt from an excellent *Wired* magazine investigative piece:

> [I]n Ukraine, the quintessential cyberwar scenario has come to life. Twice. On separate occasions, invisible saboteurs have turned off the electricity to hundreds of thousands of people. Each blackout lasted a matter of hours, only as long as it took for scrambling engineers to manually switch the power on again. But as proofs of concept, the attacks set a new precedent: In Russia's shadow, the decades-old nightmare of hackers stopping the gears of modern society has become a reality.
>
> And the blackouts weren't just isolated attacks. They were part of a digital blitzkrieg that has pummeled Ukraine for the past three years—a sustained cyber-assault unlike any the world has ever seen. A hacker army has systematically undermined practically every sector of Ukraine: media, finance, transportation, military, politics, energy. Wave after wave of intrusions have deleted data, destroyed computers, and in some cases paralyzed organizations' most basic functions. "You can't really find a space in Ukraine where there hasn't been an attack," says Kenneth Geers, a NATO ambassador who focuses on cyber-security.[63]

Then, of course, there is the quiet, sub-rosa cyber-war exemplified by the Russian attacks on a variety of democracies around the world, best represented by the current conflict with the US over influence and possible collusion and conspiracy with domestic actors regarding the 2016 US presidential election. This effort increasingly seems connected to the Brexit debacle in the UK as well.

Of course, cyber-warfare is probably taking place between a number of other capable countries as well, and while it is not something that is widely publicized, a mere look at the live cyber-threat real-time maps that several cyber-security firms maintain publicly will quickly make you think about what else might be going on.

6.3.b.iii.B CRIMINAL CYBER-ATTACKS AND CORPORATE HACKING

There are so many of these types of attacks that we have almost become inured to them except when they are really big (Equifax, for example, affecting the credit information of almost 150 million people). Even then, if it involves personal information rather than something more immediately painful (power outage, planes falling from the sky, other potentially harmful cyber-induced acts) it just seems to have become part of everyday life.

Then of course we have the "garden variety" of corporate hacks where companies or other organizations have been penetrated and some of their key data and "crown jewels" have been stolen for a variety of purposes, from criminal to espionage to something else (the proverbial fat guy with a computer in his basement).

6.3.b.iii.C PERSONAL DATA HACK

Finally, there is the personal level cyber-attack that leaves a person having to scramble to protect their checking or savings account because the information was shared on the dark web, or having to deal with malware because you clicked on the wrong email link, or having to change your tax filing identification codes because somebody hijacked your identity and has claimed tax refunds on your behalf. One could go on, but I'm sure you get the picture.

6.3.b.iv Building cyber-resilience: cyber-risk governance best practices

The bottom line for most organizations is that each organization needs to develop its own organizational cyber-resilience and forge the necessary and desirable outside partnerships with:

- Public institutions (relevant government agencies). For example, the financial sector which has been at the forefront of the cyber-insecurity edge, has forged closer relationships with US federal agencies like the FBI to keep each other informed on a real-time basis of cyber-attacks, attempts, etc.
- Private associations (either organizations similar to your own or a cross section of organizations with similar cyber-requirements). For example, associations like ISACA (the Information Systems Audit and Control Association)[64] and the ISA (the Internet Security Alliance) provide different levels of peer expertise sharing at different levels of an organization, in the former case a much larger association of information security professional, in the latter a tightly knit group of the highest level executive or CISO in member companies.
- Nonprofits (associations that can be helpful in providing support, research, benchmarking, policies, templates, etc.). In this area an example might include the associations mentioned earlier but also local educational nonprofits that produce educational research and/or training materials to help with cyber-resilience and related topics like business continuity and crisis management. An example of such an organization is Disaster Recovery Institute (DRI) International.
- Expert outside service providers, both hardcore in the sense that they have the technological solutions, and advisory – legal, strategic, governance, risk management – the skills that are necessary for and complementary to creating longer-term sustainable cyber-resilience.

Finally, cyber-resilience is something that is deliberately and consciously built by an organization into its culture. As long as there is management and/or board ignorance, resistance or willful blindness, an organization and its stakeholders are at grave risk of being unprepared for an attack at best and of falling victim to a serious and malignant (potentially existential) attack at worst.

In 2015 and 2016, I conducted two research efforts for The Conference Board on best practices in cyber-risk governance (or the art of the board and management preparing for cyber-risk) and wrote a chapter in a book published by the ISA[65] on building internal cyber-resilience. Table 6.15 provides a high-level summary with a number of updates of the key findings from these three efforts, which very much fall under the category of how to transform cyber-risk into cyber-resilience.

Table 6.15 Principal take-aways from study on strategic cyber-road map for the board[66]

1. All directors opined that the board must tackle the topic of cyber-security in a manner that is appropriate to its industry, footprint, geography, assets, and people.
2. Most directors said that the board should have either a committee, cyber-expert, or both, tackling the issue of cyber-security oversight as part of overall IT oversight. Such a technology or other committee should report to the board twice a year.
3. Most directors did not favor that the audit committee assume responsibility for cyber-security oversight and some favored a technology (and cyber-security) committee approach.
4. Several directors favored the inclusion of a cyber-security savvy or knowledgeable director on the board.
5. Most directors felt that at least some board members engage in cyber-security preparedness education and training.

Source: Andrea Bonime-Blanc. The Conference Board.

The ten key take-aways from the original "Emerging Practices in Cyber-Risk Governance" have stood the test of time, and I am happy to share them here as part of a well-rounded package of measures that leaders of organizations – whether executive management or board – should consider on behalf of their stakeholders. They are the following:

#1 DEVELOP A TRIANGULAR GOVERNANCE APPROACH TO CYBER-RISK GOVERNANCE

We discuss this triangular approach to governance in Chapter 7 as part of our discussion of the eight elements of organizational resilience, but suffice it to say here that especially for something as potentially existentially threatening as a cyber-attack an organization's board, executive management and functional experts need to be on the same, highly coordinated strategic and tactical page about cyber-management.

#2 UNDERSTAND THE REPUTATION RISK CONSEQUENCES TO STRATEGIC CYBER-RISK MANAGEMENT GONE WRONG

As we have all witnessed multiple times, the reputation risk consequences (which can certainly add to the financial consequences) can magnify and exacerbate the impact of a cyber-event on the organization and its stakeholders.

In the research project I conducted for The Conference Board in late 2015, I looked at four publicly traded companies that had suffered significant cyber-attacks – Target, JP Morgan, SONY and Anthem. Figure 6.2 presents a summary overview showing the reputation risk profile for each company through three lenses – the highest RepRisk Index (RRI) over a period of two years, the Peak RRI (as per RepRisk) after the attack, a moment in time (July 2015) and against their peer group of companies at that specific moment in time (July 2015).[67]

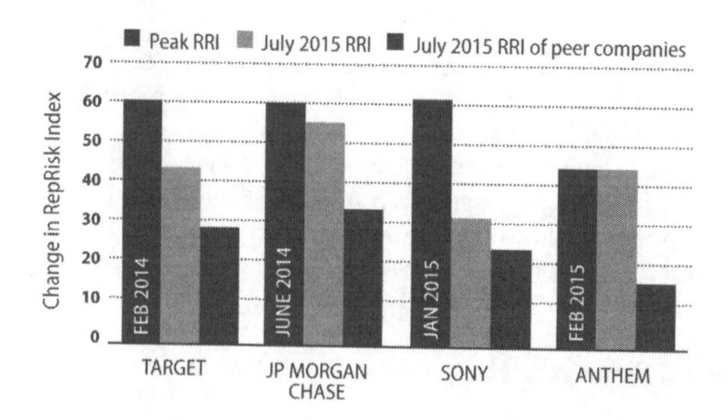

Figure 6.2 Comparison of RepRisk company profiles for four cyber-attacks.

Source: RepRisk and The Conference Board.

Significantly, the RepRisk Index was higher both right after the attack for each of these companies and after significant periods of time after the attack, and in both situations always higher than that of a peer group of companies each company was compared to.

A similar set of observations can be made regarding the Equifax cyber-disaster. In the category of a "pictures speak louder than words", Figures 6.3 and 6.4 show the RepRisk Index Trend and stock market effect of the Equifax situation.

Figure 6.3 shows the RepRisk Index going from an average of 25 (low risk) out of 100 right before the incident was revealed to 79 (extremely high risk) shortly after the company disclosed the incident involving the cyberhack of personal information of almost 150 million Americans and non-Americans – an appalling situation given that the business of Equifax is to provide credit risk analytical services and be a repository of deeply personal protected data.

Note how the RepRisk Index at the time of this writing is at 40, still significantly higher than before the disclosure, which I would argue reflects a continuing lack of trust from stakeholders. The fallout from this case included the resignations of the CEO, CISO, CIO, appointment of new executives and other measures as well as an unprecedented US$700 million settlement with the US Federal Trade Commission in the summer of 2019. But the revelations and the egregious violation of trust with not only consumers but also non-consumers – people whose data was stored on Equifax's servers without their knowledge or consent – caused a whole new level of outrage that may lead to attempts at stronger legislation to regulate the likes of Equifax.

Figures 6.4 and 6.5 also speak volumes, showing the one-year and five-year stock performance for Equifax and the deep chasm in the stock price when the breach announcement was made on September 1, 2017 (several months after it actually occurred, which added to the shock in the marketplace), going from an average of $140 per share before the announcement to a low of $89 after the breach became known. It took Equifax over a year to almost make it back to the level they were in the summer of 2017.

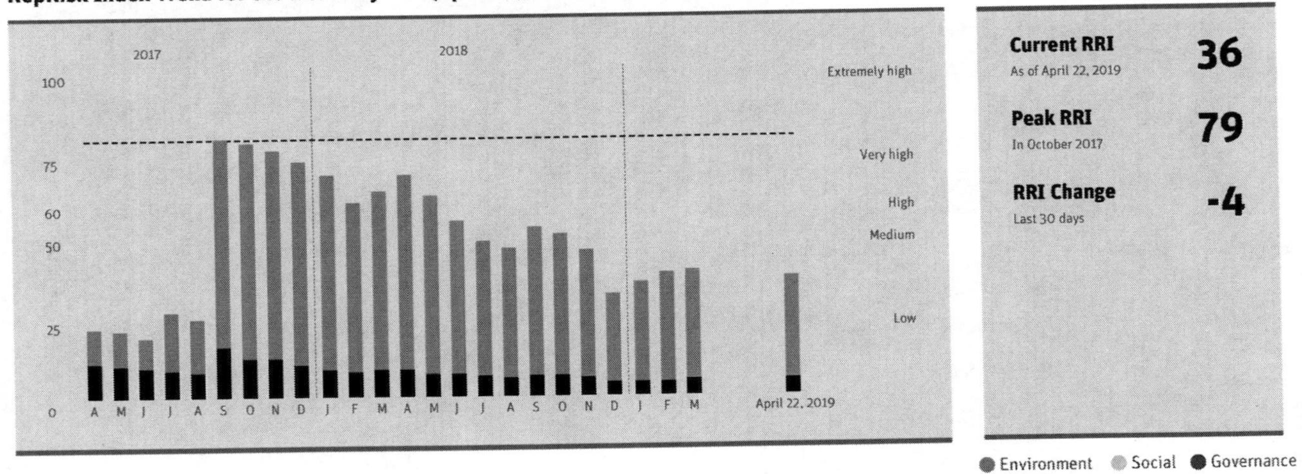

Figure 6.3 Equifax RepRisk Index Trend 2017–2019.

Source: RepRisk.

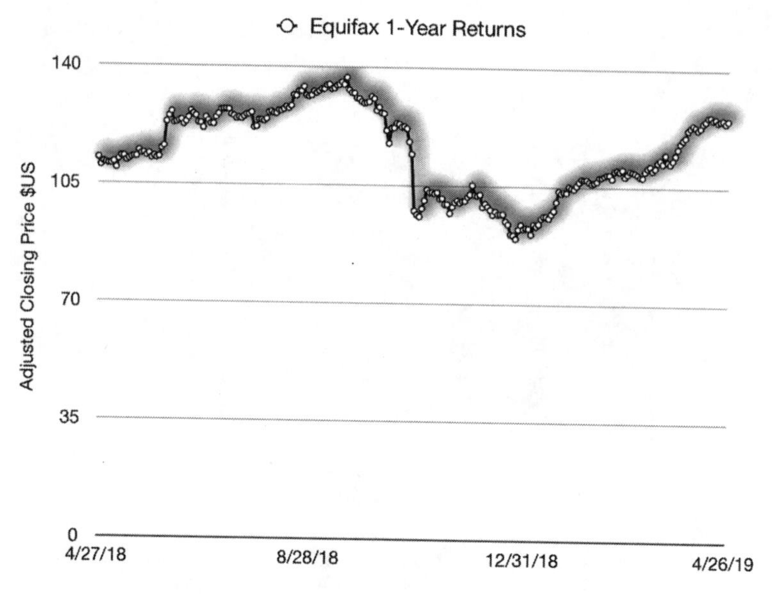

Figure 6.4 Equifax stock market performance – one year.

Source: Steel City Re.

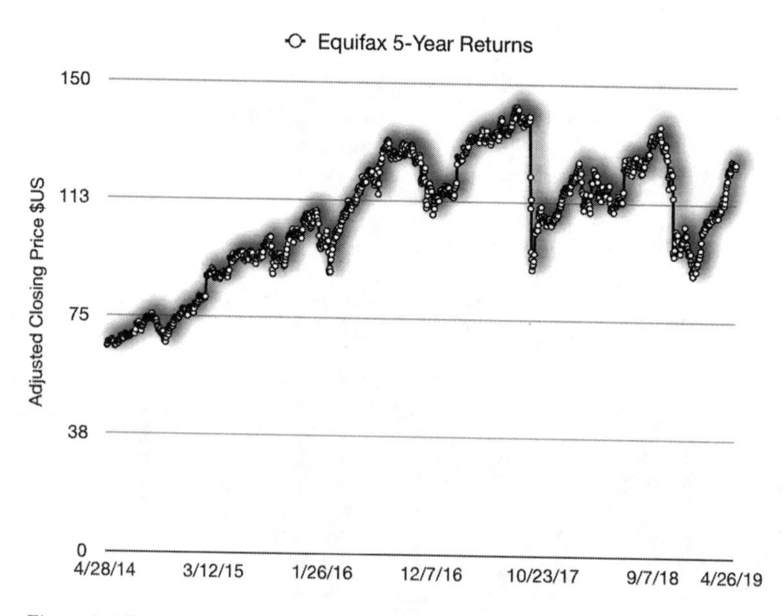

Figure 6.5 Equifax stock market performance – five years.

Source: Steel City Re.

#3 KNOW WHO YOUR CYBER-RISK ACTORS AND STAKEHOLDERS ARE

It's crucial for an organization to understand who the cyber-actors who might want to mess with them are and to understand where they will be coming from and after what (see the point on "crown jewels" later, in #4). Moreover, an understanding of your key stakeholders is also absolutely critical – it wasn't until the Cambridge Analytica scandal that Facebook appeared to care at all about protecting its users' private information. Indeed, Facebook users are arguably its most critical stakeholders without which Facebook would not exist because of the rich data extraction opportunity they represent to Facebook. And yet users were (and continue to be to a great extent) treated with complete disregard or worse as valuable free goods. This led in 2018 to a major drop in trust and in stock value for Facebook and the beginning of a series of major challenges from several other stakeholders – like the Congress and eventually the regulator. Table 6.16 provides an overview of key internal and external reputational stakeholders.

Table 6.16 Organizational reputation stakeholders[68]

Internal	*External*

• Owners/shareholders/investors
  • Family
  • Private
  • Public
  • Government
  • Institutional
  • Activist hedge funds
• Boards of directors, trustees or supervisors
• Board committees and chairs
• Council of advisors
• Employees
• Temporary and contract workers
• Labor unions
• Workers councils
	• Customers, purchasers and clients
	• Users of products and services
	• Prospective owners, shareholders, investors
	• Partners and suppliers
	• Non-governmental organizations
	• Prospective employees
	• Prospective temporary and contract workers
	• Government agencies, regulators, legislatures and law enforcement:
	  • Local
	  • Provincial/state
	  • National/federal
	  • Regional/international
	• Media and social media

Source: Author.

There are a variety of cyber-actors and it's important for an organization to understand which ones might be a threat to them. Of course, one never knows where a threat of this nature might come from, as SONY Pictures found out in the most unpleasant way after its release of a movie that caricatured Kim Jong Un and led the North Korean government to cyberhack their emails and intellectual property – something they were completely unprepared for.

However, a common mistake many organizations make is to not realize how their greatest threat comes from within – employees or contractors and third parties who are authorized and have access to the entity's information technology infrastructure and either innocently or maliciously exploit that access. This is why internal cyber-resilience, including training, testing and cyber-hygiene activities are so critical Table 6.17 provides an overview of the universe of possible cyber-actors.

Table 6.17 Universe of possible cyber-actors: insiders and outsiders[69]

Insiders	*Outsiders*
• Current employees • Former employees • Current service providers/consultants/contractors • Former service providers/consultants/contractors • Suppliers • Business partners • Customers	• Terrorists • Organized crime • Activists • Activist organizations • Hacktivists • Information brokers • Competitors • Foreign entities/organizations • Foreign nation states • Domestic intelligence service • Hackers

Source: PwC.

#4 HAVE A DEEP UNDERSTANDING OF THE ORGANIZATION'S "CROWN JEWELS"

Every organization needs to know what its crown jewels are from a cyber-standpoint and where they are located. Once that analysis has been conducted and the crown jewels have been categorized, the proper resources and prioritization of protections need to be introduced.

Whether it is personal health data, addresses, identification numbers, checking and banking account information or controls on a nuclear power plant's generation equipment, this is an exercise all organizations need to undertake and their leaders need to understand and constantly review and question.

#5 ENGAGE IN RELEVANT CYBER-RISK PUBLIC/PRIVATE PARTNERSHIPS

#6 DEVELOP A CROSS-DISCIPLINARY APPROACH TO CYBER-RISK MANAGEMENT

#7 DEVELOP A CROSS-SEGMENTAL/DIVISIONAL APPROACH TO CYBER-RISK MANAGEMENT

We discussed this earlier, but it deserves repeating – like all other technologies today, things have gotten way too complicated for any one person to understand – I don't care how technologically savvy they may be. That means that cooperation, collaboration, sharing of best practices, information and tools is absolutely essential – across sectors and with the government, and internally both from a multidisciplinary/functional standpoint as well as a cross-divisional or business unit or segment. Remember one thing: cyber-resilience is only as good as the weakest link in the chain.

#8 MAKE CYBER-RISK GOVERNANCE AN ESSENTIAL PART OF YOUR ORGANIZATION'S RESILIENCE APPROACH

Here's where it is so important to underscore how cyber-resilience is part of the broader approach to crisis management, data protection and business continuity for an organization. Again, cyber needs to be part of that broader approach integrated, adapted and customized to the needs of the organization.

#9 CHOOSE ONE OF THE THREE EFFECTIVE CYBER-RISK GOVERNANCE MODELS

Since a picture speaks louder than words, Figure 6.6 shows a typology of organizations depending on whether their leaders get it or not, on the one hand, and, on the other, whether the entity itself is on the higher or lower edge of the spectrum of cyber-impact – for example, an electric grid company is on the highest end of the spectrum while a grocery chain would be on a lower end in terms of material and severe impact on people and assets.

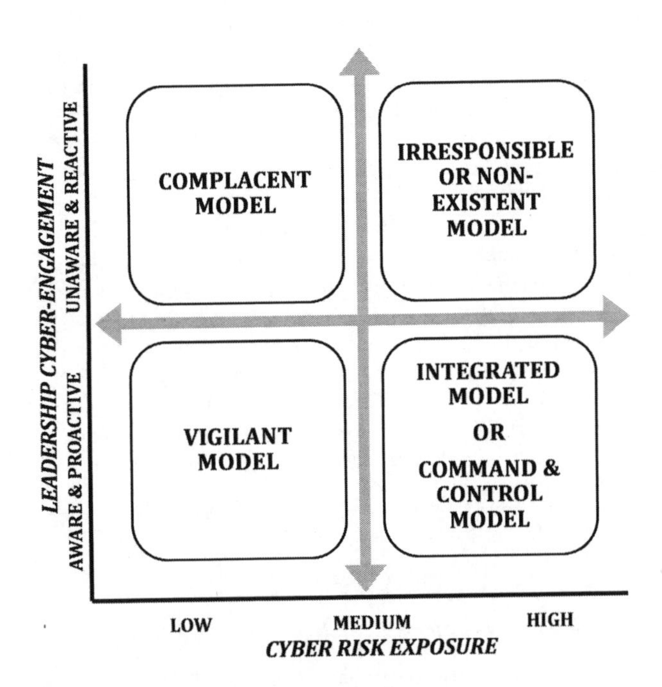

Figure 6.6 A typology of cyber-resilience models.

Source: GEC Risk Advisory.

#10 TRANSFORM EFFECTIVE CYBER-RISK GOVERNANCE INTO AN OPPORTUNITY FOR BETTER BUSINESS

This is my favorite part of everything I do and am writing about in this book: how leaders transform some of their greater challenges and risks into an opportunity to create value for their organization and stakeholders. The bottom line is this: responsible organizations and companies that take care of their own issues, risks and threats and build the appropriate programs around protecting people (stakeholders) and assets will then have the basis, foundation and trust to create, improve and deliver products and services that make use of that resilience directly or indirectly, thus adding to the tangible and intangible value of the entity. Table 6.18 summarizes two examples gleaned from the Conference Board study I authored on Emerging Practices in Cyber-Risk Governance.

Table 6.18 Two global Fortune 50 companies transforming cyber-risk into cyber-resilience and cyber-value[70]

Company	*What they did*
European Fortune 50 Insurance and Financial Company	Plans are currently in place to increase cyber-risk and security monitoring, detection, and mitigation within the company through advanced malware detection and the creation of a security operations center. Knowing that this is a risk that will continue to grow exponentially, and as an insurer with a cyber-crime insurance product, the company recognizes the need to understand this risk in real time and to have a well- informed view of what works internally within a company to combat cyber-risk and maintain healthy cyber-security. This extends to an understanding of what a systemic threat is, not only to a company but also to an entire sector or industry and to business generally.
US-Based Fortune 50 Global Technology Company	Because of the nature of its business, the company sees a large opportunity for creating a cyber-security competitive advantage in its products and services. This has served as an additional driver of board and executive attention, especially with cloud-based services, including cyber-security solutions and services and products that are now generally being delivered through the internet as opposed to shrink-wrapped boxed products. Through robust risk management practices and predictable and repeatable operational security hygiene programs, the company feels it can meet the objectives set forth in its core security priorities.

Source: Andrea Bonime-Blanc. The Conference Board.

6.3.c Building sustainable tech value: transforming tech fear into tech trust

> Facebook, Twitter and Google's YouTube have become the digital arms dealers of the modern age.
>
> Kara Swisher, Author, Journalist, Recode Founder[71]

> We see vividly – painfully – how technology can harm rather than help. Platforms and algorithms that promised to improve our lives can actually magnify our worst human tendencies.
>
> Tim Cook, CEO of Apple[72]

6.3.c.i Big tech and trust: Facebook, Google, Twitter et al.

Big tech has a trust problem. The mistrust is based heavily on one of their pervasive business models built on giving platforms "away for free" in exchange for scraping and using your personal data and then selling it without further permissions, often involving unscrupulous (and perhaps even worse) third parties as we have painfully learned from the Cambridge Analytica Facebook disaster. While this may be an oversimplification it's a core part of the business model of a variety of tech giants like Google, Facebook, Amazon and others.

One of the big-ticket issues facing corporations and individuals is data privacy and data use. The US and the EU have very different approaches to this topic which may converge in the future. Until now, however, tech companies in the US have been able to do pretty much what they want to do – the alternative for individuals has meant that an individual must specifically opt out and/or not use such ubiquitous and necessary services like Gmail, Google, Facebook, Instagram, etc.

The EU, on the other hand, recently deployed a much more comprehensive and universal set of protections of data privacy and usage because in the EU privacy of one's personal data is viewed as a fundamental human right. Their new law, GDPR, first implemented in 2018, puts the onus on commercial and other interests on how and why they can get and use such personal data.[73]

The US is seeing a slow but steady movement in a similar direction to the EU in some states – California, Vermont – and with all the tech trust issues the world is currently confronting, it is possible that the US Congress may move sooner than expected on enacting a federal law on data privacy, something the US has never done before. Given the current political climate in the US, this is bound to be a messy, fraught, ideologically clouded process that may yield something worse than what the US has now – nothing. Be careful what you wish for and stay tuned, as they say.

Big tech until recently was completely resistant to the idea of any form of self-regulation or, heaven forbid, regulatory oversight. The 2016 election in the US, and especially the Cambridge Analytica revelations of early 2018 vis-à-vis Facebook, put that all behind. Indeed, calls for responsibility, accountability, etc., were all but ignored especially by Facebook but also by Twitter and Google (re: YouTube) until then.

Now there is a rush to self-regulation or voluntary programming which will not be enough given the climate that has erupted around big tech and social media. Indeed, the climate is such that big tech will not be able to please all of the people all of the time or even a portion of them part of the time. As the news site, Axios, was smart to note in a piece on August 30, 2018:

the companies – led by Facebook and Google but with Twitter, Apple, and Amazon also in the mix – are caught in a partisan vise, between privacy-oriented critics on the left who fear further election interference and newer charges from the right of anti-conservative bias and censorship.[74]

Table 6.19 Recent Facebook "trust and ethics issues" in the news

Headline – Date, content and source	*Trust/Ethics issue*
March 17, 2018 How Trump consultants exploited the Facebook data of millions. – *The New York Times*	**Data privacy/user trust**
March 19, 2018 Facebook and Cambridge Analytica: what you need to know as fallout widens. – *The New York Times* **March 22, 2018** Facebook faces reputational meltdown. – *The Economist*	**National security** **Democracy** **Free speech** **Reputation risk**
July 11, 2018 Facebook hit with first fine over Cambridge analytica data scandal UK levies £500,000 penalty and accuses group of not protecting user information. – *The Financial Times*	**Regulatory oversight and fines**
July 13, 2018 Facebook won't ban infowars in its fake-news purge – a site that says that 9/11 was staged and the moon landing was fake. – *Business Insider*	**Content policy enforcement**
July 20, 2018 Facebook sets lobbying record amid Cambridge Analytica fallout. – *Bloomberg*	**Lobbying**
July 20, 2018 Facebook suspends analytics firm on concerns about sharing of public user-data: Facebook says it is investigating data-analytics firm crimson hexagon, which says its repository of one trillion public social-media posts is the largest." – *The Wall Street Journal*	**Third party supplier/ customer due diligence** **Data privacy**
July 26, 2018 Facebook investors wake up to era of slower growth: social network shocks those who rushed into the supercharged FAANGS group of tech stocks. – *The Financial Times*	**Investor and stakeholder trust**
July 28, 2018 A withering verdict: MPS report on Zuckerberg, Russia and Cambridge Analytica – select committee criticizes Facebook response and urges tighter Internet regulation. – *The Guardian*	**Customer care**

(*Continued*)

Table 6.19 Continued

Headline – Date, content and source	*Trust/Ethics issue*
August 6, 2016 Facebook to banks: "give us your data, we'll give you our users". Facebook has asked large US banks to share detailed financial information about customers as it seeks to boost user engagement. – *The Wall Street Journal*	**Stakeholder relations** **Stakeholder trust** **Partner relations**
August 16, 2018 Weaponized ad technology: Facebook's moneymaker gets a critical eye. – *The New York Times*	**User care** **Customer care** **Stakeholder trust**
August 16, 2018 Facebook is failing to control hate speech against the Rohingya in Myanmar, report finds. – *TIME*	**Human rights** **Hate speech**
August 22, 2018 Facebook removes data-security app from Apple Store: Apple last week informed Facebook that the Onavo app violated data-collection policies. – *The Wall Street Journal*	**Privacy** **Data security**
August 23, 2018 Facebook has suspended over 400 apps after Cambridge Analytica scandal. – *CNN Money*	**Third party due diligence and care**
August 28, 2018 It's not just Facebook: customer confidence in social media companies has deteriorated overall. – *Business Insider*	**Reputation risk** **stakeholder trust**
August 28, 2018 Dozens at Facebook unite to challenge its "intolerant" liberal culture. – *The New York Times*	**Values** **Code of conduct** **corporate culture**

Source: Author compiled from news sources cited.

Indeed, the Axios report characterizes each of the big tech companies with a snapshot take on what their stance might be toward possible regulation (certainly at the end of the summer of 2018) as follows:[75]

- Facebook – "We're at the table. We're willing to accept some regulation. We don't have all the answers". (See Table 6.19 for an overview of recent Facebook trust and ethics issues in the news.)
- Amazon – "We don't do elections. We're not a social network. We pay fair wages".
- Apple – "We don't sell your info. We don't have a social network. We're pro-privacy".
- Google – "Our algorithms have no politics". (See Table 6.20 for an overview of recent Google trust and ethics issues in the news.)
- Twitter – "We're listening to users and working with the authorities. We're being more transparent about political ads. And we're cracking down on fake accounts". (See Table 6.21 for an overview of recent Twitter trust and ethics issues in the news.)

Table 6.20 Recent Google "trust and ethics issues" in the news

Headline – Date, content and source	*Trust/Ethics issue*
August 8, 2017 Google has fired the engineer whose anti-diversity memo reflects a divided tech culture: James Damore's sexist screed indicted all of Silicon Valley. – *Vox*	**Diversity/ discrimination**
May 10, 2018 "Duplex shows Google failing at ethical and creative AI design" (Sundar Pichai, CEO, demos AI that can have "normal human conversation" on the phone raising a host of questions about trust, information sharing, disclosure and the like). – *Techcrunch*	**Ethical AI design**
May 21, 2018 Google eliminated that "don't be evil" motto". – *Digital Trends* (Referencing Google's founders original credo.)	**Code of conduct/values**
June 1, 2018 Google will not renew Pentagon contract that upset employees. – *The New York Times* (Referencing AI defense work involving AI drone deployed facial recognition work called "Project Maven".)	**Ethical use of AI/facial recognition**
June 14, 2018 Google releases first diversity report since the infamous anti-diversity memo: the company has the hardest time retaining black and brown employees. – *Techcrunch*	**Diversity/ discrimination**
August 13, 2018 Google continuously tracks people's locations: even when users turn off their location services, Google still collects and stores location information. – *US News* and *World Report*	**Privacy/personal data abuse**
August 16, 2018 Google chief tries to quell anger over China project. – *Financial Times* Google employees circulate statement: "We, the undersigned, are calling for a code yellow addressing ethics and transparency, asking leadership to work with employees to implement concrete transparency and oversight process".	**National Security** **Democracy** **Code of conduct/values**
August 29, 2018 14 powerful human rights groups write to Google demanding it kill plans to launch a China search engine. – *Business Insider*	**Human rights** **Democracy**
August 31, 2018 Google CEO Sundar Pichai takes bipartisan fire for declining Senate Intelligence Committee hearing. *Fortune Magazine*	**Transparency/ self-regulation/ disclosure**

Source: Author compiled from news sources cited.

Meanwhile, a big-tech company that has distinguished itself from a corporate responsibility standpoint is Microsoft. Though one could argue that their product and service offerings are not as social media-centric as the other five and therefore not as subject to the vicissitudes of everyday headlines, one could also observe that Microsoft has suffered its share of external critiques on a wide variety of topics while simultaneously making a concerted effort under Satya Nadella to achieve greater corporate social responsibility. Tim Cook of Apple has also been on a similar forefront on these issues.

Indeed, where Facebook, Amazon and Twitter have been more silent or reactive, Apple and Microsoft have demonstrated more proactivity and conscientiousness of some of the deeper tech threats to their stakeholders and have tried to stay ahead of the curve in managing expectations, responsibilities and rights.

One could very well argue that this preventative, proactive, informed approach has led them to have less reputation risk than the other tech companies and the corollary – greater stakeholder trust. One can also be cynical and say that these are different ways of making more money ultimately. But aligning the profit motive with what's good for a broad swath of stakeholders, not just shareholders, is not such a bad idea, right? Figure 6.7 which visualizes the RepRisk Index trends for six technology companies (Alphabet (Google), Amazon, Apple, Facebook, Microsoft and Twitter) seems to support the observation that companies minding their tech trust issues do a little better on reputation risk management.

Here is Apple's take on privacy and private data:

> Apple's philosophy and approach to customer data differs from many other companies on these important issues … We believe privacy is a fundamental human right and purposely design our products and services to minimize our collection of customer data … The customer is not our product, and our business model does not depend on collecting vast amounts of personally identifiable information to enrich targeted profiles marketed to advertisers … When we do collect data, we're transparent about it and work to disassociate it from the user.[76]

We already featured some of the statements of Microsoft President Brad Smith in an earlier part of this chapter on both the issue of AI ethics and the issue of facial recognition software. Microsoft's socially responsible and proactive stance on this issue stands in contrast to the quieter, more reactive approach that Amazon has taken on this issue so far. Here's what was reported in the summer of 2018:

> In a letter addressed to Amazon CEO Jeff Bezos in June, nearly 19 groups of shareholders expressed reservations over the company's decision to provide Rekognition to law enforcement in Orlando, Florida, and the Washington County (Oregon) Sheriff's Office, joining the American Civil Liberties Union, Amazon employees, academics, and more than 70 other groups in protest. And in July, after the ACLU demonstrated Rekognition's susceptibility to error, a trio of Democratic Congressmen raised concerns that the technology would "pose [a danger] to privacy and civil rights".

Only when attacked publicly did the company respond with some positive use cases rather than a social policy stand. This may change in the future but

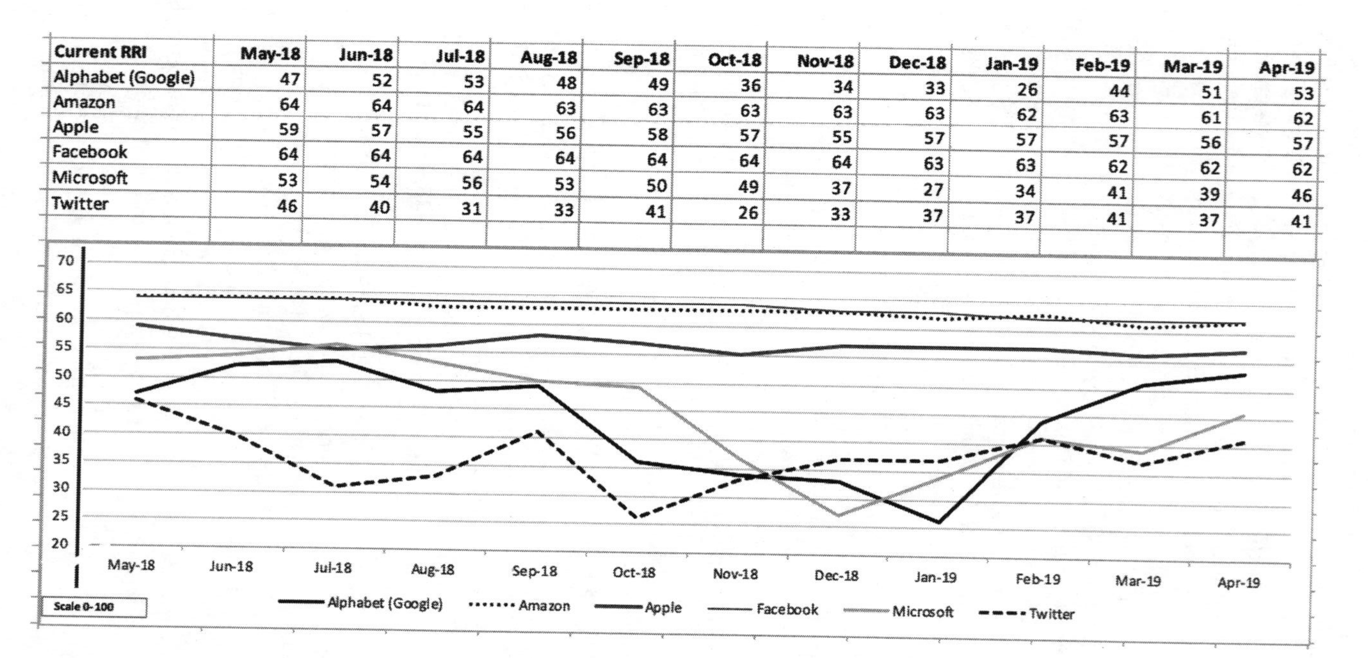

Current RRI	May-18	Jun-18	Jul-18	Aug-18	Sep-18	Oct-18	Nov-18	Dec-18	Jan-19	Feb-19	Mar-19	Apr-19
Alphabet (Google)	47	52	53	48	49	36	34	33	26	44	51	53
Amazon	64	64	64	63	63	63	63	63	62	63	61	62
Apple	59	57	55	56	58	57	55	57	57	57	56	57
Facebook	64	64	64	64	64	64	64	63	63	62	62	62
Microsoft	53	54	56	53	50	49	37	27	34	41	39	46
Twitter	46	40	31	33	41	26	33	37	37	41	37	41

Figure 6.7 RepRisk Index trends comparison for six technology companies 2018–2019.

Sourced and compiled: RepRisk.

represents a different approach to tech trust and ethics than that exhibited, for example, by Microsoft, Apple and Twitter.[77]

Meanwhile, back at Google, their CEO Sundar Pichai has come under increased pressure after a series of events most notably deriving from YouTube-related problems. Table 6.20 surveys some of their recent trust and ethics issues.

Finally, another CEO who has been visible discussing trust and ethics issues affecting his company – Jack Dorsey of Twitter – has also been making a few headlines on these issues in recent months. Table 6.21 contains some of Twitter's trust and ethics issues in the news.

Table 6.21 Recent Twitter "trust and ethics issues" in the news

Headline – Date, content and source	*Trust/Ethics issue*
January 27, 2018 The follower factory – everyone wants to be popular online. Some even pay for it. Inside social media's black market" – *The New York Times*	**Bot policing Fake account policing**
July 6, 2018 Twitter suspended more than 70 million accounts in May and June, and the pace has continued in July. – *The Washington Post*	**Fake account policing**
July 11, 2018 Battling fake accounts, Twitter to slash millions of followers. – *The New York Times*	**Fake and hate account policing**
August 8, 2018 Twitter CEO Jack Dorsey: we haven't banned Alex Jones because he hasn't violated our rules. – *Fortune Magazine*	**Rules development**
August 13, 2018 Jack Dorsey: Twitter is "thinking about values" to adopt. – *Barrons*	**Company values**
August 15, 2018 Jack Dorsey says he's rethinking the core of how Twitter works. – *The Washington Post*	**Company values and mission**
August 19, 2018 Twitter CEO Dorsey admits not being clear on company's goals amid right-wing ban controversies. – *CNBC*	**Values Humility**
August 22, 2018 It sounds like Jack Dorsey wants to massively change how you follow people on twitter. – *Business Insider*	**Strategy rethink**

Source: Author compiled from news sources cited.

Big tech has a long way to go to establish (or reestablish) trust with the public – and with their stakeholders generally. Given the burst of stories we have witnessed recently ranging from misuse and abuse of personal data by third parties to sexual harassment and discrimination and everything in between, big tech and tech generally have a lot of work to do to build back stakeholder trust. They are in luck – all they need to do is read a couple of prescriptions I offer in this chapter and our final Chapter 8 applicable to management and the board to help them become aware of and begin to take actionable steps toward restoring trust.

6.3.c.ii SDGs big data for good

We have left the best for last in this chapter. A positive and constructive example of what can be done to transform technology risk into technology value can be seen in the UN Global Pulse initiative, which is doing just that: matching big data projects with the SDGs.

Relating directly to the 17 SDGs, are examples of what the UN – together with its other public, private and nonprofit partners around the world are doing – to harness big data, data science and analytics to support and further the 17 SDGs:[78]

1. **SDG # 1 – No poverty**.
   Spending patterns on mobile phone services can provide proxy indicators of income levels.
2. **SDG #2 – Zero hunger**.
   Crowdsourcing or tracking of food prices listed online can help monitor food security in near real-time.
3. **SDG #3 – Good health and well-being.**
   Mapping the movement of mobile phone users can help predict the spread of infectious diseases.
4. **SDG #4 – Quality education.**
   Citizen reporting can reveal reasons for student drop-out rates.
5. **SDG #5 – Gender equality**.
   Analysis of financial transactions can reveal the spending patterns and different impacts of economic shocks on men and women.
6. **SDG #6 – Clean water and sanitation**.
   Sensors connected to water pumps can track access to clean water.
7. **SDG #7 – Affordable and clean energy**.
   Smart metering allows utility companies to increase or restrict the flow of electricity, gas or water to reduce waste and ensure adequate supply at peak periods.
8. **SDG #8 – Decent work and economic growth**.
   Patterns in global postal traffic can provide indicators such as economic growth, remittances, trade and GDP.
9. **SDG #9 – Industry, innovation and infrastructure**.
   Data from GPS devices can be used for traffic control and to improve public transport.
10. **SDG #10 – Reduced inequality**.
    Speech to text analytics on local radio content can reveal discrimination concerns and support policy response.

11. **SDG #11 – Sustainable cities and communities.**
    Satellite remote sensing can track encroachment on public land or spaces such as parks and forests.
12. **SDG #12 – Responsible consumption and production.**
    Online search patterns or e-commerce transactions can reveal the pace of transition to energy efficient products.
13. **SDG #13 – Climate action.**
    Combining satellite imagery, crowd-sourced witness accounts and open data can help track deforestation.
14. **SDG #14 – Life below water.**
    Maritime vessel tracking data can reveal illegal, unregulated and unreported fishing activities.
15. **SDG #15 – Life on land.**
    Social media monitoring can support disaster management with real-time information on victim location, effects and strength of forest fires or haze.
16. **SDG #16 – Peace, justice and strong institutions.**
    Sentiment analysis of social media can reveal public opinion on effective governance, public service delivery or human rights.
17. **SDG #17 – Partnerships for the goals.**
    Partnerships to enable the combining of statistics, mobile and internet data can provide a better and real-time understanding of today's hyper-connected world.

6.3.c.iii Blockchain … to the rescue?

What is Blockchain?

> You (a "node") have a file of transactions on your computer (a "ledger"). Two government accountants (let's call them "miners") have the same file on theirs (so it's "distributed"). As you make a transaction, your computer sends an email to each accountant to inform them. Each accountant rushes to be the first to check whether you can afford it (and be paid their salary "Bitcoins"). The first to check and validate hits "reply all", attaching their logic for verifying the transaction ("proof of work"). If the other accountant agrees, everyone updates their file.[79]

Blockchain and cryptocurrency (like bitcoin) are often spoken about in the same breath, as bitcoin and its crypto-brethren have been around for about a decade. The history of crypto-currencies is something above and beyond the scope of this book (and the paygrade of this author) but suffice it to say that they are deeply intertwined, as many experts and others have noted sometimes quite negatively. Witness this account from a World Economic Forum Agenda article in early 2018:

> The world has been captivated by the drama surrounding Bitcoin's meteoric rise and subsequent collapse. In December, the price of a single coin reached nearly $20,000 before plummeting to below $6,000 in

early February. Matters seemed decidedly grim when economist Nouriel Roubini slammed the "melting bitcoins" of the crashing cryptocurrency. Several days later, the general manager of the Bank for International Settlements joined in, likening the mania around Bitcoin to "a combination of a bubble, a Ponzi scheme and an environmental disaster".[80]

Author and risk authority extraordinaire, Dante Alighieri Disparte, has done excellent early work on blockchain and cryptocurrency. His views are that blockchain should not be lumped in with cryptocurrency:

> Blockchain, like the early days of the internet, is in its thousand flowers blooming phase. The market picks winners, technologists are merely the gardeners.[81]

The essence of Disparte's argument is that blockchain is both incredibly promising and at its earliest stages of development – only first achieving beta stage in 2017. He downplays the naysayers (Nouriel Roubini, for example, has blasted blockchain as just another scam), and makes the case that:

> in addition to the pilot projects being carried out by the 50 largest companies in the world (with some industries opting for "coopetition"), there is a growing cadre of blockchain-based projects gaining serious global recognition for their potential to change the fundamental nature for how economies and essential services are organized.[82]

On that positive note, Table 6.22 summarizes six non-crypto use cases for blockchain which put even further positive light on this still very early-stage but exceptionally promising technology.

Table 6.22 6 non-crypto use cases for blockchain[83]

SMART CONTRACTS	**Overview:** smart contracts are self-executing programs that automatically check the rules of the transaction, verify and process the transaction, and, in some cases, enforce the obligations of the parties. This type of automation can dramatically increase productivity and lower costs. **Use cases:** companies like Slock.it use smart contracts to automate payments for renting electric vehicle charging stations, while Fizzy tracks flight delays and automatically refunds passengers when flights take off more than two hours late.
TOKENIZATION OF CONTENT	**Overview:** tokenization allows you to take an asset and fractionalize its ownership by creating digital tokens. Each token represents a percentage of ownership in the asset, and the use of blockchain makes the chain of custody and proof of ownership immutable. In practice, you can do this with both digital and physical assets, but in the absence of laws or regulations, claiming title to physical property without proper government records tends to yield unexpected results. That said, there are quite a few tokenization projects in the works.

(*Continued*)

Table 6.22 Continued

	Use cases: tokenizing content allowed messaging app Kik to have a "reverse ICO (initial coin offering)" (in which the company decentralized itself by selling tokens, like stocks, to interested investors) to raise $100 million and continue to grow its platform. Simple Token empowers companies to easily create branded tokens without having to worry about regulatory issues or creating an ICO.
ELIMINATING COUNTERFIT PRODUCTS	**Overview:** because blockchains are permanent, immutable ledgers, it's easy to identify and trace the chain of ownership of assets. Storing serial numbers or other product identity information on a blockchain allows all parties (manufacturers, distributors, retailers, and consumers) to verify that the item in question is authentic. **Use cases:** Blockverify uses blockchain technology to boost anti-counterfeit measures by helping to identify counterfeits and prevent counterfeit duplication of products and by enabling companies to verify their products and monitor their supply chains. The world's largest diamond producer, De Beers, is working with blockchain technology to create an immutable and permanent digital record for every registered diamond – and cut down on conflict ("blood") diamonds.
SUPPLY CHAIN IMPROVEMENTS	**Overview:** by identifying production components and processes and storing that information on a blockchain, you can monitor (and optimize) your supply chain from raw materials to finished goods. **Use cases:** Walmart uses blockchain to allow its employees to scan goods (like fruit) in the store's app and track it along every step of the journey from harvest to the store floor. The world's largest shipping company, Maersk, uses blockchain to monitor its cargo ships, while British Airways uses blockchain to ensure the information it shares on its site, in its apps, and on airport displays is up-to-date and correct.
DIGITAL TWINS	**Overview:** a digital twin is a virtual representation of a physical asset. Through sensor data, artificial intelligence, and human input, digital twins mirror their real world counterparts and create value by allowing for training, maintenance, troubleshooting, simulation, and more. **Use cases:** Deloitte uses digital twins to "detect physical issues sooner, predict outcomes more accurately, and build better products," while GE uses digital twins to optimize its wind farms, leading to an increase of up to 20% in annual energy production.
ENCRYPTED MESSAGING	**Overview:** encrypted messaging has become "table stakes" for business communication. There are many traditional solutions for end-to-end encryption, but blockchain has inspired a new approach that leverages decentralization. Using blockchain, messages can be anonymous (even IP addresses can be masked). Work is generally done locally, so no private user data is transferred. And some blockchain-based encrypted messaging solutions include anonymous crypto-currency-style payment options as well. **Use cases:** there are quite a few organizations working on blockchain-powered encrypted messaging platforms, including Matrix, Crypviser, and ADAMANT.

Source: Shelly Palmer.

6.3.c.iv Building tech trust: a mini-blueprint

As Brad Smith, president of Microsoft, has put it: "We're going to have to develop these ethical principles, and we're going to have to work through the details that sometimes will be difficult." He goes on to say that the reason for this is "because the ultimate question is whether we want to live in a future of AI where only ethical people create ethical AI, or whether we want to live in a world where, at least to some degree, ethical AI is required and assured for all of us. There's only one way to do that, and that is with a new generation of laws".[84]

What follows is a mini-blueprint of useful and constructive things management (especially in tech companies) can do *today* to ensure that products and services are being developed – from their inception – in an ethical, diverse and socially responsible way.

6.3.c.iv.A DEPLOY A DIVERSITY OF TECH AND NON-TECH PEOPLE AND SKILLS

Developing new technologies should be diverse at its inception. It should be a multidisciplinary and multifunctional endeavor in almost every case involving people with technology skills but also people with more traditional skills like transport, security, health care, agricultural and food expertise, forestry, economics, governance, shipping, transportation and many more. The main point here being that in order to pull this all together a great variety of complementary skills are needed.

6.3.c.iv.B HARNESS INTERSECTING TECHNOLOGIES

While there is a thread that runs through all of these initiatives – the use, deployment, collection, analysis etc. of big data – it is also significant to see so many different other technologies intertwined with the big data piece, such as mobile telephony, satellite imagery, internet-based crowdsourcing and tracking, social media. See JT Kostman's piece about the five most important technologies that are driving the future, summarized in Table 1.10 in Chapter 1. While many of these technologies are moving in distinct silos, they are also totally interconnected and together they will completely change the face of humankind in the near future.

6.3.c.iv.C EMBRACE MULTISECTOR ENGAGEMENT

Which brings me to my third major point: to achieve the goals outlined in the SDGs, there are enormous and necessary opportunities for public, private, NGO, academic and other sector cross-fertilization and resourcing. These are opportunities for a win/win/win where:

- The public sector (be it a national, state or local government) serves the needs of its main stakeholder: getting the products or services to the population in need or affected.

- The private sector (be it a global corporation, national company or local service or product provider) serves the needs of its main stakeholders: owners who are looking to make a profit, customers who are looking to acquire the products or services, employees who are providing the products or services.
- The NGOs (be they small local charitable institutions or much larger global nonprofits devoted to specific missions – health, food, water, etc.) serve the needs of their most important stakeholders – beneficiaries - helping to deliver or provide products and services to them while serving the mission and purpose of the entity and its donors.
- The International Agencies (be it UN agencies such as the World Food Programme or the World Health Organization or another institution like the World Bank or IMF) serve the needs of their principal stakeholders, namely the ultimate beneficiaries, the donor nations that have come together to provide resources and funds and private donors.

6.3.c.iv.D SERVE A GREATER PURPOSE

Finally, a theme running throughout this book (if you haven't already noticed) is that regardless of who you are, what organization you work for or represent, we all play a part in making the world a better place. Every single aspect of what we do in our corporations, nonprofits, international or domestic government agencies – especially when it involves the development or application of new technologies – requires us to put on our global citizen hat and think about and even verbalize how that technology and that work can help the globe – even in its most microcosmic or local representation.

6.4 Endnote: a technology issue exercise and a technology issue (incomplete) inventory

As we wrap up this chapter, I will leave my readers with a couple of practical bits and pieces. First up, in Table 6.23 you will find a hypothetical technology/cyber-threat scenario exercise which you can deploy at your organization for purposes of cyber-preparedness. Nothing like starting now to prepare for a threat you know you have already experienced (although you may not know it) and experience again in the future.

Table 6.23 A technology vignette: your executive team has been cyber-breached – what now?

Hypothetical Scenario	**Fast Analysis**
• Executive management and board personnel private data housed in your company's human resources department has been hacked.	• This is a cyber-attack. • Social media and media implications. • Privacy violations. • Reputation risk implications. • Possible personal security implications.

(*Continued*)

Table 6.23 Continued

• Among these personnel are former us military officials. • Also, among your board members are very high net worth individuals with high visibility in media. • Leading newspapers and Twitter are rapidly socializing this discovery on media and social media. • You are the C-Suite – what do you do now?	**Immediate Actions** • Assemble and deploy crisis management team. • Deploy IT department and outside expert cyber-firm to analyze what happened and close cyber-security holes asap. • Start an investigation under general counsel and possible outside law firm to preserve attorney client privilege. • Disclose breach to affected parties. • Ensure all government disclosure requirements are attended to immediately.
What about the Long-Term Implications? • What do you as leaders of the company and board oversight do for the longer-term? • Were you prepared for this event? • If not, consider doing a root cause analysis.	

Source: GEC Risk Advisory.

And, for good measure and sure to be incomplete but intended nevertheless to get you thinking about your own organization(s), Table 6.24 provides an inventory of technology issues that may help you think about your organization's technology issues and to which you can freely add new ones that may come up periodically or maybe even more often than that.

Table 6.24 A technology issues (incomplete) inventory

- Artificial intelligence (AI) and the military (robot warfare)
- AI and geopolitics (cold war/hot digital war)
- AI and labor
- AI and corporate responsibility
- AI ethics
- AI, robotics and universal basic income
- Augmented reality
- Autonomous vehicles
- Bioengineering
- Biotechnology
- Conflicts of interest and technology
- Crispr (bioengineering)
- Cyber-espionage
- Cyber-governance
- Cyber-resilience
- Cyber-risk
- Cyber-security
- Cyber-war
- DNA and genetic engineering
  1. Genetic engineering of animals
  2. Genetically modified food

(*Continued*)

Table 6.24 Continued

3. Crispr challenges 4. Genome editing 5. Synthetic life forms • Data privacy and protection • Data privacy v. security 1. Geolocation/GPS 2. Internet of things 3. Wearable technologies • Digital copyright 1. Theft 2. Protection of artists and authors rights • Drones • Fake information (news, audio, video) and deep fakes • Geoengineering • Health and safety: 1. New products – ecigarettes 2. Wearables 3. Implants 4. Sex robots • Internet of things • Nanotechnology • Quantum computing • Robotics • Social media, media and the rise of the deep fakes • Technology company collaboration with military: 1. Of your own country 2. Of another country 3. Of an enemy country • Technology governance • The singularity • Virtual reality • Wearable technologies

Source: Author.

Notes

1 Zoe Henry. "Mark Zuckerberg's 10 Best Quotes Ever". October 14, 2014. *Inc Magazine*. https://www.inc.com/zoe-henry/mark-zuckerberg-move-fast-and-break-things.html. Accessed on September 3, 2018.

2 Catherine Clifford. "Hundreds of AI Experts Echo Elon Musk and Stephen Hawking in Call for a Ban on Killer Robots". November 8, 2017. CNBC.com. https://www.cnbc.com/2017/11/08/ai-experts-join-elon-musk-stephen-hawking-call-for-killer-robot-ban.html. Accessed on November 25, 2018.

3 Bernard Marr. "28 Best Quotes on Artificial Intelligence". *Forbes*. July 25, 2017. https://www.forbes.com/sites/bernardmarr/2017/07/25/28-best-quotes-about-artificial-intelligence/#48618e3a4a6f. Accessed on August 25, 2018.

4 Klaus Schwab. "How Can We Embrace the Opportunities of the Fourth Industrial Revolution?". World Economic Forum Agenda. January 15, 2016. https://www.wefo-

rum.org/agenda/2016/01/how-can-we-embrace-the-opportunities-of-the-fourth-industrial-revolution/. Accessed on November 25, 2018.
5 University of Cambridge. "Synthetic Biology and Bioengineering: Risks and Opportunities". *Science Daily*. November 20, 2017. https://www.sciencedaily.com/releases/2017/11/171121121422.htm. Accessed on November 24, 2018.
6 University of Cambridge. "Synthetic Biology and Bioengineering: Risks and Opportunities". *Science Daily*. November 20, 2017. https://www.sciencedaily.com/releases/2017/11/171121121422.htm. Accessed on November 24, 2018.
7 Christianna Reedy. "Kurzweil Claims the Singularity will Happen by 2045". Futurism.com. October 5, 2017. https://futurism.com/kurzweil-claims-that-the-singularity-will-happen-by-2045. Accessed on August 18, 2019.
8 Scott Hartley. *The Fuzzy and the Techie: Why the Liberal Arts Will Rule the Digital World.* Houghton Mifflin Harcourt 2017; and Susan Crawford. "Why Universities Need Pubic Interest Technology Courses". *Wired*. August 22, 2018. https://www.wired.com/story/universities-public-interest-technology-courses-programs. Accessed on September 1, 2018.
9 "Technology". Wikipedia. https://en.wikipedia.org/wiki/Technology#cite_note-98. Accessed on August 26, 2018.
10 *The Business Dictionary*. "Technology". http://www.businessdictionary.com/definition/technology.html. Accessed on August 25, 2018.
11 Lego Engineering. "What Is Technology?" http://www.legoengineering.com/what-is-technology/. Accessed on August 26, 2018.
12 Jake Kanter. "Academics Created a Periodic Table of Mind-Blowing Tech and It's a Handy Guide to How the World Will Change Forever". *Business Insider*. August 4, 2018. https://amp-businessinsider-com.cdn.ampproject.org/c/s/amp.businessinsider.com/imperial-college-london-table-of-disruptive-tech-will-blow-your-mind-2018-7. Accessed on August 25, 2018.
13 Cambridge Centre for Existential Risk. https://www.cser.ac.uk/. Accessed on July 29, 2018.
14 World Economic Forum. *Global Risks Report 2019*. http://www3.weforum.org/docs/WEF_Global_Risks_Report_2019.pdf. Accessed on April 12, 2019.
15 UN Sustainable Development Goals. https://sustainabledevelopment.un.org/?menu=1300. Accessed on July 29, 2018.
16 Kasey Panetta. "5 Trends Emerge in the Gartner Hype Cycle for Emerging Technologies". Gartner. August 16, 2018. https://www.gartner.com/smarterwithgartner/5-trends-emerge-in-gartner-hype-cycle-for-emerging-technologies-2018/. Accessed on August 27, 2018.
17 Plug and Play Tech Center. https://www.plugandplaytechcenter.com. Accessed on August 30, 2018.
18 Allianz. *The Allianz Risk Barometer 2019*. https://www.agcs.allianz.com/content/dam/onemarketing/agcs/agcs/reports/Allianz-Risk-Barometer-2019.pdf. Accessed on April 12, 2019.
19 RepRisk. https://www.reprisk.com/our-approach. Accessed on July 17, 2018.
20 UN Global Compact. https://www.unglobalcompact.org/. Accessed on July 29, 2018.
21 Jake Kanter. "Academics Created a Periodic Table of Mind-Blowing Tech and It's a Handy Guide to How the World Will Change Forever". *Business Insider*. August 4, 2018. https://amp-businessinsider-com.cdn.ampproject.org/c/s/amp.businessinsider.com/imperial-college-london-table-of-disruptive-tech-will-blow-your-mind-2018-7. Accessed on August 25, 2018.
22 Jake Kanter. "Academics Created a Periodic Table of Mind-Blowing Tech and It's a Handy Guide to How the World Will Change Forever". *Business Insider*. August 4, 2018. https://amp-businessinsider-com.cdn.ampproject.org/c/s/amp.businessinsider.com/imperial-college-london-table-of-disruptive-tech-will-blow-your-mind-2018-7. Accessed on August 25, 2018.

23 Jake Kanter. "Academics Created a Periodic Table of Mind-Blowing Tech and It's a Handy Guide to How the World will Change Forever". *Business Insider*. August 4, 2018. https://amp-businessinsider-com.cdn.ampproject.org/c/s/amp.businessinsider.com/imperial-college-london-table-of-disruptive-tech-will-blow-your-mind-2018-7. Accessed on August 25, 2018.
24 Cambridge Centre for Existential Risk. https://www.cser.ac.uk/. Accessed on July 29, 2018.
25 UN Global Pulse. https://www.unglobalpulse.org/about-new. Accessed on August 30, 2018.
26 Kasey Panetta. "5 Trends Emerge in the Gartner Hype Cycle for Emerging Technologies". Gartner. August 16, 2018. https://www.gartner.com/smarterwithgartner/5-trends-emerge-in-gartner-hype-cycle-for-emerging-technologies-2018/. Accessed on August 27, 2018.
27 Plug and Play Tech Center. https://www.plugandplaytechcenter.com. Accessed on August 30, 2018.
28 University of Cambridge. "Synthetic Biology and Bioengineering: Risks and Opportunities". *Science Daily*. November 20, 2017. https://www.sciencedaily.com/releases/2017/11/171121121422.htm. Accessed on November 24, 2018.
29 Allianz. *The Allianz Risk Barometer 2019*. https://www.agcs.allianz.com/content/dam/onemarketing/agcs/agcs/reports/Allianz-Risk-Barometer-2019.pdf. Accessed on April 12, 2019.
30 RepRisk . https://www.reprisk.com/our-approach. Accessed July 17, 2018.
31 World Economic Forum. Global Risks Report 2019. http://www3.weforum.org/docs/WEF_Global_Risks_Report_2019.pdf. Accessed on April 12, 2019.
32 United Nations Global Compact. https://www.unglobalcompact.org/what-is-gc/mission/principles. Accessed on August 7, 2018.
33 David Stupples. "What Is Information Warfare?" World Economic Forum. December 3, 2015. https://www.weforum.org/agenda/2015/12/what-is-information-warfare/. Accessed on September 1, 2018.
34 US Department of Justice. "United States v. Internet Research Agency et al.". February 16, 2018. https://www.justice.gov/file/1035477/download. Accessed on November 25, 2018; and US Department of Justice. "United States of America v. Netyksho et al. Indictment". July 13, 2018. https://www.justice.gov/file/1080281/download. Accessed on September 2, 2018.
35 Michael Isikoff and David Corn. *Russian Roulette*. Twelve 2018; Malcom Nance. *The Plot to Destroy Democracy*. Hachette Books 2018; Malcolm Nance. *The Plot to Hack America*. Skyhourse Publishing 2016; and Clint Watts. *Messing with the Enemy*. Harper 2018.
36 Jack Stubbs, Christopher Bing. "Exclusive: Iran-Based Political Influence Operation – Bigger, Persistent, Global". August 28, 2018. Reuters. https://www.reuters.com/article/us-usa-iran-facebook-exclusive/exclusive-iran-based-political-influence-operation-bigger-persistent-global-idUSKCN1LD2R9; Axios. Why China Hasn't Followed Russia Yet". https://www.axios.com/china-and-russia-online-disinformation-election-meddling-255d7196-41da-4f1b-9fdf-2f185218cb2c.html?utm_source=newsletter&utm_med; and Holger Roonemaa and Inga Springe. "This Is How Russian Propaganda Actually Works on the 21st Century". August 31, 2018. Buzzfeed. https://www.buzzfeednews.com/article/holgerroonemaa/russia-propaganda-baltics-baltnews. All accessed on September 2, 2018.
37 Katie Benner, Mark Mazetti, et al. "Saudi's Image Makers: A Troll Army and a Twitter Insider". *The New York Times*. October 20, 2018. https://www.nytimes.com/2018/10/20/us/politics/saudi-image-campaign-twitter.html. Accessed on October 21, 2018.
38 Amanda Robb. "Anatomy of a Fake News Scandal". *Rolling Stone Magazine*. November 16, 2017. https://www.rollingstone.com/politics/politics-news/anatomy-of-a-fake-news-scandal-125877/. Accessed on November 25, 2018.

39 Laignee Barron. "Facebook Is Failing to Control Hate Speech Against the Rohingya in Myanmar, Report Finds". August 16, 2018. http://time.com/5368709/facebook-hate-speech-myanmar-report-rohingya/. Accessed on September 1, 2018.
40 Shivam Vij. "Rumors on WhatsApp Are Leading to Deaths in India. The Messaging Service Must Act". *The Washington Post*. June 19, 2018. https://www.washingtonpost.com/news/global-opinions/wp/2018/06/19/rumors-on-whatsapp-are-leading-to-deaths-in-india-the-messaging-service-must-act/?utm_term=.50e8480e8d04. Accessed on September 1, 2018.
41 Glenn Fleishman. "Elon Musk Triples Down on Pedophilia Claims Against British Diver". *Fortune*. September 4, 2018. http://fortune.com/2018/09/04/elon-musk-pedo-tweet-pedophilia-accusations-diver/. Accessed on November 25, 2018.
42 *Oxford Dictionaries*. https://en.oxforddictionaries.com/definition/post-truth. Accessed on October 14, 2018.
43 Robert Chesney and Danielle Citron. "Deep Fakes: A Looming Challenge for Privacy, Democracy, and National Security". July 14, 2018. 107 *California Law Review* (2019, Forthcoming); U of Texas Law, Public Law Research Paper No. 692; U of Maryland Legal Studies Research Paper No. 2018-21. http://dx.doi.org/10.2139/ssrn.3213954. Accessed on September 1, 2018.
44 See Chesney and Citron. "Deep Fakes" for more examples.
45 George Dvorsky. "Deepfake Videos Are Getting Impossibly Good". *Gizmodo*. June 12, 2018. https://gizmodo.com/deepfake-videos-are-getting-impossibly-good-1826759848. Accessed on September 1, 2018.
46 Sheera Frenkel, Nicholas Confessore, Cecilia Kang, Matthew Rosenberg and Jack Nicas. "Delay, Deny, Deflect: How Facebook's Leaders Fought Through Crisis". November 14, 2018. *The New York Times*. https://www.nytimes.com/2018/11/14/technology/facebook-data-russia-election-racism.html. Accessed on November 25, 2018.
47 Emily Dreyfuss. "Facebook's Fight Against Fake News Keeps Raising Questions". *Wired*. July 20, 2018. https://www.wired.com/story/facebook-fight-against-fake-news-keeps-raising-questions/. Accessed on September 2, 2018.
48 Betsy Morris, Deepa Seetharaman and Robert McMillan. "Sheryl Sandberg's New Job Is to Fix Facebook's Reputation and Her Own". *The Wall Street Journal*. September 4, 2018. https://www.wsj.com/articles/sheryl-sandberg-leans-into-a-gale-of-bad-news-at-facebook-1536085230. September 9, 2018.
49 Nicholas Confessore and Gabriel J. X. Dance. "Battling Fake Accounts, Twitter to Slash Millions of Followers". *The New York Times*. July 11, 2018. https://www.nytimes.com/2018/07/11/technology/twitter-fake-followers.html?emc=edit_na_20180711&nl=breaking-news&nlid=48081158ing-news&ref=headline. Accessed on September 1, 2018.
50 Deborah Bach. "5 Takeaways from Brad Smith's Speech at the RISE Conference". *Microsoft News*. August 13, 2018. https://news.microsoft.com/on-the-issues/2018/08/13/5-takeaways-from-brad-smiths-speech-at-the-rise-conference/. Accessed on August 27, 2018.
51 Brad Smith. "Facial Recognition Technology: The Need for Public Regulation and Corporate Responsibility". Microsoft Blog. July 13, 2018. https://blogs.microsoft.com/on-the-issues/2018/07/13/facial-recognition-technology-the-need-for-public-regulation-and-corporate-responsibility/. Accessed on September 1, 2018.
52 Natasha Singer. "Microsoft Urges Congress to Regulate the Use of Facial Recognition". *The New York Times*. July 13, 2018. https://www.nytimes.com/2018/07/13/technology/microsoft-facial-recognition.html. Accessed on September 1, 2018.
52 NewsGuard. https://www.newsguardtech.com/. Accessed on September 1, 2018.
54 Sara Fischer. "NewsGuard Launches First News Product with Help from Microsoft". Axios. August 23, 2018. https://www.axios.com/newsguard-launches-first-product-2143fc9e-470f-44b6-b8f1-6006646d26db.html. Accessed on September 1, 2018.

55 Sara Fischer. "How Post-Trust Initiatives Are Taking Over the Internet". Axios. October 14, 2018. https://www.axios.com/fake-news-initiatives-fact-checking-dfa6ab56-3295-4f1a-9b38-e61ca47e849f.html?utm_source=newsletter&utm_medium=email&utm_campaign=newsletter_axiosam&stream=top. Accessed on October 14, 2018.
56 Institute of Risk Management. https://www.theirm.org/knowledge-and-resources/thought-leadership/cyber-risk/. Accessed on August 30, 2018.
57 Cisco. What Is Cyber-Security? https://www.cisco.com/c/en/us/products/security/what-is-cybersecurity.html. Accessed on August 30, 2018.
58 Chris Moschovitis. Cybersecurity Program Development for Business: The Essential Planning Guide. Wiley, 2018.
59 Andrea Bonime-Blanc. "Emerging Practices in Cyber-Risk Governance". The Conference Board. October 2015. https://gecrisk.com/download-pdf-cyber-risk-governance-full-report/. Accessed on August 28, 2018.
60 Andrea Bonime-Blanc. "Deploying a Voluntary Cyber-Resilience Program: A Strategic Imperative". Chapter 15 in The Internet Security Alliance, The Cybersecurity Social Contract. 2016. https://gecrisk.com/wp-content/uploads/2016/09/ABonimeBlanc-Cyber-Resilience-Chapter-15-in-ISA-CyberSecurity-Social-Contract-2016-.pdf. Accessed on August 30, 2018.
61 Gideon Rose. "World War Web Issue Introduction". Foreign Affairs. September/October 2018. https://www.foreignaffairs.com/articles/world-war-web. Accessed on August 31, 2018.
62 Cisco. What is Cyber-Security? https://www.cisco.com/c/en/us/products/security/what-is-cybersecurity.html. Accessed on August 30, 2018.
63 Andy Greenberg. "How an Entire Nation Became Russia's Test lab for Cyberwar". *Wired*. June 20, 2017. https://www.wired.com/story/russian-hackers-attack-ukraine/. Accessed on August 28, 2018.
64 ISACA. https://www.isaca.org/about-isaca/Pages/default.aspx. Accessed on August 28, 2018.
65 ISA. https://isalliance.org/. Accessed on August 28, 2018.
66 Andrea Bonime-Blanc. "A Strategic Cyber-Roadmap for the Board". The Conference Board. November 2016. https://gecrisk.com/wp-content/uploads/2016/11/ABonimeBlanc-Strategic-Cyber-Roadmap-for-the-Board-November-17-2016.pdf. Accessed on August 28, 2018.
67 Andrea Bonime-Blanc. "Emerging Practices in Cyber-Risk Governance". The Conference Board. October 2015. https://gecrisk.com/download-pdf-cyber-risk-governance-full-report/. Accessed on August 28, 2018.
68 Andrea Bonime-Blanc. *The Reputation Risk Handbook*. Greenleaf 2014.
69 Adapted from PwC table used in Andrea Bonime-Blanc. "Emerging Practices in Cyber-Risk Governance". The Conference Board. October 2015. https://gecrisk.com/download-pdf-cyber-risk-governance-full-report/. Accessed on August 28, 2018.
70 Andrea Bonime-Blanc. "Emerging Practices in Cyber-Risk Governance". The Conference Board. October 2015. https://gecrisk.com/download-pdf-cyber-risk-governance-full-report/. Accessed on August 28, 2018.
71 Kara Swisher. Mark Zuckerberg wanted Facebook to change the world. And it has – but not for the better. Think. August 5, 2018. https://www.nbcnews.com/think/amp/ncna897701. Accessed on August 6, 2018.
72 Ina Fried. "Tim Cook Says Tech's Dark Side Is Real". Axios. October 24, 2018. https://www.axios.com/newsletters/axios-login-cb71a615-7d62-47de-a1f9-cd230d91ddce.html?utm_source=newsletter&utm_medium=email&utm_campaign=newsletter_axioslogin&stream=top. Accessed on October 24, 2018.
73 EU General Data Protection Regulation. May 25, 2018. https://ec.europa.eu/commission/priorities/justice-and-fundamental-rights/data-protection/2018-reform-eu-data-protection-rules_en. Accessed on September 3, 2018.

74 Ina Fried. "1 Big Thing: Tech's Make or Break Two Months". Axios. August 31, 2018. https://www.axios.com/newsletters/axios-login-934fca91-08d3-48fc-b95d-592e512ce95c.html. Accessed on August 31, 2018.

75 Ina Fried. "1 Big Thing: Tech's Make or Break Two Months". Axios. August 31, 2018. https://www.axios.com/newsletters/axios-login-934fca91-08d3-48fc-b95d-592e512ce95c.html. Accessed on August 31, 2018.

76 Devin Coldewey. "Apple's Response to Congressional Privacy Inquiry is Mercifully Free of Horrifying Revelations". TechCrunch. August 7, 2018. https://techcrunch.com/2018/08/07/apples-response-to-congressional-privacy-inquiry-is-mercifully-free-of-horrifying-revelations/. Accessed on September 3, 2018.

77 "Amazon Counters Recognition Facial ID Backlash with Positive Use Cases". Venturebeat. August 9, 2018. https://venturebeat.com/2018/08/09/amazon-counters-rekognition-facial-id-backlash-by-citing-positive-use-cases/. Accessed on September 2, 2018.

78 UN Global Pulse. https://www.unglobalpulse.org/about-new. Accessed on August 30, 2018.

79 Margaret Leigh Sinrod. "Still Don't Understand Blockchain? This Explainer Will Help". March 28, 2018. World Economic Forum Agenda. https://www.weforum.org/agenda/2018/03/blockchain-bitcoin-explainer-shiller-roubini/. Accessed on November 24, 2018.

80 Margaret Leigh Sinrod. "Still Don't Understand Blockchain? This Explainer Will Help". March 28, 2018. World Economic Forum Agenda. https://www.weforum.org/agenda/2018/03/blockchain-bitcoin-explainer-shiller-roubini/. Accessed on November 24, 2018.

81 Dante A. Disparte. "To Blockchain or Not to Blockchain?". Forbes. November 14, 2018. https://www.forbes.com/sites/dantedisparte/2018/11/12/to-blockchain-or-not-to-blockchain/#32681b7e73cb. Accessed on November 24, 2018.

82 Dante A. Disparte. "To Blockchain or Not to Blockchain?". Forbes. November 14, 2018. https://www.forbes.com/sites/dantedisparte/2018/11/12/to-blockchain-or-not-to-blockchain/#32681b7e73cb. Accessed on November 24, 2018.

83 Shelly Palmer. "Blockchain: 6 Great Non-Cryptocoin Use Cases". Shelly Palmer. July 22, 2018. https://www.shellypalmer.com/2018/07/blockchain-6-great-non-crypto-coin-use-cases/?utm_source=Daily+Email&utm_campaign=4c5a26ba89-EMAIL_CAMPAIGN. Accessed on August 31, 2018.

84 Deborah Bach. "5 Takeaways from Brad Smith's Speech at the RISE Conference". Microsoft News. August 13, 2018. https://news.microsoft.com/on-the-issues/2018/08/13/5-takeaways-from-brad-smiths-speech-at-the-rise-conference/. Accessed on August 27, 2018.

Part III

Boom

Achieving sustainable resilience and value

7 Metamorphosis

Achieving organizational resilience

Organizational resilience: risk and opportunity

Franz Kafka, the great 19th-century author of dark and disturbing stories and novels, wrote in one his most famous pieces "The Metamorphosis": "I only fear danger where I want to fear it". In an earlier, oft-quoted saying that most of us know, and which illustrates that kind of hope that follows from difficulty or even tragedy, or that positivity that flows from adversity, first attributed to the 17th-century travel writer Thomas Fuller, is another powerful encapsulation of the human condition: "It is always darkest before the day dawneth". Or, as Crosby, Stills & Nash sang around 1969: "But you know the darkest hour is always, always just before the dawn".[1]

In keeping with one of the key themes of this book, let us transport ourselves to a place of resilience by equipping, preparing and metamorphizing our organizations to reach that better place of value creation, opportunity building and positivity that resilient entities have reached. While he has been quoted controversially by some, the quintessence of this theme is captured in 19th-century German philosopher Friedrich Nietzsche's famous quote "That which does not kill us, makes us stronger", which he also embellished with "To live is to suffer, to survive is to find some meaning in the suffering". Or, as the revered 19th-century US President, Abraham Lincoln, put it powerfully and succinctly: "I will prepare and someday my chance will come".[2]

Chapter 7: Metamorphosis – summary overview

7.1 Chapter overview

In this chapter we address the issue of organizational resilience. Why? Because resilience is a core aspect of survival and hence a key element of sustainability and value creation, which we address more directly in the final chapter – Chapter 8 "Boom". But why is organizational resilience part of a discussion of transforming ESGT risk into value? Because when an organization does not have the internal equipment to incorporate, deploy and monetize ESGT issues, risks and opportunities, that organization cannot exhibit holistic resilience. Without resilience, there is no long-term sustainability, and without either there is little new value creation or there is chaotic, fragile or unstructured value creation that can lead to value erosion and even destruction. We address those value issues more specifically in Chapter 8.

We have called this chapter "Metamorphosis" both tongue-in-cheek and seriously, because the core belief here is that organizations must undergo sufficient change to be able to continuously recognize, understand, incorporate and deploy their entire suite of ESGT issues. How do they do that? By having the internal organizational fortitude and resilience that we outline in this chapter. Many organizations don't have the equipment to achieve resilience or their equipment is faulty, or they are missing major components of that equipment, etc.

This chapter tries to remedy that by presenting a holistic picture of the eight key components of organizational resilience – what does a company, a government agency, a nonprofit or a university, any kind of organization – need to be a resilient, a well-oiled machine, with all the sufficient parts in place, well lubricated and coordinated to tackle their core mission, vision, values and strategy seamlessly and with their ESGT issues fully incorporated into that activity?

In this chapter we cover the following:

- We take a look at the *eight elements of organizational resilience* – eight critical components that should exist within any type of organization to help leaders build long-term, sustainable, organizational resilience for the benefit of their most important stakeholders.
- Drawing on those eight elements, we present *four types of resilience lifecycle* – from the stellar and strong (the Virtuous and Robust Resilience Lifecycles) to the precarious (the Fragile Resilience Lifecycle) and downright broken (the Vicious Resilience Lifecycle).
- We conclude the chapter with a diagnostic tool for leaders to gauge where their organizations might be on the spectrum of organizational resilience.

7.2 What is "resilience" and "organizational resilience"?

7.2.a Resilience definitions and approaches

"Resilience" is a powerful word that can apply to a variety of different levels and contexts – from the individual, psychological and biological to the organizational and geographic. For example, while it no longer exists in its original form, the nonprofit 100 Resilient Cities was created to support and develop climatological and other risk-resilience at leading urban centers around the world. Their work was focused as follows:

> Cities in the 100RC network are provided with the resources necessary to develop a roadmap to resilience along four main pathways:
>
> 1. Financial and logistical guidance for establishing an innovative new position in city government, a Chief Resilience Officer, who will lead the city's resilience efforts.

2. Expert support for development of a robust Resilience Strategy.
3. Access to solutions, service providers, and partners from the private, public and NGO sectors who can help them develop and implement their Resilience Strategies.
4. Membership of a global network of member cities who can learn from and help each other.[3]

They define "urban resilience" as follows:

> Urban resilience is the capacity of individuals, communities, institutions, businesses, and systems within a city to survive, adapt, and grow no matter what kinds of chronic stresses and acute shocks they experience.[4]

Another excellent resilience definition comes from the Stockholm Resilience Centre:[5]

> Resilience is the capacity of a system, be it an individual, a forest, a city or an economy, to deal with change and continue to develop. It is about how humans and nature can use shocks and disturbances, like a financial crisis or climate change, to spur renewal and innovative thinking.

Getting down to the individual, more personal level, *Psychology Today* defines personal resilience as follows:

> Resilience is that ineffable quality that allows some people to be knocked down by life and come back stronger than ever. Rather than letting failure overcome them and drain their resolve, they find a way to rise from the ashes. Psychologists have identified some of the factors that make someone resilient, among them a positive attitude, optimism, the ability to regulate emotions, and the ability to see failure as a form of helpful feedback. Even after misfortune, resilient people are blessed with such an outlook that they are able to change course and soldier on.[6]

Wikipedia has an entry for "Psychological Resilience" as follows:

> Psychological resilience is the ability to successfully cope with a crisis and to return to pre-crisis status quickly. Resilience exists when the person uses "mental processes and behaviors in promoting personal assets and protecting an individual from the potential negative effects of stressors." In simpler terms, psychological resilience exists in people who develop psychological and behavioral capabilities that allow them to remain calm during crises/chaos and to move on from the incident without long-term negative consequences. Psychological resilience is an evolutionary advantage that most people have and use to manage normal stressors.[7]

Getting a little closer to the concepts we are covering in this book, let's turn to the concept of resilience in organizations. Here is a definition/description from a leading study on the subject:

> Organizational Resilience is the ability of an organization to anticipate, prepare for, respond and adapt to incremental change and sudden disruptions in order to survive and prosper.[8]

The World Economic Forum (WEF), as always, has done good work in this space, and the following are excerpts of the work of Roland Kupers for WEF that provide further insight into the subject of "Resilience in complex organizations", using a resilience lens for purposes of assessing risk management. They developed nine lenses based on an effort spearheaded by Exxon called the "Resilience Action Initiative" (RAI):

> which … resulted in a set of resilience tools and approaches informed by complexity theory but grounded in practice. One critical application is enterprise resilience: the capacity of a company or other organization to adapt and prosper in the face of high-impact, low-probability risks.
>
> The RAI work led to nine resilience lenses, grouped into the following three categories to provide the agenda for a fat-tail risk conversation:
>
> - "Structural resilience" considers the systemic dynamics within the organization itself.
> - "Integrative resilience" underlines complex interconnections with the external context.
> - "Transformative resilience" responds to the fact that mitigating some risks requires transformation.[9]

On a more direct and granular level, another source for the meaning of "resilience" comes from the association of disaster recovery professionals – the Disaster Recovery Institute (DRI) International. In their 2018 Fourth Annual International Risk and Resilience Trends Report, they found the following to be among the most important trends in resilience:[10]

- A significant concern is that many C-Level executives have no direct experience of managing a major disruptive event.
- Across all regions and sectors, it is a concern that over 30 percent of professionals believe that senior management doesn't understand their resilience role.
- Professionals must have a common base of skills to manage IT disruptions and physical interruptions, such as natural or man-made disaster.
- Resilience professionals often have the best overview of how a business functions as well as the risks posed by dependencies and disruptions.

To conclude this section, I would like to offer my own definition from the "Resilience" chapter in *The Sage Encyclopedia of Corporate Reputation*:

> Organizational resilience is the ability of an organization to provide and maintain an acceptable level of operation, service, and performance in the face of challenging conditions, disruptions, risks, and crises and to bounce back and recover quickly from them with minimal impact to the organization including to its reputation.

In that same chapter I offer a counterpart definition for the "brittle organization" as follows:

> In contrast, brittle organizations view the absence of failure as an indication that hazards are not present or that the countermeasures designed are effective. As a result, they can be overwhelmed by discrete shocks, disruptions, minor interruptions, or deviations from standard operating procedures that resilient organizations are able to absorb and handle.[11]

With the benefit of the various individual and organizational resilience definitions set forth here, let's now turn our focus to what the internal dynamics of a resilient organization look like. Based on three decades of dealing with this subject in one form or another, I offer my take on what makes for organizational resilience with eight elements of the resilient organization and four types of resilience lifecycle that an organization might exhibit for better or for worse depending on how present and well embedded the eight elements of resilience are. Our focus, as always, is on how leaders build and maintain organizational resilience, not only to survive but also to thrive – for the purpose of long-term value preservation, creation and sustainability for the benefit of all key stakeholders.

7.3 The eight elements of organizational resilience

A question readers may be asking themselves just about now is the following: why is the author suddenly talking about organizational resilience when most of this book has been focused on ESGT issues? Two answers:

- First, while the core focus of this book until now has been on understanding and teasing out each of the major issues, risks and opportunities relating to the four main categories under ESGT, that was never the stand-alone purpose of this book. Indeed, the whole point of reviewing those topics in depth was to get leaders of all types of organization to focus on those often-forgotten topics that they might not consider to be core to their mission and strategy but are indeed essential to their organization's resilience, sustainability and ability to protect and create value for stakeholders.
- Second, resilience at its core is about surviving and thriving. By examining the eight elements of organizational resilience, we will gain a deeper

understanding of the types of organizations that exist based on concepts of resilience and resilience lifecycle. We will also gain a lens into how organizations can develop a roadmap to move to a higher, more resilient, sustainable and valuable form that benefits the greatest swath of relevant stakeholders (as we explore in our next and final Chapter "Boom").

Indeed, readers should think about an organization as a machine or an old-fashioned watch full of gears that should be synchronizing at all times to be able to provide the right time. The resilience questions are these:

- Are all the necessary and desirable internal components of this organization – as driven and supported by its leadership – in place and working in synchronicity?
- Are the components maintained regularly and properly, and do they get the "lubrication" they need from leaders and their culture to function smoothly and efficiently?

Or is the contrary true?:

- Are things missing, or working out of sync or even against each other, in unsupportive and unintegrated ways that lead to abrasive and destructive tendencies that might harm the very viability and sustainability of the organization and thereby the value proposition for stakeholders?

Organizations are more or less well managed, more or less resilient and more or less capable of understanding, properly managing and even extracting value from their ESGT issues. Understanding the big picture of organizational resiliency that is needed to have a robust and effective management of ESGT is absolutely critical for leaders of any type of organization – whether a business, an NGO, a university or a government agency.

The eight infrastructure elements are the following:

1. Governance
2. Culture
3. Stakeholders
4. Risk
5. Strategy
6. Performance
7. Crisis
8. Improvement

Figure 7.1 is a generic representation of the resilience lifecycle that these eight elements represent. As we will see in our discussion, each of these elements are not theoretical or conceptual topics; they are hard-core components, programs and practices that should be taking place in any type of organization fueled and driven by that organization's leadership in response to that organization's stakeholders, their expectations and requirements.

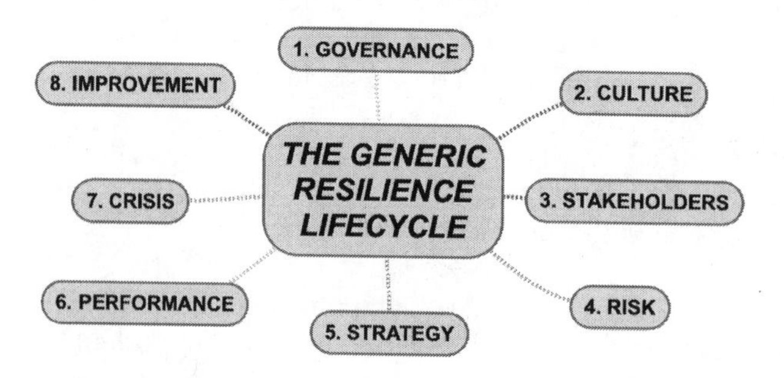

Figure 7.1 The Generic Resilience Lifecycle.

Source: Author and GEC Risk Advisory.

An important principle to keep in mind as you read this chapter and the best practices in creating organizational resilience and achieving what we call the "Virtuous Resilience Lifecycle" is that the eight elements are organized in descending order of criticality – not importance, because all of them are important. But as you read through these elements, keep in mind that the effectivity of later elements depends almost entirely on the existence and effectivity of the earlier ones. So, for example, without proactive, lean-in governance, it is questionable whether any of the following seven elements will be as strong and effective as they could or should be.

This is why I often say that if there is only one thing that I could understand about an organization to judge it, it would be the first element – governance – closely followed by culture (the second element) because the quality and character of those two elements will say volumes about the rest of the elements and the organization as a whole in terms of resilience, sustainability and value creation (or destruction).

Another observation: without the thorough integration of ESGT issues, risks and opportunities into an organization's strategy, how can an analysis of stakeholders be complete? And how can risk management succeed without a thorough understanding of those ESGT issues and stakeholders since so many risks – and opportunities – come out of a comprehensive and informed knowledge of ESGT issues that an organization faces?

Looking at the "Generic Resilience Lifecycle" in Figure 7.1, the essence of what we are discussing in the following pages is that ultimately, the most resilient, best managed, well-oiled organizational machines will have these eight elements in place. Not only that, but the elements will also display a certain kind of edge in a positive and constructive sense. Figure 7.2 shows the best possible organizational resilience lifecycle – the Virtuous Resilience Lifecycle – consisting of variations on the theme of the eight elements of organizational resilience:

1. Lean-In Governance.
2. A Culture of Empowered Integrity.
3. Stakeholder Emotional Intelligence or EQ.
4. Risk Intelligence.
5. Strategic ESGT.
6. Performance Equilibrium.
7. Crisis Readiness.
8. Innovation Ethos.

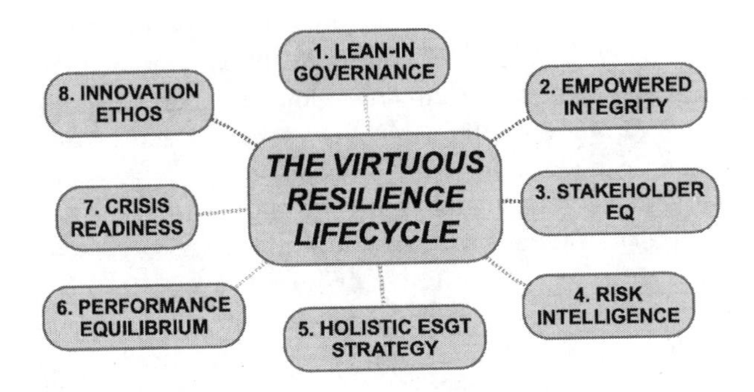

Figure 7.2 The Virtuous Resilience Lifecycle.

Source: Author and GEC Risk Advisory.

In each of the following eight sections that discuss the eight elements of organizational resilience, we focus on what the gold standard should look like in each of these areas. Each of the eight sections will be headlined by what we consider to be the best practices illustrated in Figure 7.2. We will also break down what that means in each instance, knowing full well, however, that every organization is different and clearly the kind of organizational resilience that might be needed at a ministry is different from that of a small local NGO, etc. Most of the focus of this book and of this chapter has been on the business sector, but I'd like my readers to always think about how such a gold standard might apply to their specific organization – regardless of sector, purpose or size. How could or should the resilience lifecycle be customized and tailored to be effective, even "virtuous", for your specific organization and your specific key stakeholders?

Having worked in one way or another for 30 years in the corporate, nonprofit and government agency world (as a corporate executive, a nonprofit board member and as a consultant to all kinds of organizations), I can say that these concepts apply to every type of organization. The key to success, however, is for savvy leaders to adapt, design and apply the right nuances of each of these elements to their particular organization. Customize, customize, customize.

7.3.a Governance: deploying Lean-In Governance

The following discussion of governance is focused broadly on the word "governance" as it relates to leaders at the highest levels of organizational oversight. That would typically include the board of directors or trustees or even another form of oversight body, for example, a congressional subcommittee that might have oversight responsibilities for a government agency, etc.

Regardless of the type of entity, "Lean-In Governance" means two distinct things:

- First, that there is a board or other oversight body (in the case of non-business entities) that is properly populated by highly qualified, experienced and diverse board members who bring directly relevant experience and knowledge to the organization, take their jobs very seriously and proactively and bring their best game to the oversight table.
- Second, "Lean-In Governance" means having the right attitude, taking full responsibility for duties, roles and responsibilities vis-à-vis executive management, including oversight of the CEO or president or executive director over whom the board exercises oversight. We are talking about lean-in governance, not "sit-back", "friends and family" style governance. It is about diversity of views, experiences and, yes, background, histories and personal characteristics of the board members. It's about the board reflecting the makeup of the shareholders, stakeholders and beneficiaries they serve. It's about exercising proactive oversight over strategy, risk and opportunity where ESGT issues have as big a seat at the proverbial strategy table as other considerations that are core to the mission of the organization. Yes, this is about integrating ESGT into strategy by exercising proactive, lean-in governance.

7.3.a.i Lean-In Governance: core practices

Employing Lean-In Governance means taking various actions which may be more or less palatable, such as:

- Exercising proactive oversight over every one of the other seven elements of organizational resilience, as follows:
  - Exercising proactive oversight of organizational culture and making sure it is one of high integrity.
  - Knowing who the organization's principal stakeholders are and prioritizing them.
  - Exercising strategic risk governance – understanding the most important (likely and/or high-impact) strategic risk profile of the organization.
  - Making sure organizational strategy is holistic – including relevant ESGT issues and strategic risk oversight.

  - Ensuring executive performance incentive alignment and equilibrium, especially at the CEO and executive team level by tying their performance to ESGT metrics in addition to the typical financial and operational metrics.
  - Overseeing, understanding and periodically participating in crisis management planning and exercises.
  - Ensuring the organization is proactively learning lessons from failures and mistakes and is transforming its lessons learned and knowledge into continuous improvement and value creation.
- Exercising proactive governance self-management which may mean:
  - Regular board self-evaluations.
  - Having independent board members and/or a lead independent board member or chair.
  - Removing board members who do not contribute to the diversity or talent needed.
  - Removing board members who aren't qualified because they are "friends and family" or simply rubber stamps for the all-powerful CEO or president.
  - Periodically bringing in new board members with diverse experiences, talents and backgrounds relevant to the organization.
  - Questioning and even removing the CEO when and if necessary.
  - Having board members who understand ESGT on the board.
  - Bringing in outside experts to educate and update the board on ESGT periodically.
- Questioning everything management does – in a respectful way, of course.

The best governance happens when the board itself has all the right people, objectives, tools and diversity needed to oversee the mission, vision, values and strategy of the organization in a proactive, lean-in kind of a way.

The worst kind of governance of course is the absence of governance, though there are two runners-up:

- "Absentee landlord" or "Potemkin Village" governance – where luminaries are brought in by a founder to look the part simply to rubber-stamp everything (Theranos, Massey Mining and the Weinstein Company boards come to mind); and
- The "dysfunctional" form of governance where an overly powerful CEO (usually founder as well) is also chair of the board and has packed the board with friends and family who are either well paid or controlled or both and look the other way while the CEO/founder pretty much does whatever he/she wants to do – at least for a while. Tesla, Wynn Resorts and Uber (prior to founder Kalanick stepping down from being Chairman and CEO) come to mind. For a variety of dysfunctional director types, see this excellent piece in *Directorship Magazine* from the summer 2018 where a variety of prototypes of dysfunctional board member are profiled.[12]

The point here is to underscore what it means to have good, lean-in governance, and a visual I created a few years ago for an *Ethical Boardroom* article on governance is replicated in Figure 7.3. In essence, the "Lean-In Triangular Governance" consists of the following three maxims about lean-in governance:

1. The board exercises lean-in proactive oversight over the CEO, the strategy and ESGT.
2. The executive team/C-Suite/CEO develop a holistic strategy that includes ESGT.
3. The executive team and their functional and operational teams exercise strategic and tactical collaboration, integration and implementation of the holistic strategy.

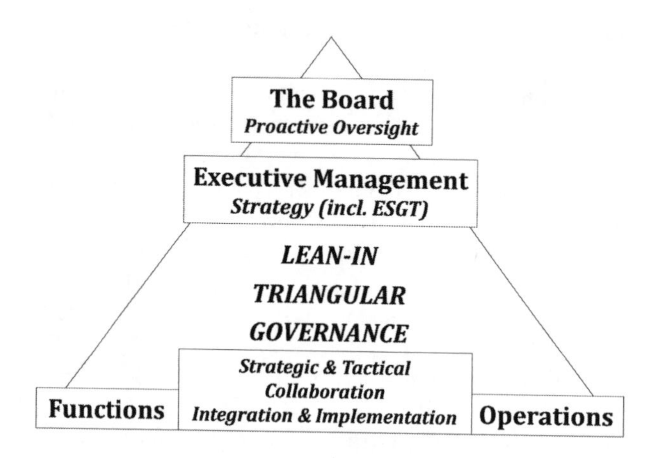

Figure 7.3 Lean-In Triangular Governance.

Source: Author and GEC Risk Advisory.

7.3.a.ii *The Resilient Leadership Manifesto: The 10 Commandments of Resilient Leaders*

In Table 7.1 I have assembled a somewhat humorously titled list of the "10 Commandments of Resilient Leadership" which I believe summarizes some of the top behaviors and qualities of resilient leaders of resilient organizations, applicable to both management and the board.

Table 7.1 The Resilient Leadership Manifesto: the 10 Commandments of Resilient Leaders

1. Resilient Leaders integrate ESGT into their organization's strategy.
2. Resilient Leaders empower and support a cross-functional team of ESGT experts to run an ESGT Value Chain approach to issue, risk, opportunity and value creation.
3. Resilient Leaders "walk the talk" supporting ESGT not only verbally but substantively – through proper budget and resources.

(*Continued*)

Table 7.1 Continued

4. Resilient Leaders aspire to become at least "Responsible Leaders" and at best "Enlightened Leaders". (See Chapters 2 and 8 for more on this leadership typology.)
5. Resilient Leaders focus on building and institutionalizing a culture of integrity.
6. Resilient Leaders remain attuned to and savvy about global megatrends that affect their organizations.
7. Resilient Leaders understand and cater to their most important stakeholders – not one most important one but the top 3–5.
8. Resilient Leaders allow for mistakes to be made and for a speak-up/listen-up culture to grow in furtherance of continuous improvement, innovation and sustainability.
9. Resilient Leaders look at the long-term horizon as well as mind the store in the short term.
10. Resilient Leaders welcome ESGT oversight – whether from a board of directors, governors, trustees or other governmental entity.

Source: Author.

7.3.b Culture: embracing Empowered Integrity

What do we mean by "Empowered Integrity"? Generally, it denotes an organizational culture that is productive, focused, diverse, transparent, collaborative – all the things you would want to enable successful implementation of an organizational strategy.

Empowered Integrity also means that there is a solid infrastructure within an organization to handle issues, concerns, complaints, allegations, etc., what some call a "speak-up/listen-up" culture where all employees and relevant third parties, regardless of rank, function, location, status, seniority, are able to voice these issues without fear of retribution or retaliation. But it does not end at the "speak-up" part of the equation – there is also a "listen-up" feature and that means that leaders – at every function, level and location – are also taking in the concerns, problems, and ideas, not just pretending to hear them. This requires a tone from the top that is unfortunately not always there.

Empowered Integrity would also include a predictable and professional organizational approach or system, that is well-advertised and transparent to examine, analyze, investigate and/or resolve these concerns, allegations and problems, with the concomitant resolution yielding, when necessary, some form of discipline that is applied evenly to employees involved regardless of rank, and seniority, all the way up to the CEO and board members. In some circles this is sometimes referred to as "organizational justice" or "procedural justice".

Besides creating this inclusive and empowered culture of speak-up/listen-up, "Empowered Integrity" means quite a bit more. It means achieving not only the letter but the spirit of such a culture through a solid and clear leadership tone from the top – from both the CEO and the board – and their sponsorship of an internal ethics and compliance or corporate integrity program and ethos, as well as an externally facing corporate responsibility program (that isn't just marketing) that lubricates the system with that speak-up/listen-up culture and much more. It all begins and ends with leadership tone from the top – especially the CEO (or equivalent).

The Ethics and Compliance Initiative (ECI) put together a Blue-Ribbon Panel in 2016 to develop the "Principles of a High-Quality Ethics and Compliance Program" made up of five core principles (HQP Principles). I had the privilege of being a founding faculty member at the ECI where I helped to develop these principles and taught them to many organizations – from the corporate world to NGOs and governmental agencies. These principles, when properly implemented, are truly on the cutting edge of what it means to have Empowered Integrity.

Table 7.2 summarizes the five HQP Principles:

Table 7.2 Ethics and Compliance Initiative HQP standards

Principle 1	Ethics and compliance are central to business strategy.
Principle 2	Ethics and compliance risks are identified, owned, managed and mitigated.
Principle 3	Leaders at all levels across the organization build and sustain a culture of integrity.
Principle 4	The organization encourages, protects, and values the reporting of concerns and suspected wrongdoing.
Principle 5	The organization takes action and holds itself accountable when wrongdoing occurs.

Source: Ethics and Compliance Initiative.

Table 2.14 in Chapter 2 is a comparison of the standard best practices that we have all grown accustomed to in the ethics and compliance field over the past 30 years (the 8–9 best practices originally derived from the Defense Industry Initiative of 1985, US Federal Sentencing Commission and the OECD Best Practices as discussed in detail in Chapter 2) and the HQP Principles.

The difference between the HQP Principles presented in Table 7.2 and the practices outlined in Table 2.14 in Chapter 2 are fairly strong, with the most important underlying theme being that ethics and compliance cannot be a separate department or activity undertaken within an organization – it needs to be part and parcel and fully integrated with the central strategy of the organization and deeply intertwined with and inextricably part of leadership and culture.

7.3.b.i The World Food Programme: integrating ethics into strategy

A leading example of an organization that has decided that their ethics program needs to be embedded and become an integrated part of its long-term business strategy comes from an unexpected place – the World Food Programme (an agency of the United Nations), which in 2016 undertook a unique initiative (certainly within the UN but more broadly, as well) to integrate their already existing and well-developed ethics program into their just-adopted five-year business strategy inspired entirely by the SDGs. Table 7.3 shows the main strategic initiatives developed by the ethics program to integrate with this long-term business strategy.

Table 7.3 World Food Programme Ethics Program: Strategic Integration Initiative

UN SDG	*WFP Strategy objectives*	*WFP Ethics strategy*
#2 Support countries to achieve zero hunger. **#17 Partner to support implementation of SDGs.**	1. End hunger by protecting access to food. 2. Improve nutrition. 3. Achieve food security. 4. Support SDG implementation. 5. Partner for SDG results.	1. Align and implement ethics office charter and other administrative issuances. 2. Facilitate regional and country ethics networks. 3. Facilitate cross-cutting/cross-functional network. 4. Use communications to enhance leadership of WFPs overall ethical culture and culture of accountability. 5. Deepen internal and external stakeholder relationships.

Source: The World Food Programme.

But there's more than having a high-quality ethics and compliance program within an organization. Empowered Integrity does not only depend on the CEO, leadership and executive team. though in the better-run more enlightened companies that may be sufficient to ensure that the right kind of culture is indeed embedded in everyday working life. What is needed above and beyond that is board and governance commitment to Empowered Integrity.

7.3.b.ii Empowering culture governance

What does board and governance commitment to "Empowered Integrity" mean? Very simply that the board is equipped to deal with ethics, compliance, risk and culture in one form or another but with at least the following characteristics present:

- Board committee and/or full board that hear periodic reports on ethics, compliance, CSR, risk and culture.
- Board members who have experience and/or informed interest in these topics and know what questions to ask.

- Off the record, executive sessions with key management representatives with deep knowledge of ethics, compliance, CSR, risk and culture issues.

As we discussed in greater detail in Chapter 2, the board must proactively oversee ethics, compliance, CSR, risk and culture, Table 7.4 provides a summary of some of the tools they can deploy.

Table 7.4 The three roles and seven tools of "culture governance"

Role	*Tools*
The Culturally Equipped Board The board is sensitive to and informed on issues of workplace culture.	a. The board is educated on the culture big picture by experts. b. Every board has at least one independent director with broad and deep ESGT/culture expertise.
The Culturally Tuned-In Board The board thoroughly understands the culture of its organization.	c. The board insists on getting the right members of management to report on culture to the board periodically. d. The board regularly reviews, is updated on, and asks questions based on a customized "culture" dashboard. e. The board thoroughly understands the culture of the organization it oversees.
The Culturally Conscientious Board The board is the guardian of a healthy culture and an instigator of culture change when necessary.	f. The board exercises proactive CEO ESGT/culture accountability oversight. g. The board enforces CEO and management ESGT/culture accountability including discipline.

Source: Author and GEC Risk Advisory.

7.3.c Stakeholders: achieving Stakeholder EQ

What is Stakeholder EQ? Very simply, it's a shorthand way of saying "stakeholder emotional intelligence". Organizations need to ask themselves whether they have a comprehensive and effective way of knowing who their most important stakeholders are? In other words: does your organization have Stakeholder EQ?

7.3.c.i Know your stakeholders: plural

In today's increasingly complex and chaotic world, a world suffused with social media and other technological changes that impact us daily, maybe even hourly or minute-by-minute during a crisis, we personally and as part of an organization need to understand who the most important stakeholders of our organization are. Why? Because they have expectations, and if the organization ignores or outright works against those expectations and interests, stakeholders may turn against the organization, depriving it of its otherwise good reputation and having an impact on its financial and reputational viability and well-being. Herein lies the nexus between stakeholders, reputation risk and reputation opportunity:

- When an organization meets or exceeds the expectations of its stakeholders, it will be creating reputation opportunity, the opportunity certainly to protect and preserve value and maybe also to increase the value of the organization in the eyes of those stakeholders and others as well.
- However, when an organization is unable to meet stakeholder expectations or downright blows those expectations, the entity can suffer reputation risk, which in turn may erode and or destroy value not only for that particular stakeholder but for others as well both now and in the future.[13]

A holistic and healthy assessment of who an organization's most important stakeholders are begins at the core – who is your most important stakeholder, the one that is driving your mission, vision and values, the one that is central to everything you do?

- If you're a business, it is your shareholders and owners.
- If you're an NGO or a nonprofit, it would be the beneficiaries for which your charity was created.
- If you're a domestic government agency, your citizens and residents as well as transients coming through your jurisdiction are your prime stakeholders.
- If you're a university or educational institution, you would say it's the students.
- If you're an international governmental agency like the World Food Programme, for example, your prime stakeholders are the beneficiaries of your work – people in need of food because of famine or war or both.

But what about other interested parties who may have a stake in your organization in some form or another? Figure 7.4 and Table 7.5 list a variety of possible internal and external organizational stakeholders that should be considered. It is critically important for every organization to have a clear sense of who the other strategically important stakeholders are and to constantly analyze their expectations and requirements. This is what the most resilient and prepared organizations do.

Figure 7.4 A universe of potential stakeholders.

Source: Author and GEC Risk Advisory.

Table 7.5 Potential internal and external organizational stakeholders

Internal	*External*

• Owners, shareholders, investors:
  • Family
  • Private
  • Public
  • Government
  • Institutional
  • Activist hedge funds
• Boards of directors, trustees, supervisors
• Boards or councils of advisors
• Employees (and their families)
• Temporary and contract workers
• Labor and trade unions
• Workers councils
	• Customers, purchasers, clients, consumers
	• Users of products and services
	• Prospective owners, shareholders, investors
	• Partners, supplies
	• Communities
	• Non-governmental organizations
	• Foundations, donors, sponsors, contributors
	• Prospective employees
	• Facility/office visitors
	• Government agencies, regulators, enforcers:
	  • Local, municipal
	  • Provincial, state, regional
	  • National
	  • International
	• Media and social media

Source: Author and GEC Risk Advisory.

In considering the views and expectations of stakeholders, I like to use the perspective (or multiple perspectives) that the great movie director Akira Kurosawa deployed in his powerful classic movie *Rashomon.* I call it the Stakeholder Rashomon Effect or "Stakeholder Rashomon 360" and it means that for each major ESGT issue, an organization and its leaders should consider engaging in a Stakeholder Rashomon 360 exercise to accomplish the following:

- Identify and understand who the most important stakeholders are for that issue.
- Understand what the perspective and stake of each major stakeholder might be on that issue.
- Analyze what the stakeholders' principal expectations of the organization are on that issue.
- Qualitatively (and quantitatively if possible) determine the downside (reputation and financial risk) or upside (reputation or financial opportunity) that your organization's actions on that issue may have on that and other stakeholders.

7.3.c.ii The "Stakeholder Rashomon 360": BP Deepwater Horizon

Figure 7.5 presents a generic visualization of the Stakeholder Rashomon Effect. Table 7.6 presents a "Stakeholder Rashomon 360" exercise regarding the BP Deepwater Horizon disaster – reverberations of which continue to this day. Did BP systematically look at their ESGT issues, their impact on strategy and on key stakeholders beyond shareholders? Something tells me they probably did not – at least prior to that disaster.[14]

Table 7.6 Applying the "Stakeholder Rashomon 360" to BP Deepwater Horizon

Stakeholders	*Expectations*	*Damage/Impact*
Employees and Their Families	Protection of people and assets. Effective environmental, health and safety (EHS) programs.	Loss of life, injuries. Loss of livelihoods.
Shareholders, Investors	Protection of people and assets. Effective EHS programs. Return on investment. Good reputation.	Share price and other economic losses. Asset destruction. Long-term reputation risk hit.
Partners	Proper division and implementation of EHS responsibilities and duties.	Litigation to assign culpability. Share price and other economic losses. Asset destruction. Long-term reputation risk hit.
Community	Preservation and augmentation of economic value of offshore drilling platform near community, benefiting local businesses and communities.	Local natural habitat destruction. Destruction of small dependent businesses, employment and livelihoods.

(*Continued*)

Table 7.6 Continued

Stakeholders	*Expectations*	*Damage/Impact*
Protected Habitats	Protection and preservation of natural habitats including endangered species and habitats protected by applicable laws.	Destruction and/or degradation of vast swaths of coastland, fisheries, animal population.
Consumers	Patronizing a responsible corporation with a sense of corporate responsibility, CSR, ESG and stakeholder sensitivity.	Disappointed consumers boycotted company. Reputation risk hit for company with global consumers.
Regulators	That proper EHS and other necessary and desirable internal ESGT programs are in place to prevent EHS incidents, including injury, loss of life, environmental disasters like oil spills.	Heavy weight of a multitude of local, state and federal regulators, investigators and law enforcement descended on BP. Consequences for how regulators and government entities elsewhere perceive BP as a corporate citizen (or not) – reputation risk contagion.

Source: Author.

Indeed, if BP had effective issues-spotting in place before the tragedy of Deepwater Horizon – perhaps akin to the ESG heatmap from RepRisk represented in Table 7.7 – perhaps the stakeholder analysis would have been more robust, although it is always possible for a company to be aware of stakeholders and their perspectives and still not make the investment of time and budget into addressing those issues effectively.

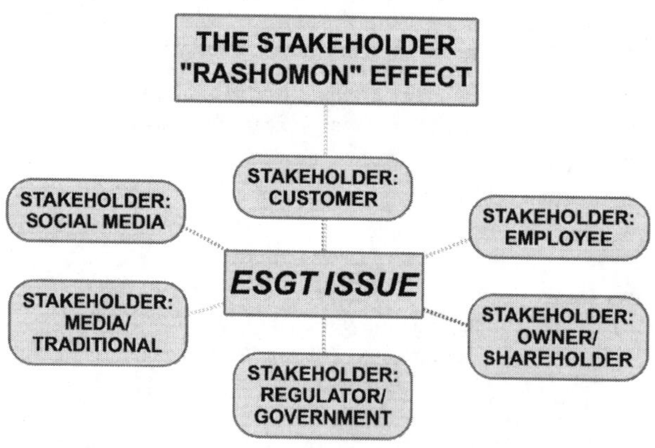

Figure 7.5 The Stakeholder Rashomon Effect.

Source: Author and GEC Risk Advisory.

Table 7.7 shows the variety of ESG issues of concern captured by RepRisk for BP for the past 10 years. It is clear that even before Deepwater Horizon, BP had at the highest end of the spectrum a variety of environmental, health and safety issues with actual or potential impacts on a variety of stakeholders beyond shareholders, including employees, customers and communities.

Table 7.7 RepRisk ESG issues heat map for BP for the ten years (April 16, 2009 to April 16, 2019)[15]

Environmental Environmental Footprint	*Social Community Relations*	*Social Employee Relations*	*Governance Corporate Governance*
Climate change, GHG emissions and global pollution	**Human rights abuses, corporate complicity**	**Forced labor**	**Corruption, bribery, extortion, money laundering**
Local pollution	**Impacts on communities**	**Child labor**	**Executive compensation issues**
Impacts on landscapes, ecosystems and biodiversity	**Local participation issues**	**Freedom of association and collective bargaining**	**Misleading communication, e.g., "greenwashing"**
Overuse and wasting of resources	**Social discrimination**	**Discrimination in employment**	**Fraud**
Waste issues		**Occupational health and safety issues**	**Tax evasion**
Animal mistreatment		**Poor employment conditions**	**Tax optimization**
			Anti-competitive practices
Cross-cutting issues – *always in combination with an ESG issue*			
Controversial products and services			
Products (health and environmental issues)			
Violation of international standards			
Violation of national legislation			
Supply chain issues			
Key to shades of gray:			
Low risk	**Medium risk**	**High risk**	**Very high risk**

Source: RepRisk.

In the case of BP, take employees as a category of stakeholder. An oil rig employee's expectation is that their company will properly train and equip them for health and safety protection as well as provide avenues to speak up and report health and safety concerns without fear of retaliation. If that doesn't happen or is done in a lackluster or ineffective way (or worse, there is a culture of fear that prevents employees from reporting problems) and there is an accident that injures or kills the employee, that employee and family will have had their expectation of safety obliterated, and the consequences both reputational and financial to the organization might be dire. The BP

Deepwater Horizon liability tally since 2010 is $86 billion and counting, and one of the post-event investigation conclusions had a lot to do with the issues described in Table 7.7.

7.3.c.iii *Achieving Stakeholder EQ*

The point of having Stakeholder EQ is that organizations and their leaders must do something to understand their issues and how those issues affect their most important stakeholders. Doing a scenario exercise from time to time to understand the full implications of an issue, who the most important stakeholders in that issue are, what their expectations might be and the impact both reputational and financial of an adverse event is a resilience building and important step organizations and their leaders should take.

Figure 7.6, for example, provides a generic consideration of primary stakeholders of a technology company. Did Facebook, for example, do any serious exercises – at least pre-Cambridge Analytica – that would have helped them glean the issues and stakes of some of these stakeholders and the impacts on stakeholder expectations of Facebook?

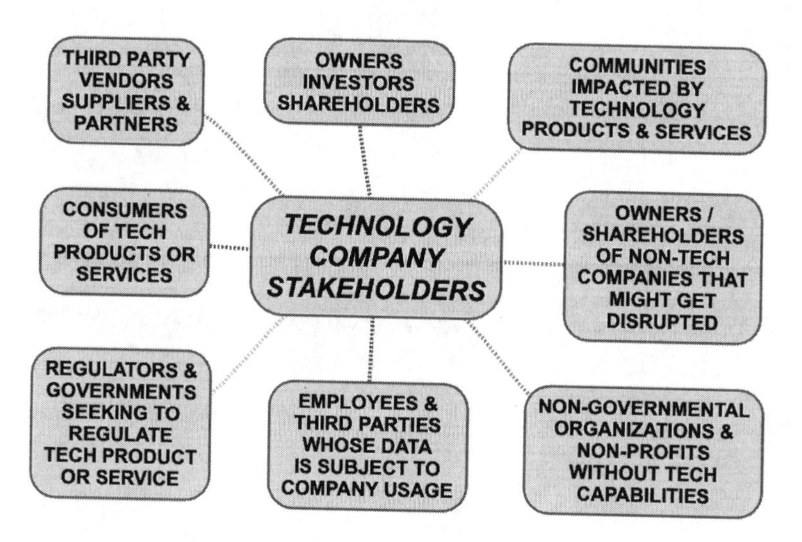

Figure 7.6 Technology company stakeholders.

Source: Author and GEC Risk Advisory.

7.3.d Risk: implementing Risk Intelligence

The point of having "Risk Intelligence" is that a full and appropriately customized suite of risk management and oversight practices are in place at an entity –

neither overdone nor underdone. in other words, done in an intelligent way that is useful and fully customized to the organization in question, hence our term "Risk Intelligence".

This can mean different things to different types of organizations, but it never means having no risk management. Figure 7.7 provides a big-picture snapshot of different stages of risk management maturity in which two of the stages (the first two) are unacceptable regardless of the type of organization. The bottom line is that every type of organization requires some form of risk management.

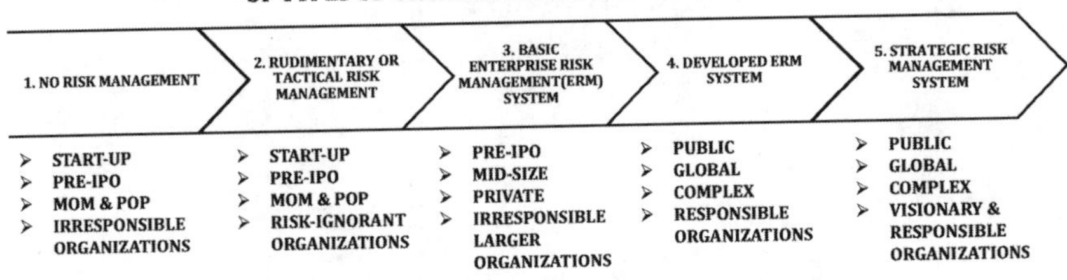

Figure 7.7 An evolutionary view of types of organizational risk management.

Source: Author and GEC Risk Advisory.

7.3.d.i Getting to the right risk architecture for your organization

What does it mean to have the right risk management approach for your entity – whether a company, an NGO, a government agency or some other form of organization? Very simply, it means having the right degree of sophistication in identifying, managing, mitigating, monitoring and constantly and periodically reviewing and renewing the risk management process and content. There are some excellent books and experts in the marketplace that have tackled the highly sophisticated nature of what risk management should look like, and I refer my readers to them, as there is much to learn on this subject and we have limited space to give to this topic in this book.[16]

Figure 7.8 provides a simple visual representation of the phases of what an effective risk management approach might look like for any type of organization. Moreover, Figure 7.9 shows a representation of what the sources of information and buckets of risk might look like in a typical enterprise risk management program. Both of these figures try to convey in a simple visual way what an organization should consider having in place in terms of Risk Intelligence, customized to its particular strategy, mission and vision as well as the needs and characteristics of the entity and its stakeholders.

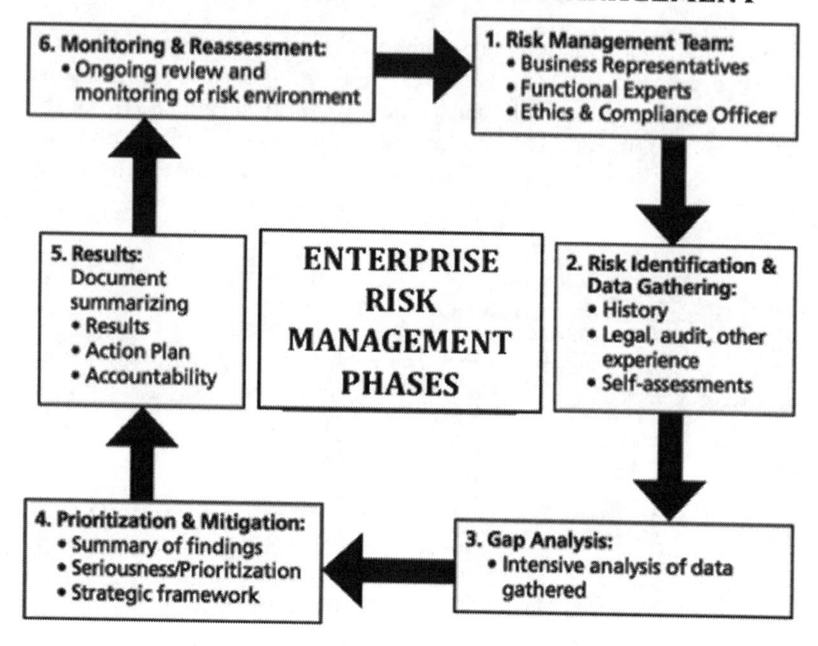

Figure 7.8 Enterprise risk management phases.

Source: Author and GEC Risk Advisory.

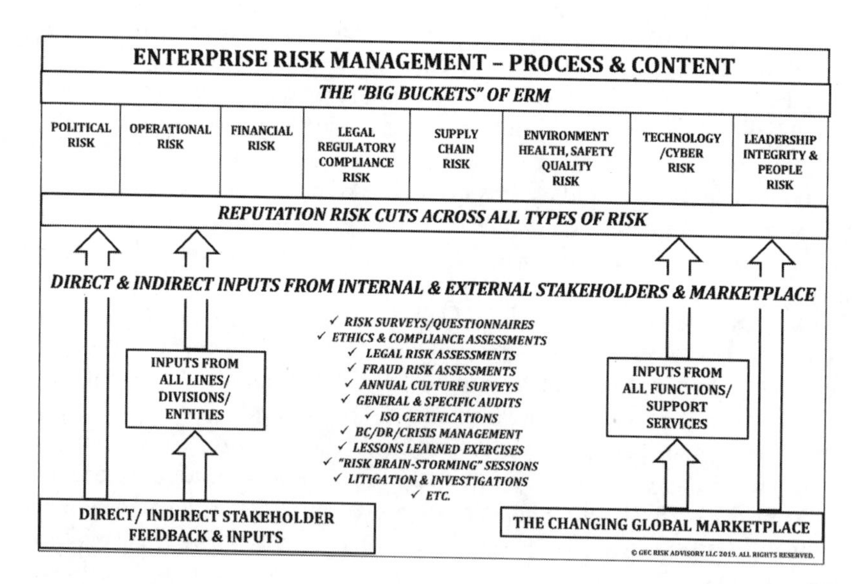

Figure 7.9 Risk architecture: a big-picture visualization of enterprise risk management.

Source: Author and GEC Risk Advisory.

Many companies and entities that have formal risk management programs in place may or may not be effective in actually proactively and successfully identifying, managing and mitigating their risks. This can happen even in the more sophisticated global companies. Sometimes sophisticated companies have an evolved form of risk management and produce significant risk data and information. However, they are not able to prioritize and synthesize their most important, material and strategic risks for proper senior level decision-making or board oversight. I sometimes refer to this as the "drinking out of a firehose" form of risk management where there is too much data but not enough strategic judgment that is helpful to decision-makers. Sadly, in such situations, organizations and their leaders may be leaving value on the table and possibly even destroying value by not seeing the risk forest for the trees. What follows are a few tips to consider that might help in avoiding this trap and in making sure that what you have at the end of the day is Risk Intelligence and not risk stupidity or uselessness.

7.3.d.ii Deploying effective Strategic Risk Governance

It is incumbent on risk management leadership – preferably a seasoned executive with deeply relevant business or organizational and risk management expertise – to select, underline and visualize the most important information for the highest-level decision-makers. I call this "Strategic Risk Governance". There is a general progression depending on the evolution, maturity and leadership and culture of an organization from no risk management to evolved and holistic risk management.

Fully evolved and mature strategic risk governance would include the following practices at the board or governance level:[17]

Board Makeup. There are one or more board members who have a rich, diverse or highly sought-after background or experience in governance, risk and compliance relevant to the business of the organization. The board may have more than an audit committee to oversee nonfinancial risk management and has considered or created a separate risk (and/or compliance and/or CSR) committee.

Risk Oversight. There is a clear understanding of the entity's enterprise and strategic risk management profile. The right questions are being asked of management and of the heads of risk, compliance, CSR and other key risk functions. The type of information and reporting that reaches the board from risk management is robust, periodic, useful and focused primarily on the company's strategic risk.

Role of the C-Suite/CEO. The C-Suite is fully and periodically informed and engaged on enterprise and strategic risk issues and the frontline risk management leadership and team are effective, interconnected and properly resourced to deal with the risks of the enterprise. Both are interconnected with the board in a synchronized way in terms of risk reporting.

Role of the Board in Resilience Building. The board understands and requires the existence of proper business continuity, crisis management and disaster recovery programs at the company. The board participates from time to time in crisis management exercises.

Table 7.8 presents a chart with a sample of some of the critical strategic risk governance issues from 2018 that corporate directors should not only be interested in knowing about but should also be proactively preoccupied about on behalf of their organization.

Table 7.8 2018 top 10 risk concerns from the National Association of Corporate Directors, Protiviti and the North Carolina State Survey 2018[18]

NACD Survey	*Protiviti/NC State*
• Significant industry change. • Business model disruption. • Changing global economic conditions. • Cyber-security threats. • Competition for talent. • Political uncertainty in the US. • Technology disruption. • Corporate tax reform in the US. • Increased regulatory burden. • Risk in M&A.	• Rapid speed of disruptive innovations and new technologies. • Regulatory changes and scrutiny. • Resistance to change. • Cyber-security threats. • Success in challenges and competition for talent. • Operations don't meet performance expectations. • Economic conditions. • Organization's culture may not escalate risk issues. • Inability to use data analytics for market intelligence. • Limited opportunities for organic growth.

Source: NACD, Protiviti, North Carolina State.

7.3.d.iii Understanding reputation risk and opportunity management

It is important to single out reputation risk management as part of risk management. It warrants a mini-section of its own because of its pervasive interconnectedness with all other kinds of strategic risk, especially in the age of hyper-transparency, social media and fake news, to just mention a few drivers of this relatively new risk.

As I laid out in my 2014 book, *The Reputation Risk Handbook*, reputation risk is:

> an amplifier risk that layers on or attaches to other risks – especially ESG risks – adding negative or positive implications to the materiality, duration or expansion of the other risks on the affected organization, person, product or service.

Speaking more visually, one can regard reputation risk as that risk that cross-cuts all other categories, as the illustration in Figure 7.9 shows.

7.3.d.iv Conquering complex risk

A critical component of the risk identification process is to understand that most if not all risks are not stand-alone matters but are interconnected with one or more other buckets of issues and risks, as illustrated visually in the bible of global risk – the WEF Global Risks Reports. Figure 7.10 from WEF provides a visually arresting demonstration of this interconnectedness of complex risks.

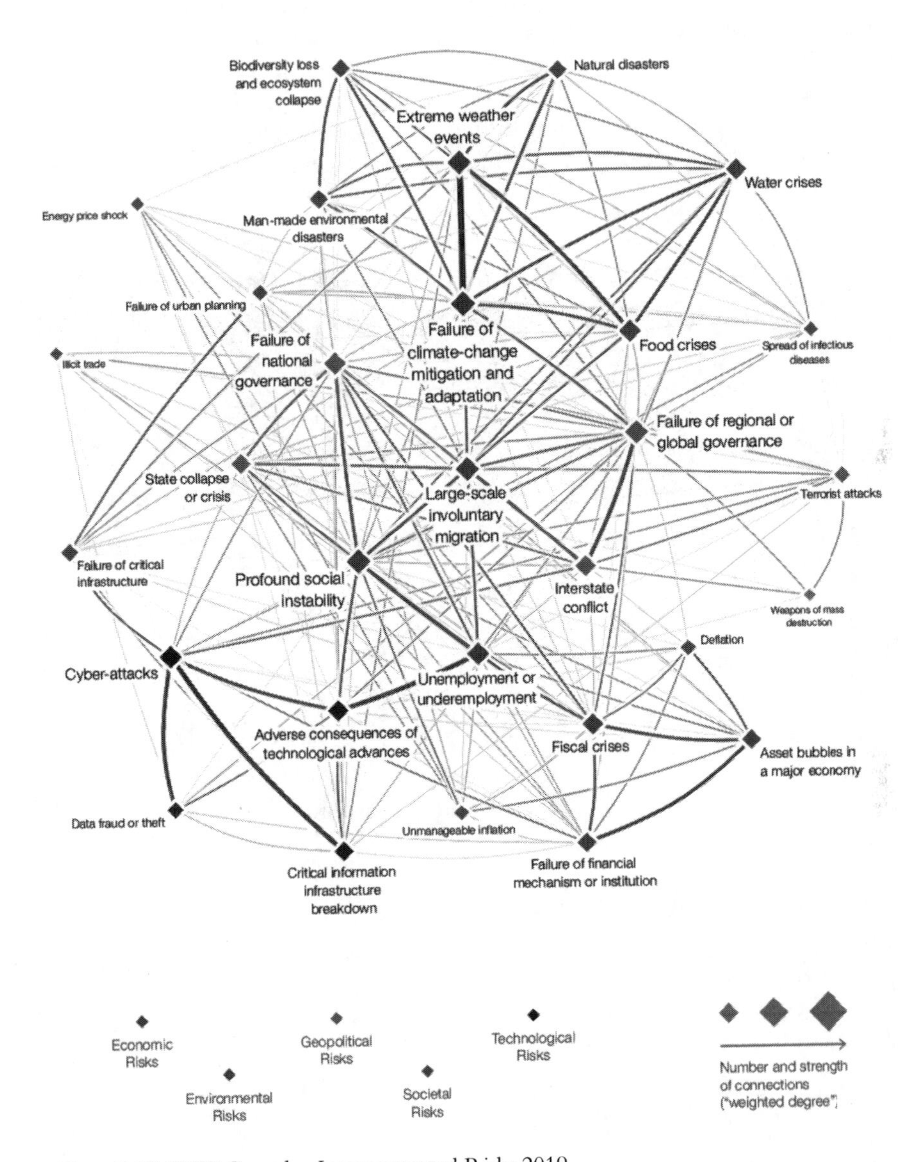

Figure 7.10 WEF Complex Interconnected Risks 2019.

Source: World Economic Forum.

7.3.d.v Deploying cross-functional risk teams

Finally, another critical component of resilient Risk Intelligence is that all relevant hands are on deck to identify, assess, evaluate, mitigate and transform risks into opportunities within the universe of risk for an entity. That could and should translate into a high-level cross-functional or cross-disciplinary team of internal (and sometimes external) experts in a wide variety of important issues and topics emanating from the company's strategy, tactics, footprint, products, services, etc. Such a team must work together periodically, consistently and systematically.

Figure 7.11 provides a visualization of what such a team might look like. The bottom line is this: every entity has risks and risk experts and owners from different parts of the entity – they should work in unison led by an appropriate risk leader to triangulate and implement Risk Intelligence customized to their organization.

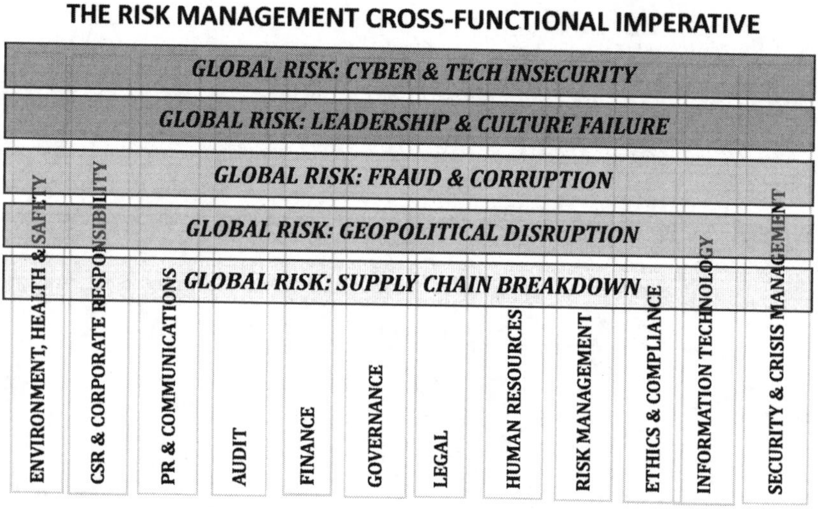

THE RISK MANAGEMENT CROSS-FUNCTIONAL IMPERATIVE

GLOBAL RISK: CYBER & TECH INSECURITY
GLOBAL RISK: LEADERSHIP & CULTURE FAILURE
GLOBAL RISK: FRAUD & CORRUPTION
GLOBAL RISK: GEOPOLITICAL DISRUPTION
GLOBAL RISK: SUPPLY CHAIN BREAKDOWN

ENVIRONMENT, HEALTH & SAFETY
CSR & CORPORATE RESPONSIBILITY
PR & COMMUNICATIONS
AUDIT
FINANCE
GOVERNANCE
LEGAL
HUMAN RESOURCES
RISK MANAGEMENT
ETHICS & COMPLIANCE
INFORMATION TECHNOLOGY
SECURITY & CRISIS MANAGEMENT

Figure 7.11 The Risk Management Cross-Functional Imperative.

Source: Author and GEC Risk Advisory.

7.3.e Strategy: developing Strategic ESGT

The resilient organization will also have a command of its strategic planning process that not only focuses on its core mission, vision and values, but also integrates seamlessly its ESGT issues, risks and opportunities into the development and implementation of that strategy.

7.3.e.i Integrating ESGT into strategy

Figure 7.12 presents a visualization of what such a high-functioning strategy incorporating strategic ESGT would look like.

Figure 7.12 An ESGT integrated organizational strategy.

Source: Author and GEC Risk Advisory.

How does an organization that hasn't already incorporated a system for understanding, deploying and monetizing its ESGT issues actually do this? Table 7.9 presents a step-by-step approach any organization can use to develop an appropriate and customized approach to its own customized ESGT issue management. I call it the "ESGT Issue Value Chain" – a visual tool for leaders to use to understand the evolution or metamorphosis of their ESGT issues into "value". Simply having issues management at the front end (via a public relations or external communications function) is insufficient in today's complex world.

7.3.e.ii The ESGT Issue Value Chain

The ESGT Issue Value Chain is the series of steps that an organization and its leaders should take to create a holistic and integrated approach to identifying their most important ESGT issues and then understand how those issues are manifested within the organization – as PR issues, risks, opportunities, crises, etc., and how they might ultimately impact value – innovation creation, preservation or destruction (see Figure 7.13).

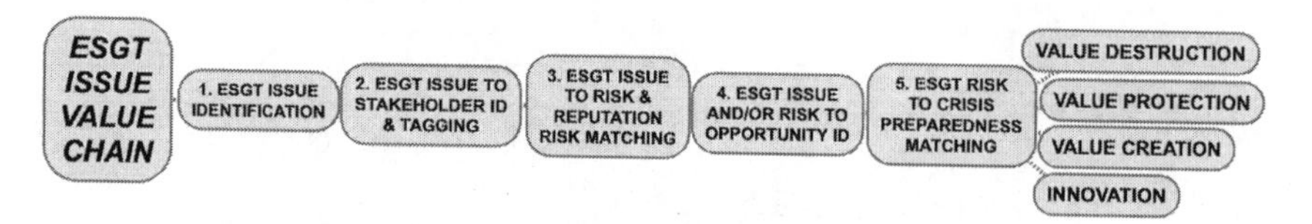

Figure 7.13 The ESGT Issue Value Chain.

Source: Author and GEC Risk Advisory.

The most important, material or strategic ESGT issues should be considered from several key perspectives along the way – from a public relations and communications standpoint (issues management), a risk management and oversight standpoint (enterprise risk management and strategic risk governance), as part of crisis preparedness (crisis management and business continuity) and in terms of developing new or enhanced products and services (strategy and innovation). Here are some of the qualities of a healthy and holistic ESGT Issue Value Chain approach.

1. **ESGT Issue Identification and Situational Awareness.** Every entity needs to know its key ESGT issues; doing this requires good situational awareness in the form of strategic, communications and public relations personnel who are trained to identify and understand the current important issues for the organization, including having tools for the rapid collection of issues popping up in social media in its many, multiplying and evolving incarnations.
2. **Stakeholder Identification.** Closely related to knowing what your issues are is understanding who the stakeholders in those specific issues are above and beyond the key ones that you know you must cater to. This means being able to go beyond the core stakeholders which are in the case of a corporation, shareholders/owners, in the case of a nonprofit, its beneficiaries, and in the case of a government agency, citizens and residents impacted by their policies.
3. **Risk Identification.** An entity needs to have good risk management – in which a person and team of people have the wherewithal to understand what issues might become risks and have some form of risk management and mitigation strategy in place – preferably some form of enterprise risk management (ERM) customized, of course, to the size, mission, strategy and footprint of the organization.
4. **Opportunity Identification.** From every issue, risk and crisis there is an opportunity to learn important lessons and convert them into value creation, such as better processes, more efficiency, liability avoidance, cost savings and

even new value creation. Only a few organizations consciously and systematically look at issue and risk management as also opportunity management and are thus able to create new value by exercising such continuous improvement.

5. **Crisis Preparedness.** It is critically important for there to be a thoughtful understanding of how an issue (or an already identified risk) can translate into a crisis.
6. **Value: Protection, Destruction or Creation?** Those entities that either fail to identify their important ESGT issues (situational awareness) or to apply robust and agile risk management to their portfolio will likely descend into value destruction. However, entities that take issues, stakeholders, risks, opportunities and crisis management seriously are more likely to preserve and protect existing value and actually create new value.

Table 7.9 provides an overview of the principal steps of the ESGT Issue Value Chain.

Table 7.9 The ESGT Issue Value Chain: actionable steps

Step 1 – ESGT issue identification – EST "situational awareness".
Develop a customized ESGT issue identification framework that provides full ESGT situational awareness for the organization.
Step 2 – ESGT issue to stakeholder identification and tagging.
Manage ESGT issues, identify and prioritize key stakeholders and understand their expectations.
Steps 3 – ESGT issue to risk and reputation risk matching.
Proactively funnel ESGT issues into risk management or enterprise risk management (ERM) frameworks and ensure that those that may be risk issues are properly incorporated into ERM or other risk management processes and exercises.
Step 4 – ESGT issue and/or risk to opportunity identification.
Proactively funnel ESGT issues and risks into strategy, business development, innovation, research & development and/or new products and services activities so that the benefit of issue and risk identification is incorporated into strategy formulation and business planning.
Step 5 – ESGT risk to crisis preparedness matching.
Ensure strategic and key tactical ESGT risks are incorporated into crisis planning and preparedness, including scenario exercises.
Outcome – one of three possible issue value chain outcomes:
1. Value protection.
2. Value creation.
3. Value destruction.

Source: Author & GEC Risk Advisory.

7.3.e.iii Hypothetical pharma company: strategic issues dashboards

Table 7.10 represents a sample ESGT issue classification dashboard that a given organization could put together – we use the example of a fictitious global pharmaceutical company, identifying what might be some of their core strategic ESGT issues.

Table 7.10 Strategic ESGT Issues Dashboard for a hypothetical global pharmaceutical company

Environmental	Social
• Climate change/global warming • Carbon footprint • Natural resource management • Waste management • Biodiversity • Animal well-being and testing • Natural & man-made disasters • Sustainability programs	• Human rights and supply chain • Labor relations • Employee diversity and inclusion • Product safety and quality • The future of work • Employee health and safety • Pandemics preparedness • Employee crisis management
Governance	**Technology**
• Fraud, corruption and bribery • Third-party management • Conflicts of interest • Board composition & diversity • Board risk governance • Reputation management • Tax and jurisdictional transparency • Technology governance	• Digital strategy • Cyber-risk and cyber-security • Data privacy and protection • Artificial intelligence ethics • Social media and deep fakes • Technology governance • Biotechnology • DNA and genetic engineering

Source: Author and GEC Risk Advisory.

A similar table, Table 7.11, once again designed for a fictional global pharmaceutical company, shows the most important internal corporate people who should have a seat at the table discussing the particular issues and who the most important stakeholders are in each such issue. This provides an example of "issue to stakeholder tagging" described in Step 2 of Table 7.9 as well as the formation of the proper cross-functional teams to triangulate and develop policies on such issues.

Table 7.11 ESGT Issue Management and Stakeholder Dashboard for a generic global pharmaceutical company

Overall issue category	*Key members of cross-functional team*	*Critical stakeholders to consider (beyond shareholders)*
ENVIRONMENTAL BIODIVERSITY/ Animal Well-Being	• EH&S • Quality • Legal and compliance • Supply chain/procurement • Risk management • Corporate responsibility • Research and development • Innovation	• Employees • Customers • Communities • Animal populations • Regulators

(*Continued*)

Table 7.11 Continued

Overall issue category	*Key members of cross-functional team*	*Critical stakeholders to consider (beyond shareholders)*
SOCIAL		

Human Rights in the Supply Chain	• Legal
	• Ethics and compliance
	• Corporate responsibility
	• Supply chain/procurement
	• Risk management
	• Operations
		• Labor in the supply chain
		  • Slave labor
		  • Child labor
		• Communities
		• Consumers
		• Employees

GOVERNANCE		
Fraud, Corruption and Bribery	• Ethics and culture • Risk management • Legal and compliance • Finance • Third-party management	• Customers • Communities • Employees • Regulators
Research Conflicts of Interest	• Ethics and culture • Risk management • Legal and compliance • Finance • Corporate secretary	• Customers • Communities • Employees • Regulators
TECHNOLOGY		
Cyber-Risk and Cyber-Security	• IT, CIO, CISO • Legal and compliance • Risk management • Data privacy officer • Operations • Finance • Crisis management	• Employees • Customers • Regulators

Source: Author and GEC Risk Advisory.

At the end of the day, organizations and their leaders need to ask themselves whether they have an effective issues management value chain that leads to proper issues management throughout the value chain (risk, opportunity, crisis and value), or whether that issues value chain is ineffective or worse, nonexistent.

In today's complex world of interconnected issues, risk and opportunities, no leadership team anywhere for any kind of organization – whether business, nonprofit or governmental – should live in the blissful ignorance of their most important ESGT issues. If they do, their bliss will be short-lived, and may quickly turn into a value-threatening nightmare.

Finally, as you build your own ESGT Issue Value Chain, here are some important pointers:

Everything is interconnected. While we talk about distinct categories of issue and risk, almost all risks are interconnected. Take a look at the World Economic Forum Global Risk Report latest edition for 2018 as well as Figure 7.10 in this chapter and see how truly interconnected the various global risks confronting the planet really are.

Mix it up. As a subset of the first pointer, beg, borrow and steal concepts from completely different walks of life for your particular walk of life. Apply quality data thinking to everything. Take lessons from biology and integrate them into urban planning. Take emotional intelligence and marry it to quantitative thinking.

Everyone is interconnected. Likewise, as we talk about distinct cases and solutions – opportunities for resilience and value – we also need to think about how we achieve those opportunities together, in an interdisciplinary, collaborative manner. It takes a village (to coin a phrase) of many different stakeholders and actors to get there.

Think outside the box – create a new box – hell, don't have a box. However we get there, we need to think differently or "think different" as the great Steve Jobs once said.

EQ over IQ. As we careen into the age of AI robotics and nanobiology, remember that data isn't everything and that emotionally intelligent human beings still have the upper hand, at least for now.

Ethics matter. As data becomes king, and machines replace a lot of the manual and menial work that humans used to do, one thing remains true: as imperfect as we are we have deep-seated neurological and biological and other roots to our thinking – morally, ethically and communally. Let's focus our brains and efforts on maintaining and even improving the jobs and tasks we perform in the brave new world we have entered.

Think forward/futuristically. Because the future is here.

7.3.f Performance: achieving Performance Equilibrium

> Like political lives, many executives' careers end in failure. Those who over-extend their tenure are often prone to stumble in their last years in office. The intense focus on pay – a lightning rod for criticism of Mr. Ghosn – adds pressure for listed company executives to perform.[19]

The preceding quote is from a *Financial Times* op-ed penned at the heels of the eruption of the latest CEO abuse of power allegation – the Carlos Ghosn scandal which at the time of this writing revolves around alleged lies, cheating and possible theft of corporate financial resources by the all-powerful chairman of the auto consortium Renault/Nissan/Mitsubishi. It certainly provides an apt example and introduction to what we are talking about in this section.

"Performance equilibrium" is another way of saying having a balanced incentive and compensation performance management system in an organization starting at the very top – whether at a government agency, a university, a corporation or an NGO. It means having a bonus and compensation system that aligns people with important business and strategic goals but in such a way that it is balanced and in equilibrium with more than the bare financial or operational objectives. It means achieving equilibrium while keeping the values and the culture of the organization in mind and in sync as well.

All organizations are congregations of people hopefully working together for a common mission, vision and strategy and with common values, goals and objectives. What they should also have in common is a performance evaluation, management and incentive system that aligns employees and management with the primary business and strategic objectives of the organization.

But it isn't enough to align individuals and teams with purely financial or operational objectives – the metrics used to measure success must also include qualitative and/or quantitative metrics about *how those goals and objectives are achieved*. In other words, it is insufficient, even dangerous, to align goals with purely financial targets – witness the Wells Fargo case of "eight is great" in which salespeople lied and cheated to achieve the sale of eight forms of account to each customer. Every organization needs to find the right "equilibrium" between achieving those hard financial and/or operational targets and how those hard targets are being achieved.

7.3.f.i An illuminating case study: aligning safety with strategy in Latin America

I would like to share an "illuminating" example from my own experience, which provided firsthand experience of what I mean by "performance equilibrium" that has stayed with me for many years. This happened when I was a senior executive at PSEG Global, a major international electric power developer and operator in the 1990s and early 2000s. I served, among other things, as general counsel and head of EHS. Our company had successfully bid on the privatization of a number of previously government-owned, environmentally dirty and unsafe power generation and distribution facilities throughout Latin America.

Specifically, PSEG Global acquired facilities and operations that had completely unacceptable worker accident rates and in some cases worker deaths. Our parent company, the large US public utility PSEG, mandated that we implement a system to improve the health and safety operations in these Latin American assets so that they operated at the same health and safety success levels as their US operations did. As we profiled in Chapter 4, PSEG had a stellar EHS track record for many years after a lackluster track record back in the 1980s from which they learned lessons and built long-term EHS resilience. In other words, the parent company told us to get the Latin American operations' EHS standards to the same level as those of their award-winning EHS US operations.

To achieve this seemingly daunting task, our CEO structured our annual bonus system to align completely with achieving what seemed impossible milestones on health and safety in jurisdictions like Peru, Chile, Argentina and Brazil. Some of the facilities we had acquired had very poor worker health and safety protections in place (let alone vigilance and enforcement) and several of them had multiple worker deaths yearly.

Within one year of receiving this mandate from our parent company and developing and implementing a successful EHS strategy for these assets, we reversed the poor and even deadly track records of several operations to align with that of our US assets.

How did we do it? I'd like to think that we were all motivated by "doing the right thing", but the bottom line was that by motivating us to achieve these goals by tying our entire or a large chunk of our annual bonus to achieving specific health and safety milestones in each asset, management enhanced our focus and resourcefulness immeasurably. Yes, we were all good people wanting to do the right thing of dramatically lowering accident rates and eliminating deaths, but by having our annual bonus completely or primarily dependent on doing this we became even more laser-focused and, I would argue, successful.

The structure of our performance management system not only achieved the goals and objectives assigned to us but also created other invaluable externalities such as engendering the goodwill of the local communities and of our workforces. In fact, we literally had workers coming from our competitors telling us how much they would prefer to work for us because we cared about their health and safety. And, by the way, our EHS team won several awards for their achievements in Latin America. I would call that a win/win/win or more officially "performance equilibrium".

It is critically important to design performance management systems that integrate not only the organizational financial and operational goals and objectives but also the relevant values, culture and other stakeholder considerations (like health and safety) not only at the CEO and executive team level but all the way down into the workforce of an organization. Table 7.12 provides a variety of additional tips and ideas for doing this, and of course readers should always consult the experts on these topics all the way up and down the organizational food chain – it all begins and ends with proper CEO compensation design at the board and governance levels within the appropriate compensation committee. Only then can such a performance management structure be properly cascaded into an organization. And that's not an easy feat – if the board doesn't set the right tone from a performance equilibrium standpoint at the top with the CEO and C-Suite then all bets are off for the rest of the organization to achieve such equilibrium.

Table 7.12 Ideas to achieve performance equilibrium

- Financial performance targets can be aggressive but should not be irrational or "too good to be true".
- Individual and team sales and revenue goals should be within reach and not so difficult that they will tempt naturally risk-taking sales and business development folks to stretch and even cheat.
- What are the stated values in your organization? How can they be defined, calibrated and even measured to allow for informed judgment by performance managers?
- Organizations that understand their ESGT profile (as discussed earlier in this chapter and throughout the book) are organizations that can identify specific ESGT goals and objectives to align their workers and teams with (as in the example of PSEG global, where we identified health and safety goals).
- Similarly, organizations that have evolved sustainability and CSR programs can use organizational targets and metrics that make sense for the company or organization or a subset or team therein.

(*Continued*)

Table 7.12 Continued

- Consider the creation of what some leading companies like Merck & Co., Inc. call a "balanced scorecard" where performance equilibrium is baked into the system based on the company's strategy, business plan and starting at the very top of the organization (CEO and executive team) and percolating into every part of every individual worker's performance.
- The role of good governance is paramount and central to performance equilibrium being achieved as the board is where the buck stops – if they do not create performance equilibrium at the very top of the food chain – the CEO – the whole system will be out of whack.
- Likewise, if the board isn't vigilant to ensure that the rest of the organization is engaged in performance equilibrium overseeing management's systems even if the CEO is properly overseen, the rest of the organization may not be optimally aligned.

Source: Author and GEC Risk Advisory.

7.3.g Crisis: adopting Crisis Readiness

Once again, as someone who has created and run crisis management plans and teams for 25 years, it still amazes me how often organizations – including sizable *Fortune 1000* type companies – don't have the most rudimentary form of crisis management team or plan (let alone both) in place.

As we laid out in some alarming detail in Chapter 1 – "Gloom", in today's chaotic, unstable and tectonically changing world, not having a crisis management plan and team ready and available at all times is tantamount to dereliction of duty to the organization and its stakeholders. Indeed, in one summer, at one of the companies I worked at, we had not one, not two but three major events (an earthquake, a flood and severe weather including a nearby tornado) that required us to fine-tune our crisis management and readiness program. The good news: this was a company that already had a good plan in place and we literally weathered the storms intact.

Why is crisis management important? Because it involves protecting life and limb (employees, customers, bystanders), it involves protecting assets (real estate and other property) and it involves protecting reputation (being willing and able to put people and assets before profits).

Thus, in essence every single entity (no matter how small) should have an emergency or crisis management plan in place, and that plan should be directly associated with the entity's issues and risk management system which should directly inform and feed the crisis planning process (as per our earlier discussion of the ESGT Issue Value Chain).

One of the first things to do is to understand what constitutes a crisis and what doesn't, and each organization needs to define that for themselves. In Table 7.13 we take a stab at providing an overall framework for varying levels of potential crisis with a few examples to provide a sense of what could happen.

Table 7.13 What is a crisis? Various levels and examples

Level 1 – Imminent/Ongoing	*Level 2 – Potential*	*Level 3 – Serious*
• Imminent or actual threat or harm to life, safety or health of personnel and/or other stakeholders (visitors, customers, community, etc.). • Imminent or actual threat or harm to assets or property (buildings, transportation, other facilities). • Imminent or actual threat of severe reputational damage via social media or media.	• Potential threat or harm to life, safety or health of personnel and/or other stakeholders (visitors, customers, community, etc.). • Potential threat harm to assets or property (buildings, transportation, other facilities). • Potential threat of reputational damage via social media or media.	• Possible danger to life, safety or health of personnel and/or other stakeholders (visitors, customers, community, etc.). • Possible danger to assets or property (buildings, transportation, other facilities). • Possible reputational damage via social media or media.
Examples		
• An armed intruder appears at one of your offices.	• A receptionist receives a call from an anonymous person threatening to intrude into an office.	• A visitor engages in a verbal altercation with an office worker.
• A cyber-attack has crippled the power and operations of a hospital.	• You have received a ransomware threat.	• You have discovered a past cyber-breach of your HR department records.
• A tornado has hit a town where you have a manufacturing facility, wiping it out as well as affecting the lives of 20 of your local workers and their families.	• A #3 typhoon is heading directly to the location of your headquarters office in the next 24–48 hours.	• A dangerous thunderstorm system is heading in the direction of a facility that is in a flood zone and may hit in the next five days.
• The mainstream media reports that your CEO has been arrested for alleged sexual assault and rape of a minor in a foreign country and social media has lit a firestorm on this issue.	• An unhappy restaurant customer has taken a video of your location claiming that the location serves dirty and unsafe food and threatens to post it on Twitter.	• There are rumors that an opposition research firm will attack your NGO via social media claiming you are a front for espionage for a hostile nation-state actor.

Source: Author and GEC Risk Advisory.

For those of us who have lived through one or more crises personally and/or professionally, it is absolutely clear how important it is for an entity to have a crisis team and plan in place. Indeed, for those of us who have lived through a major crisis (for example, 911 in New York City), it is abundantly clear that you as a person and you as a member of your family and workplace, should have your own preparedness plan. There are no excuses.

So what should an entity have in place in terms of crisis management planning? Once again, it is all dependent on size, location, diversification, purpose, complexity, etc. However, there are some very basic things that every entity should have in place:

- A customized crisis management plan, no matter how simple, with details about the who, what, when, why and how of a crisis.
- A crisis management team of specified individuals, including from the highest level of the organization and including a liaison to the board or other oversight body.
- A crisis management team leader and a back-up leader as well as alternates to the main core group members.
- Regular meetings of the team (principals and alternates) to compare notes, review, update and revise protocols.
- Conduct crisis scenario exercises periodically.
- Link crisis management and facilities management.
- Link crisis management and business continuity (indeed business continuity is another topic that many companies and organizations don't seem to have under control but should – what do they otherwise do after the crisis has occurred – what is the plan to get the business or location up and running again after 24 hours, several days, a week?).
- Link crisis management to human resources and travel protocols.
- Link crisis management and IT systems support.
- Link crisis management to data protection and retrieval.
- Link crisis management to accounting and finance systems.
- Link crisis management to legal and regulatory issues, requirements and implications.

7.3.g.i The Case of the Timely Crisis Fanny Pack

I'd like to share another personal and professional perspective on this topic, as I have done a lot of work in the crisis management and preparedness arena. This "Case of the Timely Crisis Fanny Pack" illustrates how important it is not only to protect people and assets but also to create additional long-term intangible and reputational value for an organization though crisis management readiness. I do not identify the company involved, but suffice it to say that I had direct, hands-on experience with this case, as I was the instigator and leader of the crisis plan.

The background:

- In post-911 New York City, a company that owned and operated its own skyscraper office building in a very central and vulnerable location of the city with 4,000 employees working in the building at any given time did not have a crisis management team or plan in place for the 911 crisis and for a substantial period of time thereafter.

- Then a program was developed, a team was assembled, a crisis management plan was written and adapted to the issues and circumstances of the business and its location.
- The team created crisis kits in the form of fanny packs with essential information on what to do as well as basic equipment (flashlight, batteries, water bottle, whistle, a plastic instruction sheet, a list of transportation resources and important phone numbers). These fanny packs were dropped off together with crisis/emergency instructions at every one of the 4,000 employees' desks one evening.
- Then, the East Coast Blackout of 2003 happened only two weeks after the plan had been formally adopted and the fanny packs dropped off at employee desks (and after over a year of preparation and the teasing and outright mockery of the crisis fanny packs by some of the company executives).

During the crisis:

- A building back-up generator was switched on to provide limited power to power some essential operations within the building, including the use of one elevator to transport folks from the top floors down to the lobby or the second-floor cafeteria.
- Many employees walked down the internal stairwell, some for dozens of floors, and used their handy flashlights from their fanny packs to help them see through the relative dark (some of these employees also used the flashlight to walk home from the building, including myself).
- For those who couldn't find a way to get home (there was no electricity in NYC, so unless you were within walking distance from home you were stuck in the building or would have to take a chance with a taxi or bus in the dark), basic accommodations at the cafeteria floor of the building and food were made available.
- The crisis team was able to keep abreast of developments around the city and region by deploying themselves into the appointed crisis management room in the building which was fully equipped with radios, batteries, computers and other necessary information and equipment.
- Within an hour of the blackout, the crisis team was able to ascertain through radio news that this wasn't a terrorist attack on NYC (remember it was less than two years after the shock of 911), something that we were able to share with the rest of the people in the building.

What happened next? In the days that followed, the crisis management team received many thanks and kudos from individual employees who expressed gratitude to and admiration of the company for having a thoughtful crisis plan in place and caring enough about employees' and their safety and well-being. The fanny pack, by the way, was a big hit with them despite the mocking of it by some executives that took place pre-crisis.

7.3.h Improvement: unleashing Innovation Ethos

What do we mean, finally, by "Innovation Ethos"? How is it different from "Improvement" or "Continuous Improvement" which is the generic shorthand we use for this eighth category of element identified earlier in this chapter? The term "Innovation Ethos" is intended to convey something that has permeated and suffused this book – the idea of transforming risk into value. Innovation Ethos is a culture of continuous improvement on steroids. It is intended to push leaders to think more broadly and more proactively about how continuous improvement and lessons learned can improve, support and even create new products and services. It's not just about process improvement – which is great in and of itself – it's about finding the opportunity for creating better or even new products and services through a deep understanding of your ESGT issues, risks and crisis experiences.

One of the singular purposes of this book has been to expand the lens (or the aperture of the lens to be more accurate) of all leaders of all types of organizations to not only look at their core purpose (mission, vision and strategy) but to expand their focus to include essential ESGT strategic issues heretofore ignored, slighted or overlooked. Through that expanded aperture, leaders need to know and deploy their organization's ESGT issues in furtherance of the core mission and create even greater value for their organization and stakeholders.

It is my conviction that it is only the well-equipped, highly resilient organizations that can really take this additional eight step – they must have most if not all of the preceding seven elements of resilience in place (at least to a large degree), as having them in place is what makes them knowledgeable and capable of learning the lessons needed for innovation.

Next, I profile several examples of organizations that are not only engaging in systematic continuous improvement through a robust learning culture – they are engaging in outside-of-the-box new value creation through their own version of "Innovation Ethos". Let's take a look.

7.3.h.i L'Oréal: transforming diversity risk into increased revenue

In the early 2000s, L'Oréal created a global ethics program, and one of its first products was a Global Code of Conduct that was called "The Beauty of Ethics and the Ethics of Beauty". But the program was not just skin deep; it was profound. The ethics officer for L'Oréal since its inception is innovator Emmanuel Lulin, who built the program from scratch with the very visible and comprehensive support of CEO Jean-Paul Agon. We talk a little more about this extraordinary partnership within L'Oréal between chief ethics officer and CEO in Chapter 8. However, here I would like to mention two important initiatives that the company has undertaken over the past decade that demonstrate how a highly resilient and self-aware company like L'Oréal can actually create new value.

Not so long ago, L'Oréal realized that its global beauty products catered exclusively to light skinned or white people (mainly women). That was not only an issue from a perceived possible racial insensitivity or even discrimination standpoint but meant that large and untapped markets and customers were either underserved or not served at all. It was this combination of drivers that led L'Oréal to develop whole new lines of beauty products for these customers who are not white-skinned around the world, mainly in Africa, Asia and parts of Latin America. The result: a dramatic win/win for both the company and its stakeholders achieved through the formulation of a more ethical, inclusive and diverse approach to their products and the needs of the marketplace. L'Oréal's share of the beauty products market increased exponentially partly for this reason and so did its good reputation for being an inclusive, diversity-supporting and forward-thinking company. They have been the recipients of numerous awards over the years for ethics, diversity, inclusion and similar internal company initiatives.

Also a few years ago, L'Oréal initiated a scientific program to develop artificial skin to use for beauty product testing. Major beauty and pharma companies around the world use animal testing to test their products for human safety. Animal testing is widely disliked and even condemned by many stakeholders worldwide. Although it is not always possible to police what happens in all factories and manufacturing facilities around the world – especially those in challenging countries like China – L'Oréal focused earlier and more effectively than many of its competitors on eliminating this contentious issue by developing their own patented artificial skin. Most if not all of their animal testing is now done on this artificial skin.[20]

7.3.h.ii Two global Fortune 50 *companies: transforming cyber-risk into cyber-value*

A few years ago, I undertook an extensive research project for The Conference Board and produced a lengthy report called "Emerging Practices in Cyber-Risk Governance". In that study, I profiled five leading global corporations that were willing to share their cyber-risk management and governance practices with me on the condition of anonymity. Two of those companies not only had very robust and well-developed cyber-risk management and governance practices but were also able to go above and beyond that resilience and build new value for their shareholders and other stakeholders by creating cyber-products that depend greatly on their credibility and resilience as companies.

Think about what it would be like for a company that sells certain products and services that are integral to the well-being and resilience of any organization (for example, a consulting company that sells compliance advisory services) that doesn't itself have a robust compliance program? Would you want to buy compliance services from such a company? Wouldn't you as a customer feel better about a company that did both things well and had the credibility of practicing what it preaches? The two companies profiled here from the aforementioned Conference Board study do both of these things well – internal cyber-resilience and the sale of robust cyber-related products.

- A US-based Global *Fortune 50* Technology Company:
  Because of the nature of its business, the company sees a large opportunity for creating a cyber security competitive advantage in its products and services. This has served as an additional driver of board and executive attention, especially with cloud-based services, including cyber security solutions and services and products that are now generally being delivered through the internet as opposed to shrink-wrapped boxed products. Through robust risk management practices and predictable and repeatable operational security hygiene programs, the company feels it can meet the objectives set forth in its core security priorities.
- A European-based Global 50 Financial and Insurance Company:
  Plans are currently in place to increase cyber-risk and security monitoring, detection, and mitigation within the company through advanced malware detection and the creation of a Security Operations Center. Knowing that this is a risk that will continue to grow exponentially, and as an insurer with a cyber-crime insurance product, the company recognizes the need to understand this risk in real time and to have a well-informed view of what works internally within a company to combat cyber-risk and maintain healthy cyber security. This extends to an understanding of what a systemic threat is, not only to a company but also to an entire sector or industry and to business generally.[21]

What differentiates the truly resilient organization from the more fragile, less effective organization? The fully resilient organization actually learns from its mistakes, has a process embedded into its core activities that focuses on continuous learning and improvement where it's ok to make mistakes and it's ok to fail from time to time as long as you dissect and understand the lessons learned and make the improvements necessary to make the organization, its processes, products and services more robust and more resilient. In other words, the resilient organization will put a high premium on learning and on continuous improvement, and those that are truly cutting edge and transformative (as we will explore in Chapter 8) are the ones that transform their risks and their resilience into innovation – product improvement and creation.

7.4 From Vicious to Virtuous: what type of organizational resilience lifecycle does your organization have?

In this section we put the eight elements of organizational resilience into context by providing an overview of four core types of organizational resilience lifecycles an organization may fit into. The purpose of this categorization is to help the leaders of an entity understand where their organization's gaps may be in terms of creating a "Robust Resilience Lifecycle" or even "Virtuous Resilience Lifecycle" which, at the end of the day, is where a conscientious, stakeholder-centric organization – regardless of purpose or form – would hopefully want to be. For organizations that more clearly fit into a "Fragile Resilience Lifecycle" or "Vicious Resilience Lifecycle" I would say – get to work to save your organization from danger and your stakeholders from harm and value destruction.

There are four basic types of Resilience Lifecycles. We review them here from least developed or undesirable to most developed and desirable.

7.4.a *The Vicious Resilience Lifecycle*

At one end of the Resilience Lifecycle spectrum is the "Vicious Resilience Lifecyle". The hallmark of this form of mismanagement or non-management is that although it may have one or two of the eight elements of resilience in place, even then, whatever element may be present is probably compromised or flawed (see Figure 7.14).

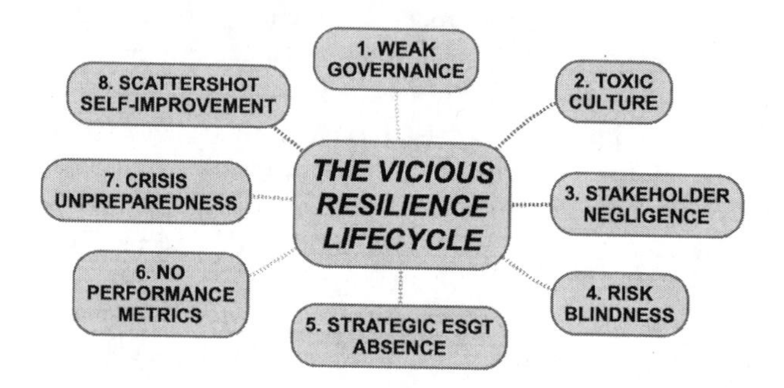

Figure 7.14 The Vicious Resilience Lifecycle.

Source: Author and GEC Risk Advisory.

The "Vicious Resilience Lifecycle" occurs mainly in organizations that have the following characteristics:

- Most, if not all, of the eight resilience elements are missing.
- Even those that may be there in some form are deeply flawed and challenged and would not qualitatively pass muster in an audit or even cursory review.
- While there may be "governance" in place, in the Vicious Resilience Lifcycle governance probably exists in name only and/or is deeply flawed and compromised.

7.4.a.i *Example: Massey Energy*

Massey Energy under CEO Don Blankenship, who ruled there with an iron fist for decades before the explosion of its West Virginia coal mine in 2010 (the Upper Big Branch mining disaster) that killed 29 miners, revealed a dark underbelly of ineffective governance, toxic culture, shoddy risk management, stakeholder ignorance or negligence, nonexistent crisis management, and misaligned performance management. Worst of all, they had a culture of health and safety gross negligence and intentional misconduct with the payment of related fines treated by the CEO and his rubber-stamping luminary board as a cost of doing business.

7.4.b The Fragile Resilience Lifecycle

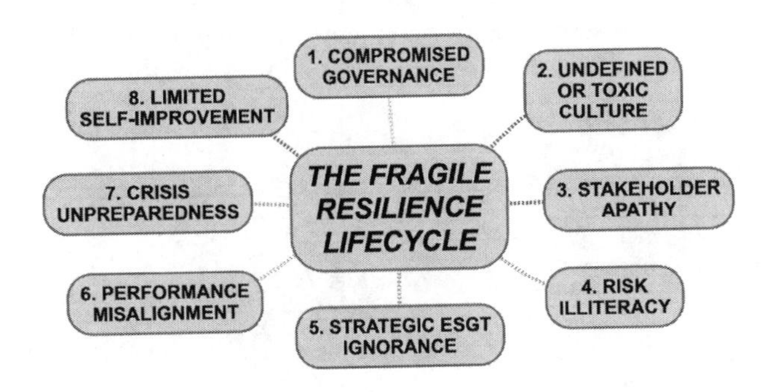

Figure 7.15 The Fragile Resilience Lifecycle.

Source: Author and GEC Risk Advisory.

What distinguishes the organization with a "Fragile Resilience Lifecycle" (see Figure 7.15) from the Vicious variety is that its leaders are not completely derelict or oblivious about having some of the elements of resilience in some shape or other. In this case, there are a few (perhaps 3–4) of the eight resilience elements present and even then, some may be flawed or compromised.

7.4.b.i Examples: Wynn Resorts/The Weinstein Company

In this category of "Fragile Resilience Lifecyle", I am lumping in two companies that have had varying degrees of problems emanating primarily from their CEOs' sexual harassment or assault allegations (ranging from bad to criminal behaviors), embedded in a more or less flawed resilience framework that shared several of the following characteristics:

- Weak, dysfunctional or passive rubber-stamping governance.
- Little or no attention paid to culture and a toxic culture emanating from the CEO.
- Ignorance or avoidance of ESGT issues or integrating them into strategy.
- A passing interest in stakeholders other than the principal stakeholder (in this case the shareholders and the controlling minority shareholder).
- Weak or compromised risk management.
- Seat-of-the-pants crisis management or solely PR-oriented crisis management.

- Ineffective or flawed performance management or performance limited exclusively to financial metrics.
- No real understanding of improvement/continuous improvement, let alone integrating lessons learned into an innovation ethos – "nothing new to learn here" approach to learning from mistake, risk or experience.

Of course there is a big difference between the Weinstein and Wynn cases – one when bankrupt while the other has survived and attempted to reinvent itself by adding some of the previously missing resilience elements to its repertoire.

7.4.c *The Robust Resilience Lifecycle*

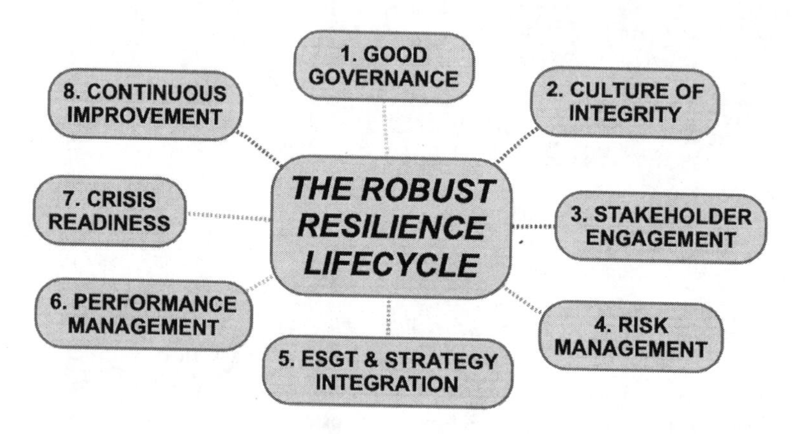

Figure 7.16 The Robust Resilience Lifecycle.

Source: Author and GEC Risk Advisory.

The "Robust Resilience Lifecycle" (see Figure 7.16) presents a completely different ball of wax – one that includes serious efforts in at least five to six elements of resilience. In the case of the "Robust Resilience Lifecycle", leaders are taking many of these elements seriously and succeeding in implementing most of them.

7.4.c.i Example: GE

General Electric (GE) presents in interesting case of a company that has been both widely admired by its stakeholders and simultaneously feared by its competitors over the years, even decades. It would probably qualify as a company that was "built to last" to coin a phrase from Jim Collins, the famous author of the *Built to Last, Good to Great* and other books. Indeed, in my own experience working for competitors of GE and as a B2B purchaser of services from GE as well as having learned a lot about some of their internal ethics, compliance, risk and other programs in some detail through professional associations, GE could have been said to not only have a "Robust Resilience Lifecycle" but perhaps a "Virtuous" one (see next section) in the past.

However, I put GE in this category because, for now, they have experienced some serious leadership, culture and operational setbacks that may point to a reversal of fortune and direction on our spectrum of resilience. Stay tuned, however, because their long-term building of internal programs for resilience may actually be a buffer and part of an arsenal of measures that may get them through their current difficult times. Some of the things we could say about their current "Robust Resilience Cycle" would include:

- Functional board governance though under duress recently.
- Attention paid to culture and the importance of having ethics, compliance and a good tone from the top. However, recent revelations about their former CEO Immelt may be clouding this claim.
- More than an awareness of ESGT issues and some integration of the perceived most important ESGT issues into organizational strategy.
- More than a passing interest in stakeholders beyond the principal stakeholder – i.e., understanding of and relationships being built with other primary stakeholders like NGOs, customers and employees.
- Risk management is taken seriously and a system is in place, even potentially a highly evolved enterprise risk management system.
- Crisis management is understood to be part of a robust approach to enterprise, people and value protection.
- Performance management is taken seriously and a more holistic approach is potentially in place, including a "balanced scorecard" approach or similar.
- This is an organization that not only implements continuous improvement but also seems to have aspects of the Innovation Ethos described earlier.

7.4.d *The Virtuous Resilience Lifecycle*

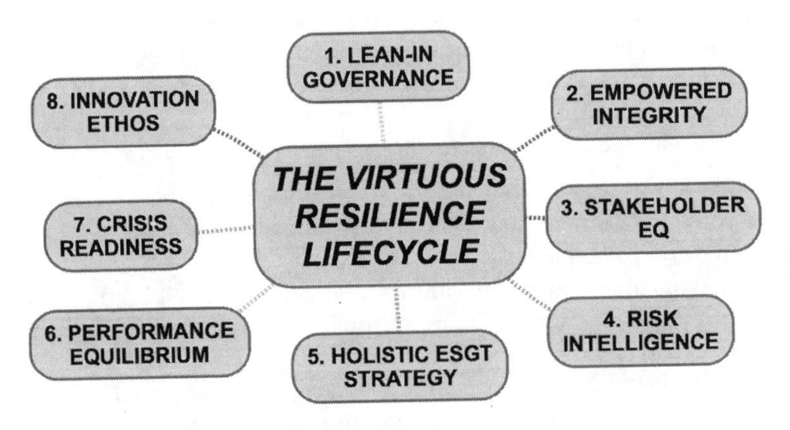

Figure 7.17 The Virtuous Resilience Lifecycle.

Source: Author and GEC Risk Advisory.

The "Virtuous Resilience Lifecycle" (see Figure 7.17) represents the epitome of resilience best practices – one that includes successful programs in all or most of the eight elements of resilience. Leaders take all of these issues seriously, provide enthusiastic support – through tone from the top and actual resources – for the eight elements which in turn are more or less successfully implemented.

The "piece de resistance" that truly distinguishes the organization that has the "Virtuous Resilience Lifecycle" from the others is the fact that they not only implement continuous improvement but go above and beyond to adopt the Innovation Ethos that permits them to create new sources of value.

7.4.d.i Example: Starbucks

Starbucks under its founder and first CEO and Chairman, Howard Schultz, represents a company with a "Virtuous Resilience Lifecycle". Starting with tone from the top – in terms of both governance and culture – and working its way through the other elements of a "Virtuous Resilience Lifecycle", under a new CEO Starbucks continues to exhibit the characteristics that Schultz built over a long period of time. This includes creating a holistic stakeholder-oriented culture that actually puts the employee first. Schultz is famous for saying that if you treat the employee really well – in terms of culture, benefits, work environment, pay – everything else follows, including customer care, shareholder value, etc. In many ways, he seems to have been right.

Indeed, a crisis that Starbucks experienced in the spring of 2018 involving a manager of one of their stores in Philadelphia who inappropriately and discriminatorily called the police on two African American men waiting in the store without purchasing a product led to an immediate and very effective crisis management response, risk management analysis and follow-up through bias training mandated for all stores in the US within weeks of the incident. Additionally, the manager was fired, and strong, timely and effective social media and media management followed.

That's what I would call a robust and even virtuous resilience response. Starbucks is also well known for not only creating its original product/service, which at the time was truly revolutionary, but for constantly innovating and looking for new products and services.

Organizations with this type of resilience lifecycle are generally characterized by:

- Governance best practices (including diversity) that are more advanced than most – perhaps even the lean-in type of governance that is most needed in today's complex, fast-changing world.
- Culture and the importance of having ethics, compliance, CSR and a great tone from the top – something that is central to this leadership's strategic and tactical goals and objectives.
- ESGT issues that are not only central to strategy but are central to new product and/or service development and value creation.
- An organizational embrace of multi-stakeholder engagement – while completely understanding the centrality of the primary stakeholder, a heavy focus on other key stakeholders as well.

- An advanced risk management system in place – both enterprise risk management and effective strategic risk governance.
- Performance equilibrium in place in the form of a "balanced scorecard" or similar approach that not only integrates nonfinancial metrics but does so from the very top of the organization (CEO) to the newest employee walking in the door.
- Crisis management that is understood to be part of a robust approach to enterprise, people and value protection that is well developed and regularly exercised.
- Continuous improvement that is the foundation and an Innovation Ethos that is the distinguishing quality of this organization where revised and improved or new products and services are created regularly.

7.4.e *The Resilience Lifecycle: key actors*

To wrap up this discussion of Resilience Lifecycles, Figure 7.18 depicts the eight elements of a "Virtuous Resilience Lifecycle" and the key actors that should be driving each of these elements within an organization. The bottom line is this: leaders must populate these elements with the right people, in the right roles, with the right cross-functional and cross-disciplinary collaboration, with the right verbal and financial support, all championed by a great tone from the top. Not a small task, but one that is eminently doable when the right people lead an organization.

Figure 7.18 The Virtuous Resilience Lifecycle: key actors.

Source: Author and GEC Risk Advisory.

Finally, Figures 7.19 and 7.20 provide a big-picture overview of the elements that would typically be present (in various degrees of development) in each of the four Resilience Lifecycles we have just reviewed.

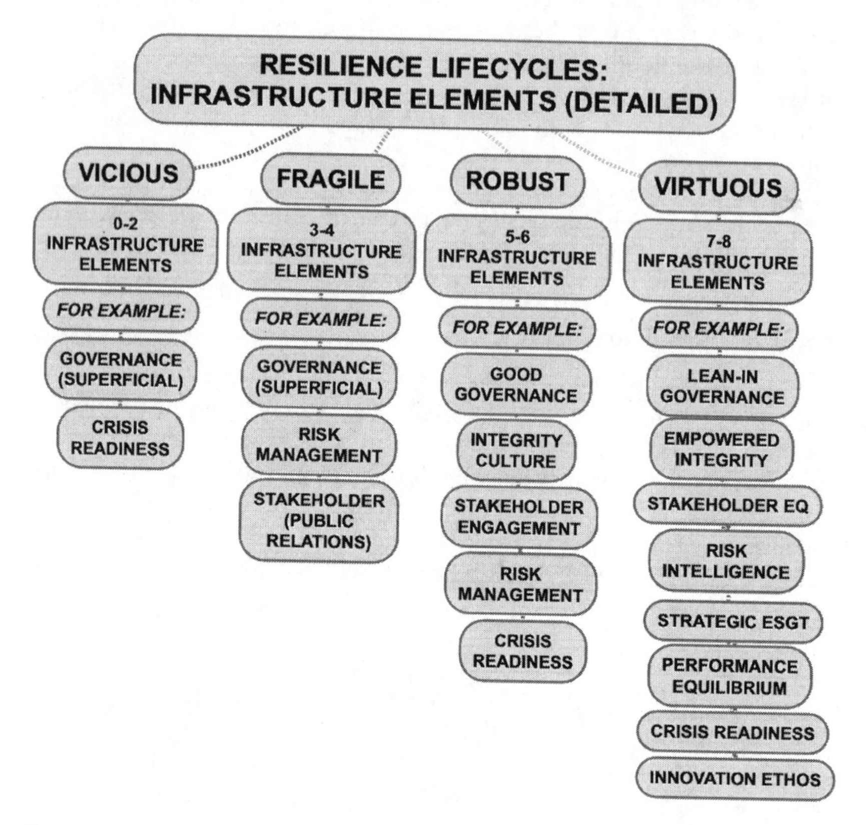

Figure 7.19 Resilience Lifecycles: infrastructure elements.

Source: Author and GEC Risk Advisory.

7.5 Determining your organizational resilience: a diagnostic tool

For those interested in undertaking a back-of-the-envelope exercise to find out where on the organizational resilience lifecycle spectrum your company or organization may lie, Table 7.14 provides a set of parameters to assess where on each of the eight elements your organization may fit (is it highly evolved [high], somewhat evolved [medium], under-evolved [low] or totally unevolved [none]) (Figure 7.21).

By reviewing each of these categories and then adding up the results you will find a key in Table 7.15 to determine whether your organization has the Vicious, Fragile, Robust or Virtuous Resilience Lifecycle. Good luck!

Figure 7.20 Organizational Resilience Lifecycles: the eight elements compared.

Source: Author and GEC Risk Advisory.

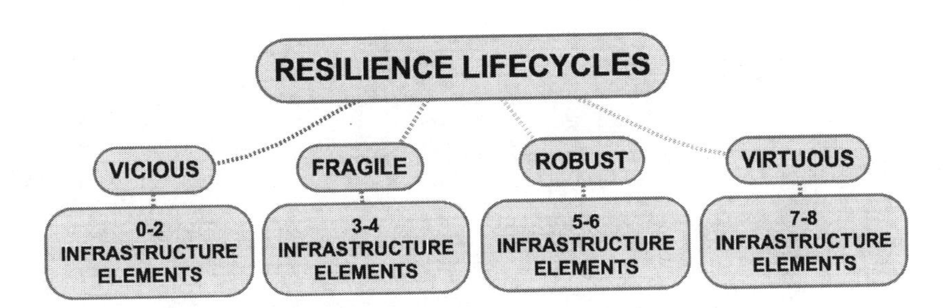

Figure 7.21 Resilience Lifecycles: diagnostic.

Source: Author and GEC Risk Advisory.

Table 7.14 The eight elements of organizational resilience: a diagnostic framework

Element	*High (3 points)*	*Medium (2 points)*	*Low (1 point)*	*None (0 points)*
1. Governance	• Diversity of board. • Lean-In Risk Oversight. • Best in class governance practices.	• Some diversity. • Risk oversight but more ad hoc. • Missing elements of best-in-class governance.	• No/bare bones diversity. • No specific risk oversight/ sit-back oversight. • Poor practices.	• No diversity. • No risk oversight. • Shoddy or no governance practices.
2. Culture	• Effective E&C Program. • Proactive Leadership. • Recognized by stakeholders.	• Some elements of an effective program. • Some leadership visibility. • No significant recognition.	• No organized program. • No leadership visibility on the subject. • No outside recognition.	• No program. • Leadership issues/ infractions. • Outside expressions of concern.
3. Stakeholder Management	• Sophisticated stakeholder management program in place, including sophisticated integrated PR/comms and risk. • Culture of catering to stakeholders through effective reputation emotional intelligence. • Robust reputation risk lens incorporated into ERM and strategy.	• Awareness of important stakeholders but no formal management program in place. • Reactive rather than proactive culture when things go wrong and affect stakeholders. • Reputation risk concerns but no systematic incorporation into risk management or strategy.	• No real stakeholder management or awareness programs/a pure PR approach. • Reactive to stakeholder related mistakes/problems. • Reputation risk lens is not incorporated into any risk or stakeholder related programs.	• No stakeholders matter other than shareholders (and even then they might be ignored). • Reputation and reputation risk are not concepts that are considered by management/leaders.
4. Risk management	• Sophisticated ERM system in place (use evolution chart). • Risk integrated into strategy. • Risk transformed into opportunity.	• Some risk management in place – not fully evolved. • Risk separated from strategy. • Risk and opportunity not connected.	• Reactive risk management derived from crisis. • No formal risk management. • No interrelationship between risk and innovation.	• Risk is not factored into anything the entity does.

5. ESGT Strategy	• Fully developed and integrated ESGT Issues Value Chain analysis, integration into overall strategy and implementation.	• Some ESGT Issues Value Chain activity, some integration into strategic considerations, some implementation.	• Limited or no ESGT Issues Value chain activity, sporadic or accidental integration of ESGT issues into strategy.	• No consideration of ESGT Issues whether through a "Value Chain" approach or otherwise.
6. Performance Management	• Fully formalized and integrated financial and nonfinancial incentive performance metrics in place from the top (CEO and C- Suite) all the way down to the entry levels of the organization • Holistic balanced scorecard approach to organizational objectives trickling down into each employee's performance objectives. • Use of outside metrics/ benchmarking to measure entity performance in sector or across other outside metrics. • Proactive lean-in board oversight of compensation.	• While formal, less stringent and less fully developed financial plus nonfinancial incentive performance metrics in place (financial may be well formed, but nonfinancial metrics are wanting). • Scorecard, if it exists, is skewed heavily to financial metrics alone. • Limited use of outside metrics/benchmarking to measure entity performance in sector or across other outside metrics. • More laissez-faire attitude by board on compensation matters.	• Mediocre to poor financial incentive performance metrics in place few if any nonfinancial metrics) – no sense of a balanced scorecard. • No use of outside metrics/ benchmarking to measure entity performance in sector or across other outside metrics. • Board "sit-back" attitude on compensation matters.	• If performance is measured, it is purely by financial targets that may often be unreasonable, unreachable or at worst incentivizing the worst kind of aggressive sales behaviors.
7. Crisis Management	• Fully formed crisis management team and plan. • Proactive scenario planning on periodic basis. • Involvement of leadership and board in crisis exercises and planning.	• More ad hoc approach to crisis management – no or static team and plan in place. • Less systematic approach to prevention. • Leadership interested but not committed.	• No formal crisis management team or plan in place. • No systematic approach to prevention. • Leadership may be interested but not committed.	• Nothing in place. • No leadership interest.

(*Continued*)

Table 7.14 Continued

Element	*High (3 points)*	*Medium (2 points)*	*Low (1 point)*	*None (0 points)*
8. Improvement	• This is an innovative organization in the first place with leaders who are constantly looking for ways to improve internal processes (six sigma), create better products and services with a systematic approach to innovation • A true "learning company" – The outside world has recognized that this is a leading-edge company through awards, best in class lists, benchmarking, etc. • Leadership has specifically asked for risk management to be developed into best in class, incorporated root cause analysis & other methods into strategy and has a proactive approach to lessons learned and constant improvement.	• Not a company that leads in innovation but is still focused on improving existing processes and attaining new and innovative products and services. • No outside recognition to speak of on innovation. • While risk management may be effective and relatively well evolved in this entity, it is not necessarily integrated with strategic product or service development or business planning. • Leadership does not see the connection between risk and innovation.	• While there may be a focus on improving existing processes and some innovation, there is no effective approach to innovation. • No outside recognition on innovation. • Risk management may be less than well evolved in this entity, and even if it is, it is unlikely to be considered or integrated with strategic or business planning • No connection between risk and innovation.	• Nothing in place. • No leadership interest in lessons learned, continuous improvement, risk management, root cause analysis and the like.

Source: Author and GEC Risk Advisory.

Table 7.15 Assessing your organization's sustainable resilience score

Point Score	*Type of Organizational Resilience Lifecycle*
20–24	Virtuous Resilience Lifecycle
14–19	Robust Resilience Lifecycle
5–13	Fragile Resilience Lifecycle
0–4	Vicious Non-Resilience Lifecycle

Source: Author and GEC Risk Advisory.

Notes

1 The English theologian and historian Thomas Fuller appears to be the first person to commit to the notion that "the darkest hour is just before the dawn" to print in his religious travelogue *A Pisgah-Sight of Palestine and The Confines Thereof*, 1650. https://www.phrases.org.uk/meanings/darkest-hour-is-just-before-the-dawn.html. Accessed on September 16, 2018; and Lyrics from the song by Crosby Stills and Nash, circa 1969, "Long Time Gone". https://genius.com/Crosby-stills-and-nash-long-time-gone-lyrics. Accessed on September 18, 2018.

2 Goalcast. Abraham Lincoln Quotes. https://www.goalcast.com/2018/01/05/abraham-lincoln-quotes/. Accessed on November 23, 2018.

3 100 Resilient Cities. http://100resilientcities.org/. Accessed on October 28, 2018.

4 100 Resilient Cities. http://100resilientcities.org/. Accessed on October 28, 2018.

5 ClimateWise. University of Cambridge Institute for Sustainability Leadership. "Investing for Resilience". December 2016. https://www.cisl.cam.ac.uk/resources/publication-pdfs/Investing-for-resilience.pdf. Accessed on April 23, 2019.

6 *Psychology Today*. "Resilience". https://www.psychologytoday.com/us/basics/resilience. Accessed on October 28, 2018.

7 Wikipedia. "Psychological Resilience". https:/en.wikipedia.org/wiki/Psychological_resilience. Accessed on October 28, 2018.

8 David Denyer. "Organizational Resilience: A summary of academic evidence, business insights and new thinking". British Standards Institution (BSI) and Cranfield University. October 2017.

9 World Economic Forum. *Global Risks Report 2018*. https://www.weforum.org/reports/the-global-risks-report-2018. Accessed on July 22, 2018.

10 DRI International. The Fourth Annual DRI International Global Risks and Resilience Trends Report. 2018.

11 Andrea Bonime-Blanc. "Resilience". The Sage Encyclopedia of Corporate Reputation. 2016. Sage.

12 Michael Pocalyko. "A Field Guide to Bad Directors". *NACD Directorship Magazine*. August 2018. https://read.nxtbook.com/nacd/directorship/july_august_2018/a_field_guide_to_bad_director.html. Accessed on October 21, 2018.

13 Andrea Bonime-Blanc and Leonard Ponzi. Understanding Reputation Risk: The Qualitative and Quantitative Imperative. *Corporate Compliance Insights*. 2016. https://gecrisk.com/wp-content/uploads/2016/12/ABonimeBlanc-LPonzi-Understanding-Reputation-Risk.pdf. Accessed on October 28, 2018.

14 Andrea Bonime-Blanc. "Volkswagen's Perfect Storm". *Ethical Boardroom*. Winter 2016. https://gecrisk.com/wp-content/uploads/2016/02/ABonimeBlanc-Volkswagens-Perfect-Storm-Ethical-Boardroom-Jan-2016.pdf. Accessed on August 22, 2018.

15 RepRisk. https://www.reprisk.com/. Accessed on April 16, 2019.

16 There are some wonderful books written by experts on risk management and in no particular order I am happy to list several of them here: James Lam. *Enterprise Risk Management.* Wiley, 2014; Daniel Wagner and Dante A. Disparte. *Global Risk Agility and Decision-Making.* Palgrave McMillan, 2016; Howard Kunreuther and Michael Useem. *Mastering Catastrophic Risk.* Oxford University Press, 2018.

17 Andrea Bonime-Blanc. "Effective Risk Governance". *NACD Directorship Magazine.* January/February 2016.

18 NACD and Protiviti.

19 Editorial Board. "Hubris is an ever-present risk for high flying executives". November 20, 2018. *The Financial Times.* https://www.ft.com/content/69343192-ebf0-11e8-8180-9cf212677a57. Accessed on November 22, 2018.

20 Bob Woods. "Companies are Making Human Skin in Labs to Curb Animal Testing of Products". May 28, 2017. CNBC.com. https://www.cnbc.com/2017/05/25/loreal-is-making-lab-produced-human-skin-to-curb-animal-testing.html. Accessed on November 26, 2018.

21 Andrea Bonime-Blanc. "Emerging Practices in Cyber-Risk Governance". November The Conference Board, 2015.

8 Boom

Transforming resilience into sustainable value

Boom: risk and opportunity

In the early 20th century, the first dean of Harvard Business School, Edwin Francis Gay, said: "The purpose of business is to 'make a decent profit, decently'—not to maximize profits at any costs".[1] An interesting comment given what has transpired in the intervening century – especially in the past 30 years since Gordon Gekko (played by Michael Douglas) famously declared that "greed was good" in the 1987 movie *Wall Street*. As a young lawyer practicing law on Wall Street at the time, that sentiment rang nasty to me then and even nastier with hindsight and the advent of the Great Recession of 2008–2009 and the income inequality that has grown precipitously globally between the top 1% (and even .001%) and the remaining 99–99.9% of humanity.

In contrast, the Dalai Lama, a recognized global leader for all the right reasons, has made countless statements about this issue, including the following recent tweets:

> In the present circumstances, no one can assume that someone else will solve their problems. Each one of us has a responsibility to help our global family in the right direction. Good wishes are not sufficient; we must become actively engaged.
>
> (September 14, 2018)[2]

> We need friends and friendship is based on trust. To earn trust, money and power aren't enough; you have to show some concern for others. You can't buy trust in the supermarket.
>
> (November 30, 2018)[3]

Other manifestations of the clash of values in today's business and society are exemplified by vociferous and successful protests by employees against their own companies – consider these quotes from Google employees upon walking out en masse in November 2018 because of steep payouts made to departing executives accused of sexual harassment:

> Happy to quit for $90M—no sexual harassment required.
>
> What do I do at Google? I work hard every day so the company can afford $90,000,000 payouts to execs who sexually harass my co-workers.[4]

And seemingly out of left field, the granddaughter of Walt Disney, Abigail Disney, made shockwaves when she tweeted and penned a sharply critical op-ed for the *Washington Post* in which she attacked the extraordinary pay of the much-admired and successful but in her opinion vastly overpaid CEO of Disney, Bob Iger, and others in similar positions of wealth and power.[5] She stated:

> It is time to call out the men and women who lead us and to draw a line in the sand about how low we are prepared to let hard-working people sink while top management takes home ever-more-outrageous sums of money. It is unreasonable to expect corporate boards to act as a check on this trend; they are almost universally made up of CEOs, former CEOs and people who long to be CEOs.

Herein lies the opportunity – the opportunity to make corrections, revisions, amendments to aspects of business, government and society that are not working and are indeed hurting larger swaths of stakeholders. In this final chapter, we pull together everything we have discussed in this book and give leaders of all kinds a road map to sustainable value creation for the most stakeholders, not only the "most important" one.

Chapter 8: Boom – summary overview

8.1 Chapter overview
8.2 A typology of organizational resilience and sustainability
- 8.2.a The generic organizational model: connecting the dots
  - 8.2.a.i Inputs: ESGT and the resilience lifecycles
  - 8.2.a.ii Enablers: leadership ESGT styles
  - 8.2.a.iii Outputs: financial and nonfinancial (ESGT)
  - 8.2.a.iv Stakeholder and value impact: expectations and reputation
  - 8.2.a.i.vWhat next?
- 8.2.b A typology of organizational resilience and sustainability
  - 8.2.b.i The Outlaw Organization
    - 8.2.b.i.A What an Outlaw Organization looks like
    - 8.2.b.i.B Silk road
    - 8.2.b.i.C 1MDB
  - 8.2.b.ii The Irresponsible Organization
    - 8.2.b.ii.A Massey Energy
    - 8.2.b.ii.B Theranos
    - 8.2.b.ii.C FIFA
    - 8.2.b.ii.D Odebrecht
  - 8.2.b.iii The Compromised Organization
    - 8.2.b.iii.A Volkswagen
    - 8.2.b.iii.B Wells Fargo
    - 8.2.b.iii.C Michigan State University
    - 8.2.b.iii.D The Catholic Church

- 8.2.b.iv The Emergent or Undervalued Organization
  - 8.2.b.iv.A Uber
  - 8.2.b.iv.B Altria
  - 8.2.b.iv.C Twitter
- 8.2.b.v The Responsible Organization
  - 8.2.b.v.A Merck & Co. Inc
  - 8.2.b.v.B Starbucks
  - 8.2.b.v.C Eli Lilly
  - 8.2.b.v.D Apple
- 8.2.b.vi The Transformative Organization
  - 8.2.b.vi.A Salesforce
  - 8.2.b.vi.B L'Oréal
  - 8.2.b.vi.C Microsoft
  - 8.2.b.vi.D Epic Theatre Ensemble

- 8.2.c A cautionary note: beware of transforming value into unacceptable risk
- 8.2.d Applying the organizational typology to governments

8.3 The value of "values": the new multi-stakeholder bottom line

- 8.3.a What is "value"? What are "values"?
- 8.3.b The value of values: the business case for ESGT, organizational resilience and sustainability
  - 8.3.b.i Some big-picture ESG "value" studies
    - 8.3.b.i.A Eccles, Ioannou and Serafeim Sustainability Studies
    - 8.3.b.i.B University of Oxford and Arabesque Partners "from stockholder to stakeholder"
    - 8.3.b.i.C Accenture's "the bottom line on trust"
    - 8.3.b.i.D Boston Consulting Group "Total Societal Impact"
    - 8.3.b.i.E BSR GlobeScan "State of Sustainability 2018"
    - 8.3.b.i.F BAML ESG investment strategy report
  - 8.3.b.ii More specific ESGT "value" data points
    - 8.3.b.ii.A McKinsey and Catalyst on the value of diversity
    - 8.3.b.ii.B National Association of Corporate Directors (NACD) "culture as an asset"
    - 8.3.b.ii.C Ethical Systems "ethics pays"
    - 8.3.b.ii.D The value of whistleblower reporting
    - 8.3.b.ii.E Ethisphere "Ethics Premium Research"
    - 8.3.b.ii.F S&P Global the value of "ESG and preparedness"
    - 8.3.b.ii.G Comparitech: the value of cyber preparedness
    - 8.3.b.ii.H Steel City Re: reputation effects on equity returns and measuring "reputation value"
    - 8.3.b.ii.I New research: resilience pays over time
  - 8.3.b.iii The value of values

8.1 Chapter overview

We have finally arrived at the "Boom" part of our journey – "boom" meant in a good way, of course! After traveling from that place of "Gloom" (Chapter 1), equipped with the leadership, culture and integrity we need to survive and thrive (Chapter 2), and having navigated through the dangers, turbulence and promise of the ESGT Scylla and Charybdis (Part II), finally arriving in Chapter 7 – Metamorphosis – to a consideration of what organizational resilience is all about, we can now finally pull all of these strands together to focus on how leadership, ESGT and resilience relate to long-term organizational sustainability and value creation.

As the reader probably knows by now, this book is focused on getting leaders to think proactively, holistically and strategically about their most important ESGT issues so that they can be prepared for the risks and the crises that will most certainly come, as well as the opportunities for value protection and creation.

In this chapter we explore the following:

- A Typology of Organizational Resilience and Sustainability that builds on everything we have discussed in this book so far and offers a typology of six kinds of organization from the Outlaw (or least desirable) to the Transformative (or most desirable) based on two key criteria – ESGT leadership quality (pulling lessons especially from Chapter 2) and organizational resilience lifecycle (which we just examined in Chapter 7).
- We zero in on a concept we have mentioned frequently but not explored in depth until now – value. What is "Value"? What are "Values"? At the end of the day, these are the things that organizational leaders are trying to create, achieve and/or preserve on behalf of their most important stakeholders. Our exploration broadens the concept from purely financial value to a broader concept of nonfinancial value and "values" and explores recent progress in this area from leading thinkers and institutions.
- We then close with a future-oriented consideration of how leaders can apply some of the lessons we have learned to a critical topic for today and tomorrow relevant to every stakeholder imaginable (all living creatures): the future-proofing of technology trust – what do organizational leaders need to do to create tech stakeholder trust? We offer a simple but hopefully useful governance perspective on what leaders can and should do right here, right now.

8.2 A typology of organizational resilience and sustainability

8.2.a The generic organizational model: connecting the dots

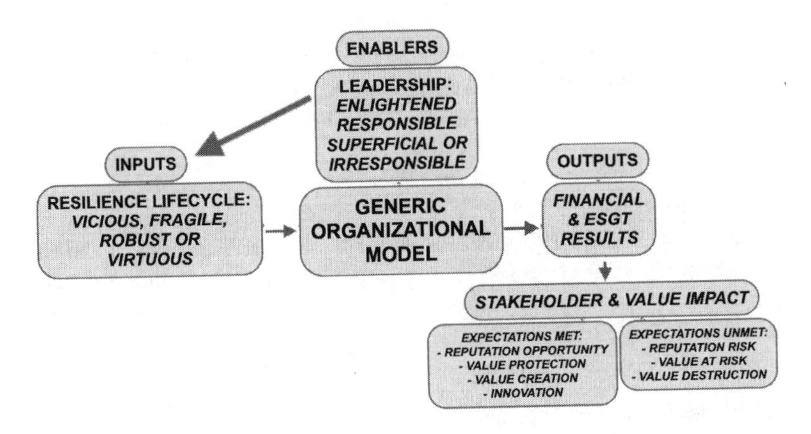

Figure 8.1 The generic organizational model.

Source: Author and GEC Risk Advisory.

Figure 8.1 illustrates in one cohesive whole a bird's-eye view of what we have been talking about throughout this book:

- The "inputs" summarized by reference to the four types of "Resilience Lifecycle" examined in Chapter 7 – vicious, fragile, robust or virtuous.
- The "enablers" or the type of ESGT leadership the organization has – enlightened, responsible, superficial or irresponsible.
- The "outputs" – both financial and nonfinancial (or ESGT) results; and
- The "stakeholder and value impacts" in the form of reputation risk or opportunity and value preservation, destruction, creation or innovation.

The purpose of bringing everything that has preceded this chapter into one simple picture is to drive home an important concept for all leaders at all levels and sectors: leaders (both management and governance) are at the center of enabling or disabling the overall dynamics of their organization – be it a corporation, a nonprofit, a university or government. All organizations are made up of people, some of whom are leaders (the board and the executive team) and some of whom populate the inner workings of the organization and have impacts both as actors and as stakeholders on the financial and nonfinancial organizational results and ultimate stakeholder well-being (or the absence thereof).

The purpose of this simple visual is to provide a bird's-eye view of our respective roles and responsibilities in making the mission, vision, values and strategy of our organizations work seamlessly and purposefully as an integrated whole.

What follows is a brief key to what each of the four main components illustrated in Figure 8.1 means.

8.2.a.i Inputs: ESGT and the resilience lifecycles

Inputs refer to all of the necessary and desirable programs, policies, controls and frameworks and ultimately strategy that an entity needs to have in place to properly manage its portfolio of issues, including ESGT issues, risks and opportunities.

More specifically, it asks the question: What form of Resilience Lifecycle does your organization have? Chapter 7 – Metamorphosis – provides guidance on the eight elements of organizational resilience to help place an organization on the spectrum of the four types of organizational resilience examined in that chapter. Are your inputs closest to the Vicious, the Fragile, the Robust or the Virtuous Resilience Lifecycle?

8.2.a.ii Enablers: leadership ESGT styles

Think of "enablers" as leadership tone from the top – from both executive management (especially the CEO) and the board (or other applicable form of governance oversight). The role of good leadership is to enable and support (both verbally and substantively) all the important issues and topics that go into creating organizational resilience (the eight elements) as well as serving the core mission, vision, values and overall strategy of the organization.

Figure 8.2 provides a quick reminder of the types of leaders we referenced in Chapter 2 based on their embrace (or lack thereof) of all things ESGT, including culture at the core.

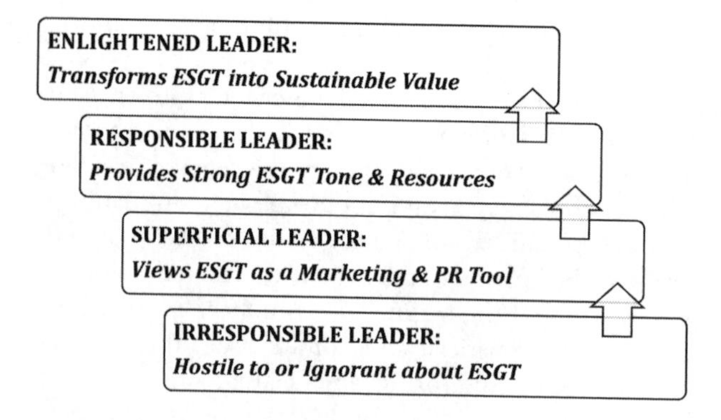

Figure 8.2 The ESGT leadership typology.

Source: Author and GEC Risk Advisory.

At the end of the day, it all depends on the type of leadership that exists at the top of an organization;

- Is it "Enlightened" or "Responsible" and thus vocal and committed from every standpoint to supporting and enabling the necessary and desirable elements of organizational resilience?
- If it is "Enlightened", it goes above and beyond "Responsible" to actually create new or improved products and services integrating and incorporating new ESGT value for the benefit of key stakeholders. Think innovation and responsibility.
- Or is leadership a bit weaker and less committed to ESGT support, as in the "Superficial" form of leadership which believes in ESGT issues only to the extent that they help with marketing and branding but don't really support most of the elements of resilience? Think "greenwashing".
- Or, worst of all, does the organization have "Irresponsible Leadership" in that there is no interest whatsoever in ESGT issues or organizational resilience, because these are viewed as either a waste of time, a distraction from the main focus of the organization (making money, e.g.), a worthless cost center and/or completely unnecessary and irrelevant to the achievement of the more important goals and objectives that leadership is seeking (namely power or money or both)? Think of the long line of disgraced CEOs and sometimes their boards that have fallen after a crisis and/or scandal.

8.2.a.iii *Outputs: financial and nonfinancial (ESGT) results*

Outputs are the results of the organization's work – whether financial or nonfinancial or ESGT – how the organization is measured in the marketplace. Depending on the type of organization, this can mean a variety of different things, including:

- *Stock performance and other financial metrics* for a publicly traded or for- profit corporation.
- *Beneficiary impact metrics* deployed at a nonprofit to measure impacts on the primary beneficiaries.
- *ESG metrics* as viewed by the investor community for a business – checking their climate (water, air, land) footprint, human rights track record and other measurables.
- *Local, state and national economic metrics* like GDP, GDP per capita, employment/unemployment numbers, etc., if the entity is governmental and its primary stakeholders are residents and citizens.
- *Ethics, compliance, sustainability and corporate responsibility metrics* used increasingly by a variety of organizations, especially larger corporations that are in the regulatory and other public spotlight and voluntarily or otherwise decide to disclose metrics, including helpline/hotline call statistics, training and communications, volunteering and philanthropy, among a wide variety of possibilities.

Both financial and nonfinancial metrics are important to value protection and value creation as well as prevention of value destruction when the results are not that good. Most organizations are well positioned to demonstrate financial and operational metrics, especially businesses that live and die by shareholder and investor expectations. But all types of organizations, including government agencies and nonprofits also need to live and die by some of their financial and operational metrics as well as other measures of value creation – their ESGT strategy and results.

The world of investment is increasingly holding businesses especially accountable for ESG results. In this book we are making the case for how organizations can beef up and deploy a broader ESGT strategy in addition to and intertwined with both financial and reputational value creation.

8.2.a.iv Stakeholder and value impact: expectations and reputation

Finally, the organizational model depicted in Figure 8.1 is completed by connecting the inputs, enablers and outputs to the principal object and purpose of the organizational activity – understanding, delivering and fulfilling the expectations of stakeholders with the following consequences:

- When the organization lives up to those expectations (i.e., both financial and nonfinancial or ESGT results), stakeholders are satisfied and reward the organization with greater trust, long-term loyalty, financial and reputational benefits and support.
- When an organization is unable to meet the expectations of one or more key stakeholders in terms or financial and/or ESGT results, stakeholders lose trust, confidence and/or support for the organization and may damage it by leaving, selling or disposing of their interest, attacking the organization in social media, etc. This leads to financial and reputational damage to the organization which can translate into value erosion or destruction.

8.2.a.v What's next?

Now that we've provided a brief tour of this generic organizational model, let's turn to a Typology of Organizational Resilience and Sustainability – the ultimate destination of our journey. Throughout our journey, we have looked at our turbulent global context, at the importance of high-integrity leadership and culture in guiding organizations through the turbulence, at a variety of strategic ESGT issues, risks and opportunities suffusing the globe today and at how organizations can become more resilient by developing the eight elements of organizational resilience and aspiring (hopefully) to have a Virtuous or Robust form of Resilience Lifecycle.

Let's turn now to where the rubber meets the road by gaining a better understanding of where and why a particular organization fits on an evolutionary spectrum of organizational resilience and sustainability. Why does this matter? Because it's all about surviving *and* thriving – not just surviving without creating value, or thriving without creating resilience, because either of those options can lead to failure or even ruin:

- *Surviving without creating value* (whatever that value might be – financial, or nonfinancial/ESGT) undercuts long-term resilience as stakeholders receive little to no benefit over the short or long term and might consider abandoning the organization, undermining its purpose and raison d'etre.
- *Thriving without resilience* may mean getting short-term fast value (mostly financial) but potentially careening out of control into one or more crises because the infrastructure of resilience (i.e., Resilience Lifecycle) is either weak, missing major elements or nonexistent (i.e., a Fragile or Vicious Lifecycle).

All of this can be (cheekily) summarized by my (feeble) attempt at distilling my vision of this book into my own personal "$E=mc^2$", as follows:

SUSTAINABILITY = LEADERSHIP + RESILIENCE

OR

"S = L + R"

8.2.b *A typology of organizational resilience and sustainability*

Our Typology of Organizational Resilience and Sustainability draws on two critical organizational must-haves – on the one hand, leadership, and, on the other, organizational resilience. When we put the concepts of good leadership and organizational resilience together on an x/y axis, we come up with a typology of organizational resilience and sustainability. Why sustainability? Because the more robust and resilient an organization is and the closer to "responsible" or "enlightened" its leadership is on ESGT, the greater the chances of not only robust value protection and creation but also sustained value creation and innovation (see Figure 8.3).

Going one step further, Figure 8.4 shows some examples (and a sneak preview of some of the cases we discuss later) of the kinds of companies and organizations that might fall under each of these types of organization along an evolutionary spectrum of resilience and sustainability – from the least evolved, "Outlaw Organization" all the way up to the most evolved and sustainable – the "Transformative Organization".

Let's take a look at each type of entity in turn, discuss the kind of leadership it is most closely associated with and explain how and why some organizations achieve resilience and sustainability (in the process maximizing organizational and stakeholder value) while others are unable to and end up mismanaging risks, experiencing crises, missing improvement and innovation opportunities, damaging stakeholder relationships and in the end eroding or destroying organizational and stakeholder value.

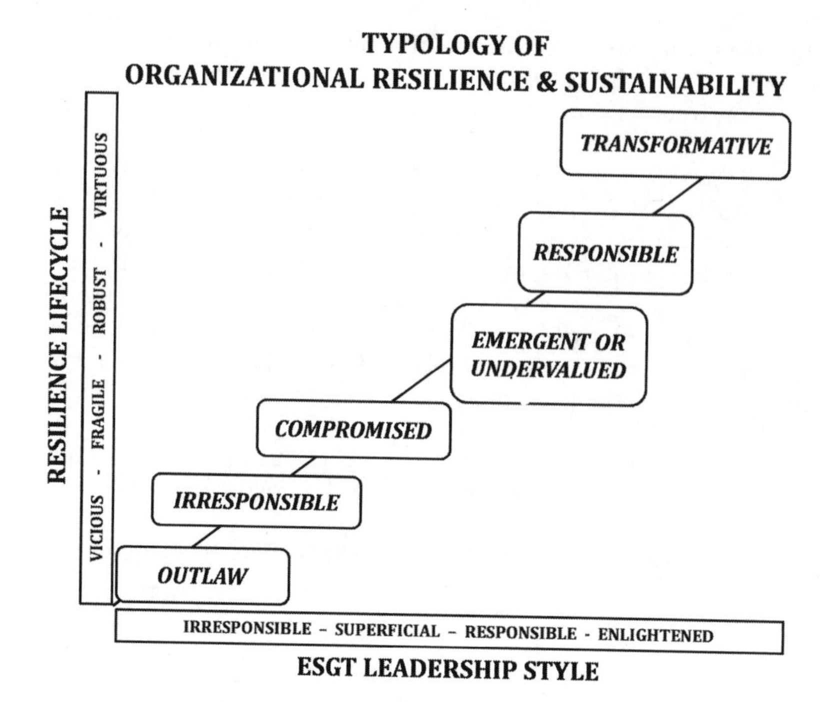

Figure 8.3 Typology of organizational resilience and sustainability.

Source: Author and GEC Risk Advisory.

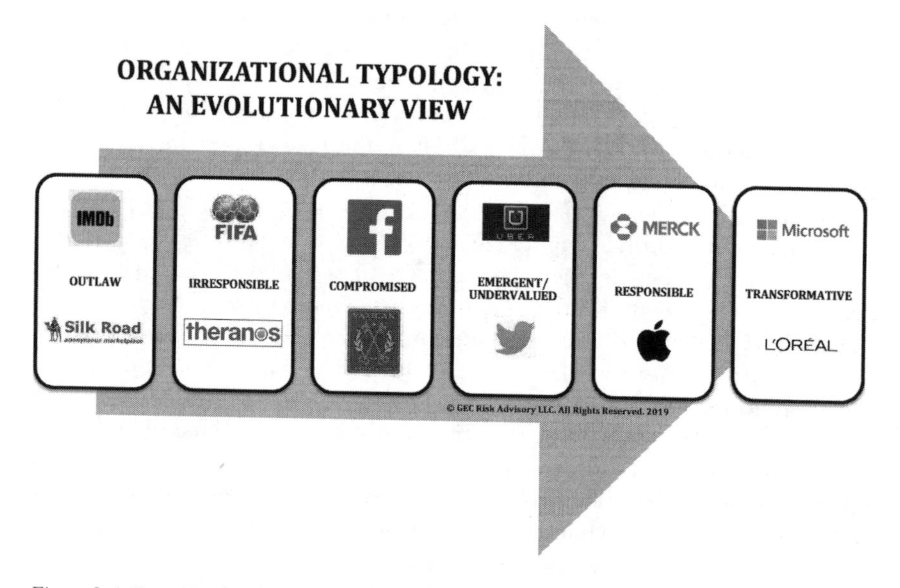

Figure 8.4 Organizational typology: an evolutionary view.

Source: Author and GEC Risk Advisory.

We start with the low end – the "Outlaw Organization" – and build our way up to the "Transformative Organization".

A note to my readers – in the following discussion of organizational types, I offer summary tables with examples of for-profit and nonprofit organizations that fit each organizational type. Somewhat more controversially perhaps, I also provide examples of current or past governments, as they too are a form of organization – in some ways the most powerful kind because of their mandate, jurisdiction and potential for severe abuse of power as well as great opportunity for good.

I'm sure some of my readers might consider this point somewhat controversial, and mind you, things change and someone we admire today may be less admirable tomorrow. In fact, I'm thinking of Prime Minister Justin Trudeau of Canada who, when I started writing this book in 2017, could do no wrong, but as I finish writing this book in mid-2019 is in quite a different spot having made some questionable ethical and policy choices that are under investigation.

That said, and relying on my PhD in political science and work on authoritarianism and democracy, I believe that it is worth thinking of specific regimes and their leaders within the context of this organizational typology. At the end of the day, each regime has:

- "Enablers" in the form of leadership (a president, prime minister, dictator) with certain leadership characteristics (good, bad or ugly as discussed in Chapter 2).
- "Inputs" in the form of national and international strategy, policies and programs (i.e., an applicable "Resilience Lifecycle").
- All of which yield national socioeconomic and political (ESGT) "results".
- Those results will either exceed, meet or miss stakeholder (citizen, resident and transient) expectations and protect, add or destroy national or local value.

Who is the most important stakeholder in a government and the ultimate arbiter of "value"? Clearly, it is the citizens who regularly vote (in democracies) or are ruled by despots (in authoritarian systems). But there are many other truly important stakeholders as well that governments whether they like it or not must consider – residents, transients, immigrants, people in neighboring countries, allies, enemies, businesses, NGOs, etc.

8.2.b.i The Outlaw Organization

At one extreme of the evolutionary spectrum of Organizational Resilience and Sustainability is the Outlaw Organization, which is pretty much what its name implies – an illegal organization, outside of the mainstream of commerce and civilized society undertaking a mission and strategy that has criminal and nefarious objectives and ends (Figure 8.5).

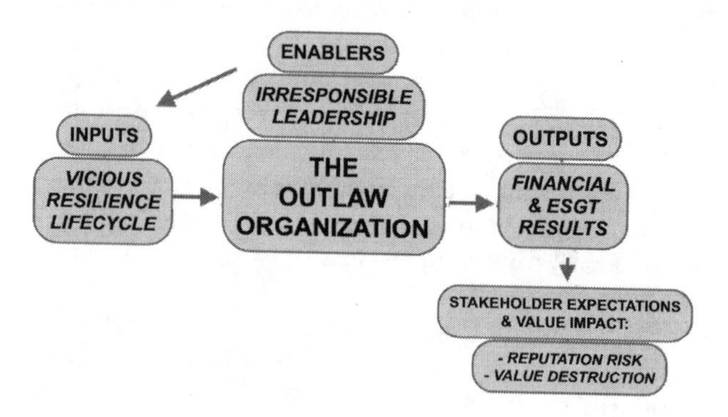

Figure 8.5 The Outlaw Organization.

Source: Author and GEC Risk Advisory.

The Outlaw Organization is based on illegal foundations, created for criminal or lawless activity without any interest in or semblance of interest in having any of the foundational, necessary or desirable elements of the Virtuous or Robust Lifecycles. Indeed, the Outlaw Organization is in the business of promoting the opposite – the negative side of ESGT: human trafficking, crime, terrorism, war, money laundering, cyber-insecurity, etc.

To illustrate what we mean by Outlaw Organizations, Table 8.1 provides examples in each of the governmental, for-profit and nonprofit sectors. We profile some of them later.

Table 8.1 Examples of Outlaw Organizations

For-Profit	• Silk Road • Mafia • The Shadow Brokers (WannaCry attack) • Cobalt crime gang[6] (internet crime gang)
Nonprofit Or Non-Governmental	• Guardians of Peace (GOP) (SONY cyber attackers) • WikiLeaks
Governmental	• MDB (Malaysian Government Development Bank) • North Korea (past and present regimes) • Syria (under Asaad) • Myanmar (Rohingya genocide)

The Outlaw Organization's leadership is clearly at one end of the spectrum of leadership styles as ESGT issues generally or systematically have nothing to do with this type of leader's world view. Indeed, the Irresponsible Leadership that is directly associated with the Outlaw Organization does not care about ESGT externalities or impacts on stakeholders unless it has something to do directly with its business model or purpose, in which case it may acknowledge the need for such ESGT issue, program or process.

8.2.b.i.A WHAT AN OUTLAW ORGANIZATION LOOKS LIKE

Looking at the elements of the "Vicious Resilience Lifecycle" and applying them to an Outlaw Organization – one could make the following observations:

- **Governance**: While governance might be tight from a control standpoint, it is certainly not robust, positive or lean-in governance that looks after the interests of all main stakeholders. Thus "weak" or "perverted" governance come to mind.
- **Culture:** Leadership dedicated to illegal or criminal enterprise would automatically qualify for "toxic" culture status.
- **Stakeholders:** Again, for leadership of an organization that targets stakeholders for exploitation rather than benefit, the automatic moniker would include stakeholder negligence, gross negligence or intentional abuse.
- **Risk:** It's possible that the Outlaw Organization (think Mafia or Silk Road) might have relatively developed risk radar as they need to circumvent, manipulate and/or subvert law enforcement, regulators and others. They may not have the standard or more holistic ESGT risk management that a *Fortune 500* company would have, but they may have reasonably good "illegal issue risk management", such as law enforcement evasion, implementing cybersecurity defensive or aggressive measures within the dark or deep web, etc.
- **Strategy:** It is unlikely that ESGT issues will make it into the strategy formulation of this type of organization.
- **Performance:** Outlaw Organizations are unlikely to have anything akin to the traditional performance management of a typical company or organization. However, they can have metrics – even sophisticated metrics – on their performance in their marketplace, whether such metrics are financial, economic or something else like measurements of cyber-subterfuge, human casualties, victims, etc., if you are talking about a cyber-hacking outfit, mercenary organization or human trafficking operation, for example.
- **Crisis:** An Outlaw Organization like the Mafia may have very well-developed crisis management capabilities since they are engaged in high-risk behavior that requires law enforcement circumvention, etc.
- **Improvement/Innovation**: Outlaw Organizations may actually be self-improving as well, engaging in continuous process improvement and even innovation in furtherance of their illegal or criminal goals and objectives.

8.2.b.i.B SILK ROAD

Silk Road, a now defunct criminal for-profit Dark Web-based enterprise focused mainly on the sale of illicit drugs, was founded by a charismatic and later criminally convicted leader, Ross Ulbricht, who is currently serving a

30-year sentence without parole. Silk Road is a good example of Irresponsible Leadership running an Outlaw Organization. It was a commercial, mostly criminal, enterprise that created a Dark Web marketplace for the sale of mostly illegal drugs and services (fake identification cards) with some legal products as well, using Tor and Bitcoin for payment. They had to use common commercial standards and practices as well as illegal ones given the purpose of the marketplace. Ultimately, because of their notoriety in the media, they became a target of an FBI sting investigation. Its founder, Ulbricht, was arrested, prosecuted and jailed, and though there were attempts to revive Silk Road (Silk Road 2.0), these failed as well.

While an Outlaw Organization can be profitable and even sustainable in that it can exist for a long time before it gets caught or embroiled in controversy, crisis or law enforcement action (witness the longevity of Mafia-type organizations), one can posit that the impact on key stakeholders will almost always be negative. For example, key stakeholders in Silk Road would be its customers, which would have included people addicted to drugs with possible health and life and death consequences. Such impacts are not only personal to the drug addict affected but also impact families, communities, health care providers and governments, all of whom must pick up the pieces of such personal and social malaise.

8.2.b.i.C 1MDB

What follows in Figure 8.6 is the RepRisk Index Trend and company profile for 1Malaysia Development Berhad (1MDB) – the Malaysian government development bank set up over a decade ago to invest in development projects in a variety of industrial sectors in Malaysia. 1MDB is currently embroiled in multiple national and international allegations and investigations about deep and extensive corruption amounting to potentially billions of dollars of theft by Malaysia's former prime minister, Najib Razak, and a Malaysian businessman, Jho Low. Multiple nations are conducting criminal investigations into the reverberations of the 1MDB scandals, including the US, whose Department of Justice, among other things, is looking at Goldman Sachs, which acted as an adviser to 1MDB a few years ago and whose top Asia executive has already pleaded guilty to criminal wrongdoing.[7]

This is clearly an ongoing situation that will take years to completely understand, but the point of including 1MDB in this section on Outlaw Organizations is simply to underscore how a government entity can also become an Outlaw Organization if its main purpose and operations appear to be criminal or illegal.

Figure 8.6 is a snapshot of the RepRisk Index Trend for 1MDB, showing how the G aspect of their ESG profile dominates the reputation risk this entity is currently under from the highest point of RepRisk Index (97, almost unheard of) to the time of this writing, April 2019, where it still hovers near 60. 1MDB's reputation risk is all about governance – the amplification of the underlying governance issues in this case – extensive corruption and fraud perpetrated primarily by government officials.

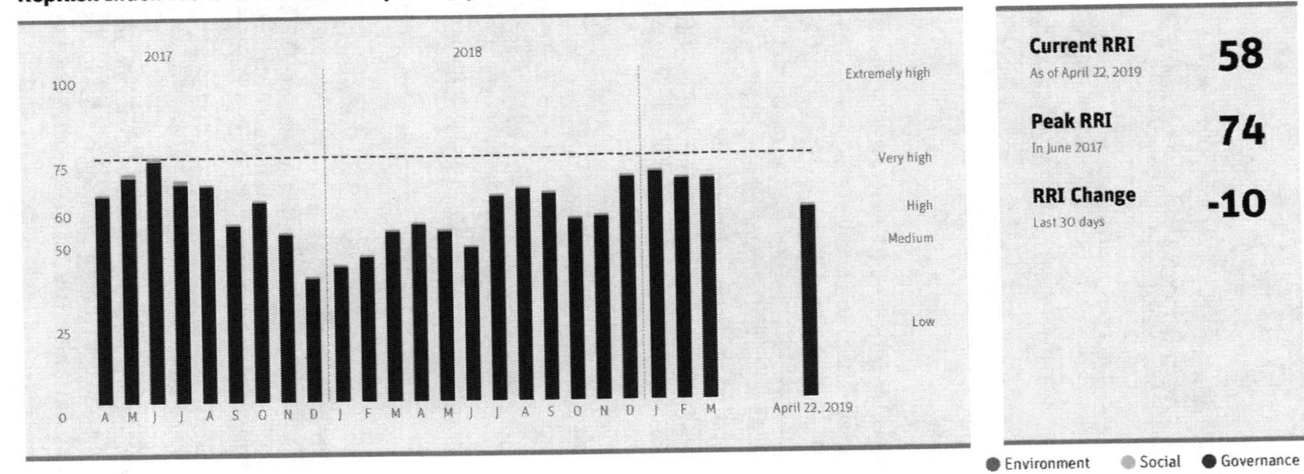

Figure 8.6 RepRisk Index Trend for IMDB 2017–2019.

Source: RepRisk.

8.2.b.ii The Irresponsible Organization

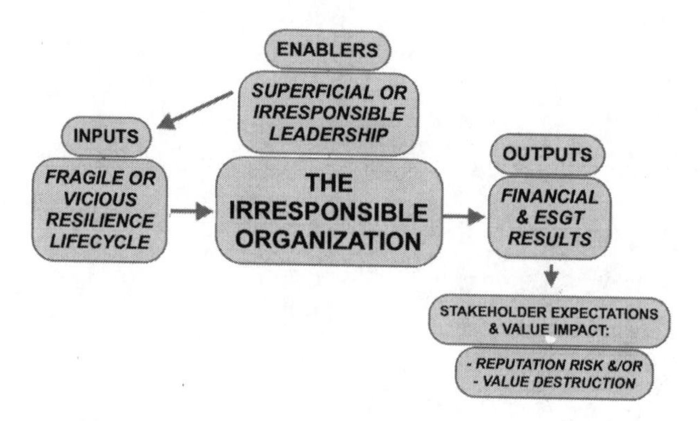

Figure 8.7 The Irresponsible Organization.

Source: Author and GEC Risk Advisory.

The Irresponsible Organization (see Figure 8.7) is run by leaders who are only focused on one thing: attaining their own goals and objectives (usually money, status, power or all three) with little or no regard for other stakeholders and with faulty, nonexistent or perverted governance that either (1) is purely paper governance, or (2) consists of "friends and family" governance that operates as a rubber stamp rather than true overseers.

Several examples come to mind, as illustrated in Table 8.2. We discuss a few of these examples next.

Table 8.2 Examples of Irresponsible Organizations

For-Profit	• Theranos (under founder Elizabeth Holmes) • Massey Mining (under Don Blankenship) • Enron (leading to bankruptcy and leadership criminal convictions) • Odebrecht (Latin America wide bribery and corruption scandal)
Nonprofit or Non-Governmental	• FIFA (under Sepp Blatter)
Governmental	• Saudi Arabian government (Mohammad Bin Salman) • Russian government (Vladimir Putin)

8.2.b.ii.A MASSEY ENERGY

Massey Energy was a West Virginia-based coal mining company with a decades-long track record of health and safety violations in their coal mining operations, run by an imperial CEO, Don Blankenship, who held court over a rubber-stamping board of luminaries who clearly didn't understand the role of the board in holding the CEO accountable. They trusted him implicitly and viewed health and safety legal violations as the cost of doing business, not a responsibility to critically important stakeholders like employees.

Then 29 miners died in a coal mine explosion, the government and regulators took a second look and the Blankenship empire imploded – he was prosecuted, convicted and jailed, the company was sold off and the board disbanded. Meanwhile, the community and families of the lost miners bore the brunt of this long-brewing scandal for which no leaders took any real responsibility to this day – even Blankenship went to jail for much less time than he could have (one year) and actually had the temerity to run as a candidate for the Republican Party nomination to the US Senate for West Virginia in 2018. The good news: he did not win his primary to become the Republican candidate.

8.2.b.ii.B THERANOS

As we discussed in greater detail in Chapter 6, Theranos was a Silicon Valley blood-testing Unicorn darling for quite some time, achieving the lofty but fraudulently estimated valuation of over $9 billion at one point. It was founded and run by an imperial CEO (Elizabeth Holmes), with a luminary board of famous, mostly retired, older white men (what I like to call "sit-back directors") who didn't ask too many questions until it was too late for the company and its many adversely affected stakeholders – customers, patients, employees, suppliers and others. The company is now bankrupt and Ms. Holmes, the founder, is slated for federal criminal trial in 2020.

8.2.b.ii.C FIFA

FIFA, the global "nonprofit" football federation, was run as an international patronage enterprise by its long-time leader, Sepp Blatter, replete with corruption, bribery, kickbacks and all manner of international intrigue and crime. He was finally deposed after a string of media exposés revealing the dark underbelly of international corruption, favoritism, conflicts of interest and bribery that suffused FIFA for decades.

The stakeholders of this organization received the short end of the stick, as is usually the case when Irresponsible Leaders commandeer an organization with ulterior personal and potentially criminal purposes, accumulating power and patronage in the process.

What about the soccer fans, youth and communities around the world who were supposed to be the beneficiaries of this nonprofit global sports institution? These stakeholders were secondary to the purpose of FIFA for many years and decades, and it will take many more years for any sense of trust and confidence in this institution to be restored (or created in the first place).

Figure 8.8 shows the RepRisk Index Trend for FIFA from 2017 to 2019, demonstrating the disproportionate swath of governance-related reputation risk that it has accumulated.

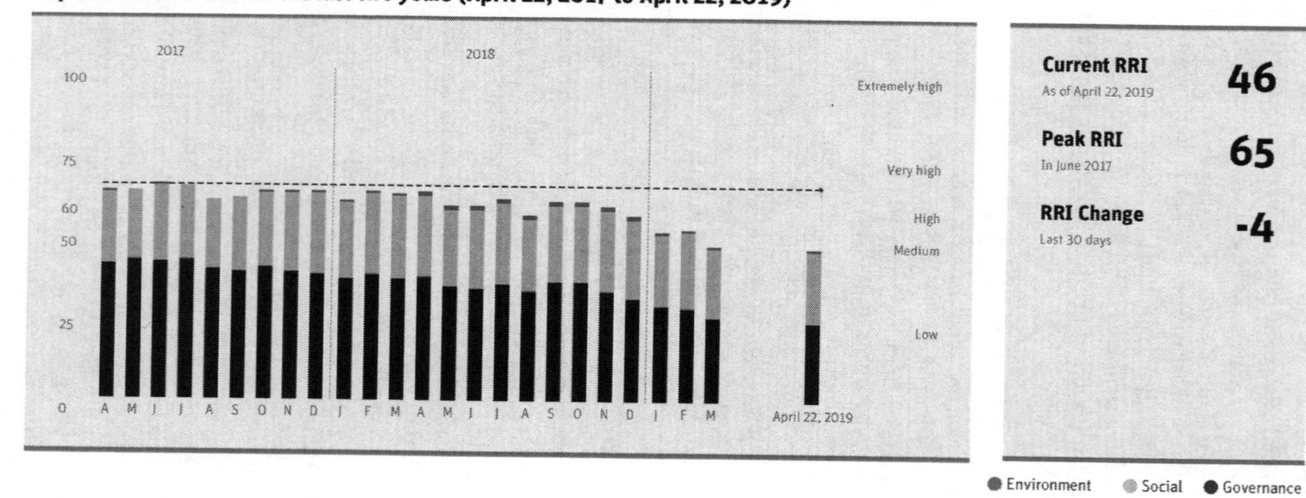

Figure 8.8 RepRisk Index Trend for FIFA 2017–2019.

Source: RepRisk.

8.2.b.ii.D ODEBRECHT

Odebrecht is the Brazil-based global construction company whose tentacles of deep and widespread corruption and bribery not only spread within their headquarters country but permeated Latin America and beyond. Odebrecht's illegal and criminal practices crept into more than a dozen countries in which they were doing business, including and up to the highest levels of government, bringing down both political and business leaders. According to *The Washington Post*:

> The largest corruption investigation in Latin America's history – revolving partly around bribes paid by the Brazilian construction giant Odebrecht to secure government contracts – has spread to 14 countries, implicating a Colombian senator, a former vice president of Ecuador, even Venezuelan President Nicolás Maduro.[8]

The Odebrecht investigation has been called by the US Department of Justice the "largest foreign bribery case in history" which has led so far to the prosecution and jailing of the company's CEO and $2.6 billion in fines relating to corruption and bribery under anti-corruption laws like the US FCPA and other similar laws under other countries' jurisdictions.[9]

And that's not all; at the time of this writing, there were two more deadly twists to this story. One in Colombia, where the possible murder of two people (father and son) took place in late 2018 because of the father's involvement as a possible whistleblower regarding the Colombian portion of the Odebrecht scandal.[10] The other one was the shocking development involving the suicide of former Peruvian President Alan Garcia, as he was about to be arrested by Peruvian authorities for alleged Odebrecht-related corruption.[11]

Figure 8.9 provides the RepRisk Index Trend for Odebrecht. Evidently, the level of risk incidents could justify categorizing this company as an Irresponsible Organization (which would otherwise qualify for Outlaw status but didn't because its ostensible business purpose – construction – is a legitimate commercial purpose).

8.2.b.iii The Compromised Organization

The Compromised Organization (see Figure 8.10) typically has leadership – both in terms of the CEO and his/her surrounding executive team and the board – that thinks of ESGT, resilience and sustainability issues as perhaps "nice to have" decorative additions to the core business but not necessarily strategic imperatives. Hence, the typical leader associated with the Compromised Organizational form is often the Superficial Leader who deploys ESGT in at best a skin-deep way – for show-and-tell but not for real.

- Example: a high-flying technology company that is growing by leaps and bounds and is in the B2B supply chain of major *Fortune 100* customers

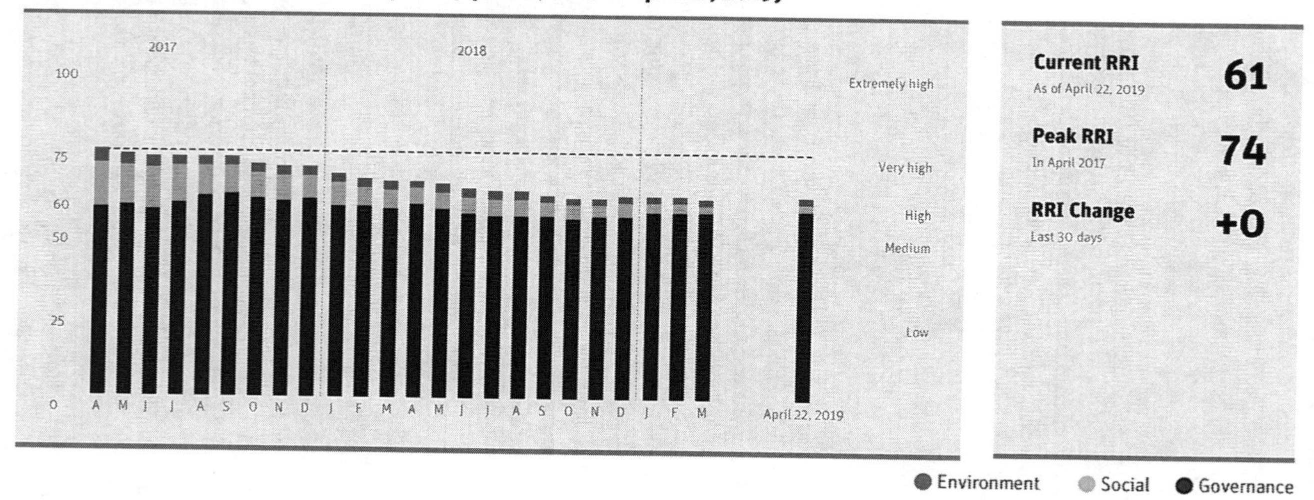

Figure 8.9 RepRisk Index Trend for Odebrecht 2017–2019.

Source: RepRisk.

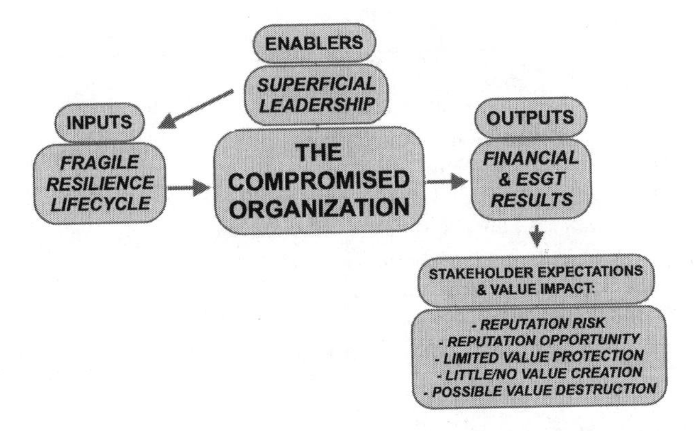

Figure 8.10 The Compromised Organization.

Source: Author and GEC Risk Advisory.

where the customers are sophisticated purchasers of services and demand that their supply chain providers have corporate responsibility programs and practices. The high-flying tech firm doesn't really have such a program in place but does a pretty good job of scraping together a few random examples of things the company does locally (annual employee volunteering, for example) and puts it all together in a glossy CSR brochure or webpage that ultimately is superficial and random rather than an effective or strategic CSR program. But to its leadership it is enough to be able to pull the wool over the eyes of the large B2B customers in their supply chain – at least for the time being.

Compromised Organizations can operate with impunity for a long time, getting away with a marketing and PR approach to ESGT issues unless and until a risk implodes or a crisis happens. Then, depending on how material and publicly noticed (social media, media) the crisis is, the Superficial Leaders will do a little patchwork and continue pretty much along the same "Superficial" path.

Sometimes, however, if the crisis event or turning point is material and severe enough, changes may happen at the leadership level. The CEO is replaced, the board becomes more proactive, and there is real change – of leadership, of resilience lifecycle elements and potentially movement toward a more evolved form of organization that can eventually propel the organization toward the Emergent/Undervalued or even Responsible type of organization.

Table 8.3 provides some examples of Compromised Organizations, some of which we explore in greater detail later.

Table 8.3 Examples of Compromised Organizations

Business	• VW (post-2015 emissions scandal) • Wells Fargo (multiple recent scandals) • BP (post-Deepwater Horizon) • Facebook (post-Cambridge Analytica)
Nonprofit or Non-Governmental	• Michigan State University (sexual abuse cases 2017–2018) • Penn State University (sexual abuse cases football program) • Catholic Church (ongoing global child abuse cases)
Governmental	• US government (under Trump) • Hungarian government (under Orban) • Brazilian government (under Bolsonaro)

Because both the Irresponsible and the Compromised types of organizations are on the more negative side of our Organizational Typology, it is useful to contrast the Irresponsible and the Compromised Organizations because there are real differences that should be noted. Think about these two types of organizations as being on the low end of the Resilience Lifecycle evolutionary ladder as well – they are organizations whose leaders don't think much about the greater good (however that might be defined for their entity). As an aside, the "greater good" in each sector could mean the following:

- For a government – the best interests of its citizens, residents and transients not just a favored minority.
- For a business – the best interests of shareholders and other major stakeholders (employees, customers, partners).
- For a nonprofit – the best interests of the intended beneficiaries as well as other major stakeholders (such as donors, the community and partners).

Table 8.4 shows a comparison of the major characteristics of the Irresponsible and Compromised Organizations:

Table 8.4 Comparison of main characteristics of the Irresponsible and the Compromised Organizations

Attribute	*Irresponsible*	*Compromised*
Leadership Style	Irresponsible or superficial	Superficial or responsible
Type of Resilience Lifecycle	Vicious or fragile	Fragile
Lifecycle Elements Comparison		
Governance	Poor or compromised	Compromised, possibly improving
Culture	Toxic or poor	Toxic, poor, possibly improving
Stakeholders	Blind to most	Cognizant of several

(Continued)

Table 8.4 Continued

Attribute	*Irresponsible*	*Compromised*
Risk	No systematic approach	Some risk responsibility (insurance)
Strategy	ESGT considerations nonexistent	Some ESGT considerations
Performance	Little alignment other than financial	Some alignment of financial, maybe considering other metrics
Crisis	Little or no preparedness	Some preparedness possibly due to crisis event already occurring
Improvement	No system for self-improvement or lessons learned	Some, maybe forced, approach to self-improvement resulting from a crisis

Superficial Leaders are "horses with blinders" at best, motivated by limited, often selfish, narcissistic, personal power and/or all-consuming financial objectives. Thus, the idea of considering resilience and ESGT issues in the mix is either anathema or at best silly, ridiculous or irrelevant to such leaders' world vision.

What distinguishes the Compromised from the Irresponsible Organization is that the latter has no concern about the law and ethics while the former may pretend to or may actually be in the process of pulling the organization out of a crisis or another negative situation.

The Compromised Organization and its leaders usually know better but don't always do what they're supposed to do (they may cheat, engage in unethical or illegal behavior as a "cost of doing business") in the belief (and perhaps past track record) that they can get away with it. They think that at worst they might have to pay some sort of social or financial penalty that is less costly, in their estimation, than not engaging in those poor behaviors to achieve their goals. Several examples follow.

8.2.b.iii.A VOLKSWAGEN

Since 2015, when VW's emissions scandal first burst onto the scene, we have witnessed investigations and revelations of a deep leadership (governance) and culture crisis that had been brewing for a long time before the exposé. While the company has taken a number of steps – to refresh its Supervisory Board, bring in more and new compliance personnel, remove some of the directly responsible leaders – it is unclear whether the company and its leadership have taken enough steps at the highest levels of the organization – especially management and the board – to qualify for an upgrade to the Emergent/ Undervalued type of organization.

While Volkswagen continues to thrive from a stock and sales standpoint, the reverberations from their scandals continue to act as a drag on both reputation and potentially on financial and economic gain that might have happened in the absence of the emissions scandal. Figure 8.11 shows the RepRisk Index Trend for VW since the beginning of the emissions scandal – right before, the RepRisk Index hovered around 40, and since the emission scandal erupted in the fall of 2015, it has ranged from a high of 75 to hovering between the mid-50s to 60s.

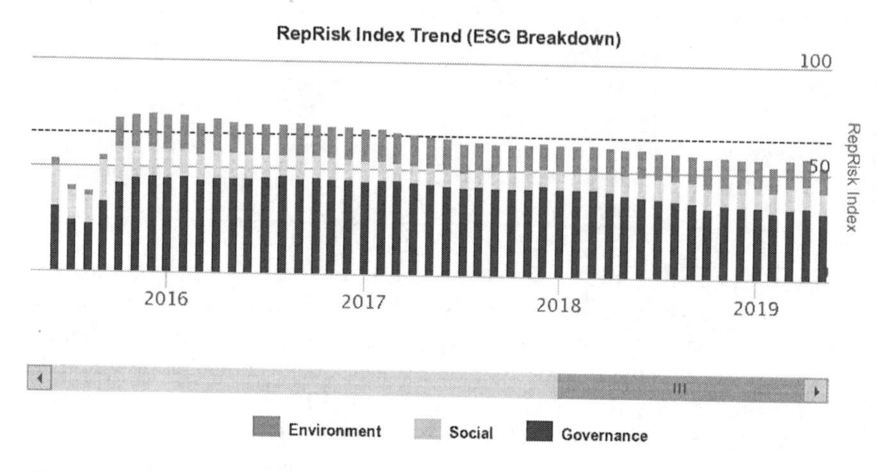

Figure 8.11 RepRisk Index Trend for Volkswagen 2015–2019.
Source: RepRisk.

8.2.b.iii.B MICHIGAN STATE UNIVERSITY

Michigan State University had a major scandal erupt over the past couple of years as accusations of sexual abuse and assault (including rape) were first slowly and then like an avalanche uncovered by the press. The accusations centered around one of their doctors, Larry Nassar, who was convicted of assaulting more than 150 young women athletes and was prosecuted and found guilty and sentenced to 40–175 years of prison.

What is worse is that the university itself was found to have known about these abuses for a long time, conducted cursory and ineffective investigations (or at worst – in cases yet to be proven – engaged in abuse cover-up). They became subject to a variety of additional civil investigations and lawsuits in addition to an already-made settlement of $500 million.

Even based on what we know so far, the extent and longevity of these accusations would certainly qualify the leadership of the university to be Irresponsible and the organization itself to be Compromised, at best. Figure 8.12 shows a RepRisk Index Trend depiction of the reputation risk exposure of the university after the major news stories broke in mid-2017.

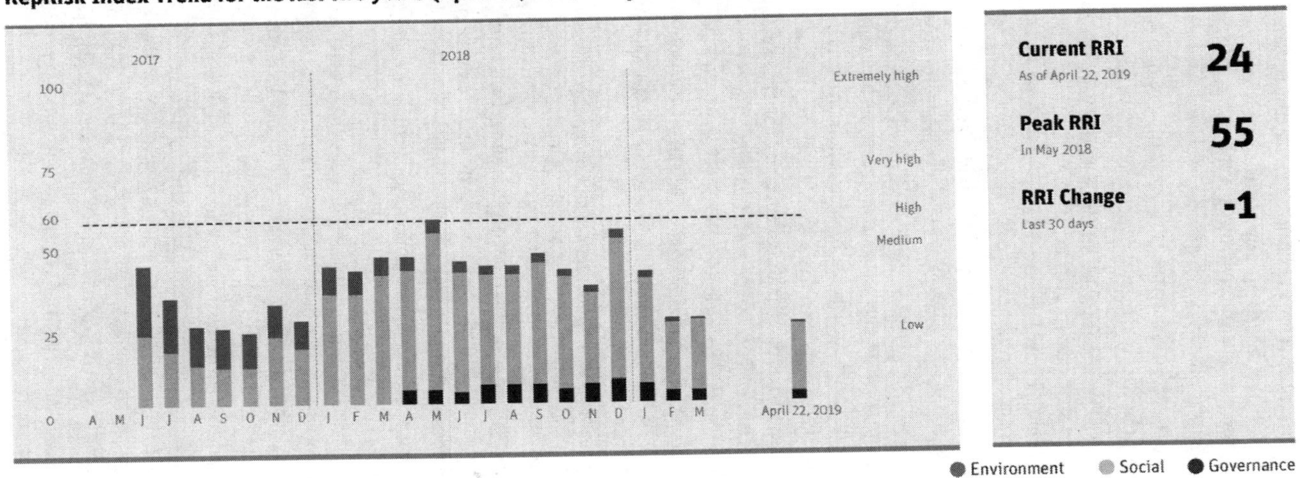

Figure 8.12 RepRisk Index Trend for Michigan State University 2017–2019.

Source: RepRisk.

8.2.b.iii.C THE CATHOLIC CHURCH

Last, but definitely not least, is the Catholic Church. It has a long-standing (decades, probably centuries-old) and constantly widening scandal of priest pedophile and sexual predation cover-ups in multiple countries around the world. What has become clearer in recent years is the internationally widespread and systematic failure of leadership and culture at the Vatican and worldwide.

Successive heads of the Catholic Church (popes) and locally (cardinals and bishops) have allowed – consciously or negligently – the systematic relocation of accused priests to other locations and the quiet paying off of some of those affected. The jury on this topic continues to be out even under the leadership of Pope Francis, lauded by so many even outside of the Catholic Church as a paragon of great leadership and culture.

This is clearly a case of at best Superficial and at worst Irresponsible Leadership and of a toxic culture of covering up unspeakable crimes that damaged the most important and delicate stakeholders anywhere – children. The only reason I have categorized the Catholic Church as a Compromised Organization and not Irresponsible or Outlaw, is that their most recent pope has exhibited signs of being a Responsible Leader, and he may yet be in a role that helps to pull the Catholic Church out from one of these negative categories into a somewhat less negative one. But the jury as of now is definitely out.

Until and unless the Catholic Church reforms itself sufficiently by adopting the eight elements of the Virtuous Resilience Lifecycle, it will be doomed to staying in a Vicious or Fragile Resilience Lifecycle category, regardless of how hard the pope tries, if he is unable to reform the inner workings of this incredibly powerful, ancient and impactful global organization.

8.2.b.iv The Emergent or Undervalued Organization

The Emergent or Undervalued Organization (see Figure 8.13) is one that is mostly on the right track and on the upswing but is in a fluid situation potentially as well, where things could go right or wrong depending on the prevailing circumstances.

There are basically two scenarios under this category:

- First, the Emergent or Undervalued Organization might be in a fragile position and have Superficial Leadership, meaning that they are not fully committed to a Robust Resilience Lifecycle. This is an organization that is teetering between evolving or devolving on ESGT issues, depending more than anything else on who its leadership is, how strong and independent their governance is and the overall responsiveness of the organization to its key stakeholders. Examples of this type of organization might include companies that pay lip service to ESGT issues superficially but really don't put enough resources or true support into the building of the eight elements of the Robust Resilience Lifecycle needed to not only survive but thrive.

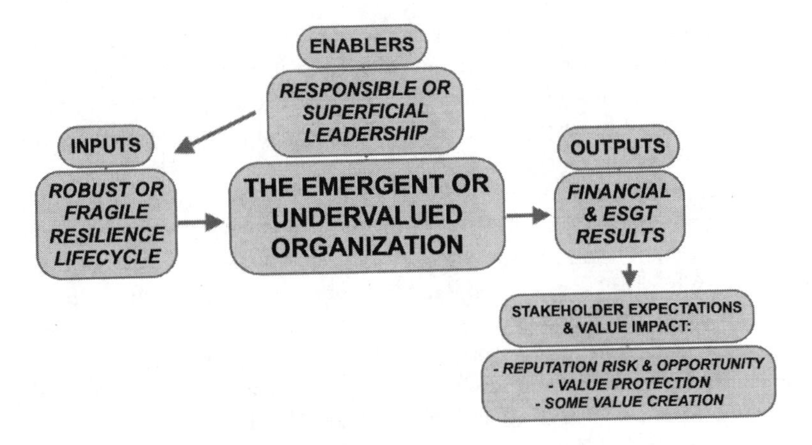

Figure 8.13 The Emergent or Undervalued Organization.

Source: Author and GEC Risk Advisory.

- Second, the Emerging or Undervalued Organization may also be pointing in an upward, evolutionary direction because it pretty much has the right kind of leadership in place (the "Responsible" variety) that is committed to building resilience and is transitioning from a more Fragile Resilience Lifecycle to a more Robust Resilience Lifecycle. What might make this variant "Undervalued" is the fact that while there are good momentum and motion toward resilience and becoming a "Responsible Organization", the leadership is unable or unwilling to properly communicate their accomplishments externally or internally not for bad reasons but because they don't have time, resources or the vision to do so.

What does this mean in the real world? Table 8.5 provides several examples.

Table 8.5 Examples of "emergent" or "undervalued" organizations

Business	• Uber • Lyft • Altria • Twitter
Nonprofit or Non-Governmental	• [No entry partly because a nonprofit in this category by definition won't be very visible/is difficult to identify]
Governmental	• Malaysia (the newly elected democratic government 2018) • Ethiopia (newly elected government 2018)

8.2.b.iv.A UBER

Uber is a fascinating company that under its founder and original CEO Travis Kalanick made many waves worldwide, not only for its disruptive technology – pretty much obviating the need for taxis in many of the major global urban centers – but because of its disruptive and some might call unethical and illegal tactics and strategy to gain market share.

Uber was embroiled for years in negative headlines on all manner of ESGT topics ranging from harassment and discrimination and privacy violations to health and safety concerns and more. After several years of getting away with it, the cup ran over when an ex-employee wrote a now famous blog on the culture of harassment and discrimination that was rampant at Uber. That seemed to tip the balance and create sufficient momentum for a turning point at the very top, first, with some changes in governance (new board members, including a well-known business woman, Ariana Huffington) and then a series of additional steps, including the eventual removal of Kalanick as chairman of the board and CEO and the hiring of a more professional corporate leader as CEO, Dara Khoshroshawi.

By most accounts, the new CEO has moved quickly and decisively to stabilize the company by creating a much more resilient set of policies, programs and culture innovation. In fact, he seems to truly understand the pivotal nature of a strong and positive culture and has devoted a great deal of time and resources to bettering the workplace environment for Uber's key stakeholders – employees (working on improving the "bro" culture and implementing important reforms to sexual harassment and discrimination); drivers (improving their benefits and protections) and customers (creating better communication and interaction).

While they are not out of the woods yet, what the new CEO is doing is most definitely moving in the right direction and propelling the company into possibly that of a Responsible Organization that is both resilient and sustainable over the long run. Of course, with Uber's IPO in 2019 it would have been malpractice (though many do it) to go public without at least a semblance of corporate responsibility on ESGT issues.

The RepRisk Index Trend for Uber shown in Figure 8.14 shows how their overall reputation risk has slowly declined from the heights in 2015–2016 at a peak RepRisk Index of 91 to hovering in the mid-50s to 60s which while not low, has been markedly better under the current CEO's leadership.

8.2.b.iv.B ALTRIA

Altria has been well known in the corporate responsibility, risk, ethics and compliance fields as a company that has built very robust and resilient programs in those spaces over the past 1–2 decades. While we can all critique the business they are in – tobacco – they are operating as a legal form of business (as long as they are obeying laws and regulations) and in doing so have decided to beef up their eight elements of resilience. Indeed, it is possible to say that, but for the industry they are in, they would probably qualify for Responsible Organization status.

However, a company like this, despite having a Robust Resilience Lifecycle and programs incorporated into its daily activities that it and its stakeholders

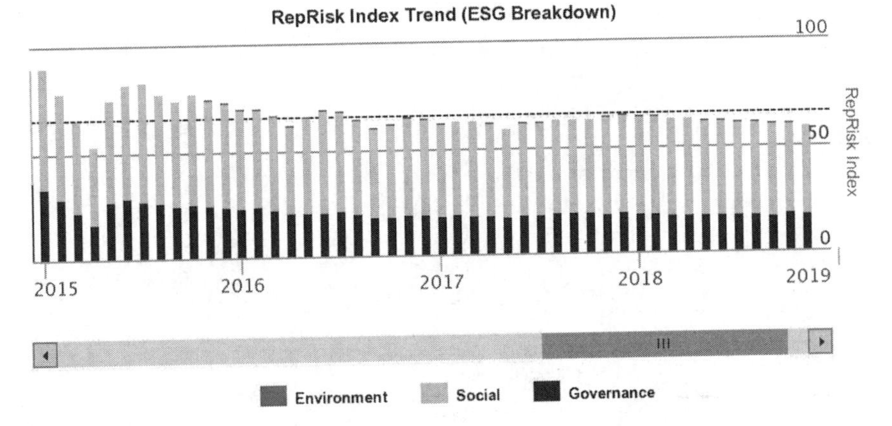

Figure 8.14 RepRisk Index Trend for Uber Technologies 2015–2019.

Source: RepRisk.

would and should be proud of, it isn't a company that is easy to praise for its internal programs given its products. Thus, Altria fits under the Undervalued/ Emergent type of Organization because of its conscious reticence about broadcasting its programs while at the same time having built appropriate resilience. Figure 8.15 showing the RepRisk Index Trend of Altria bears this out.

8.2.b.iv.C TWITTER

Twitter poses another interesting example of a relatively young but highly publicly exposed company due to the nature of its business – social media – and the fact that it is not only at the center of many of the most controversial social media challenges and crises globally today but is the actual inventor of one of the most visible and controversial social media products in the marketplace. The one and only social media used (or abused) by the current president of the US.

Twitter and its leadership are being held to an increasingly higher standard of responsibility for policing themselves and their users. It is a company that finds itself at a crossroads – on the one hand, stating that it is doing its best to learn from and manage its overwhelming challenges to balance freedom of speech against hate speech and potential violence and, on the other hand, to make money and be sustainable from a financial standpoint. I categorize Twitter as an Undervalued/Emergent Organization because, depending on how its leaders act on the very visible and complicated current and near future challenges, they could go either way – devolving into a more Compromised or even Irresponsible type of organization or growing into the more Responsible form.

Figure 8.16 shows their RepRisk Index Trend which underscores how the social (data privacy) dimension of the ESG spectrum is their greatest reputational vulnerability.

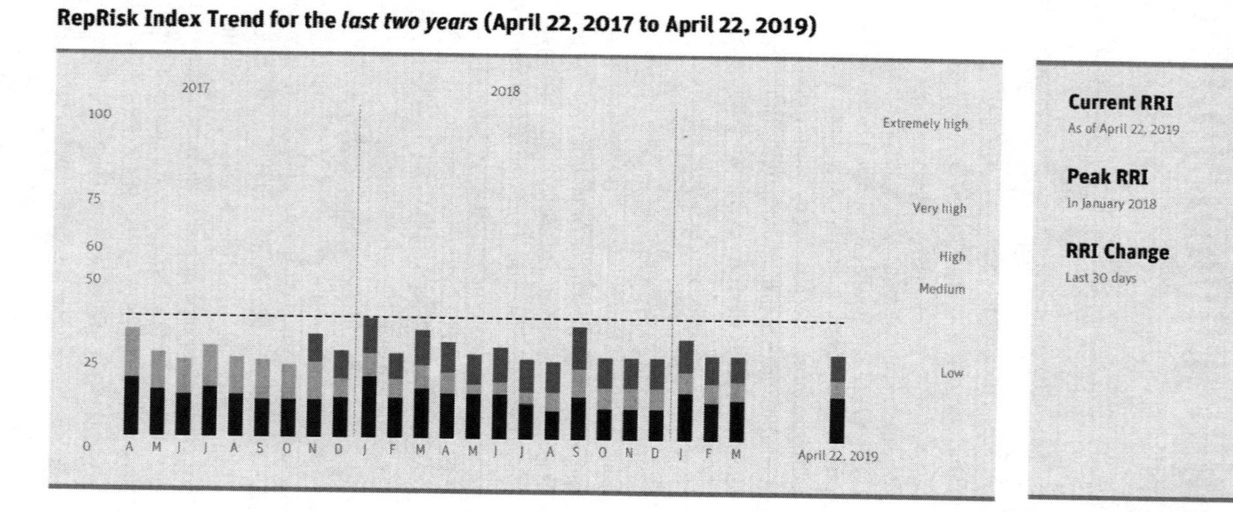

Figure 8.15 RepRisk Index Trend for Altria Group 2017–2019.

Source: RepRisk.

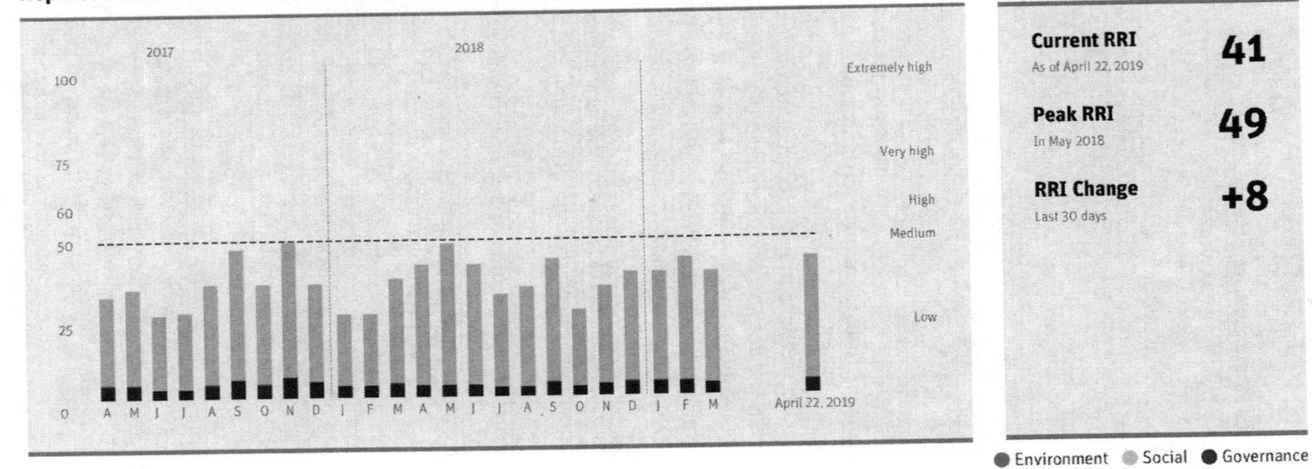

Figure 8.16 RepRisk Index Trend for Twitter 2017–2019.

Source: RepRisk.

8.2.b.v The Responsible Organization

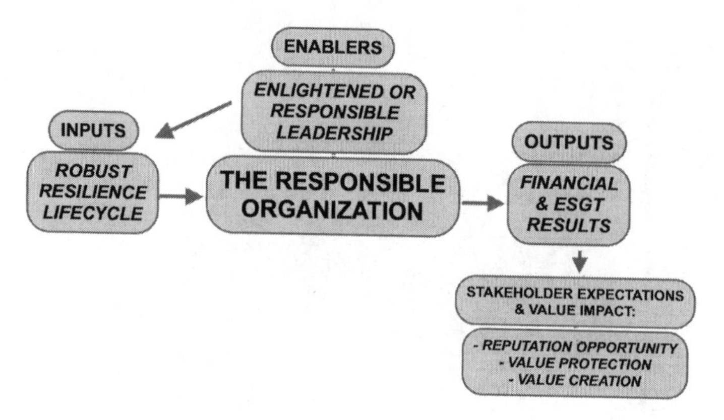

Figure 8.17 The Responsible Organization.

Source: Author and GEC Risk Advisory.

The Responsible Organization is exactly what it sounds like – it has a good balance of Responsible or even Enlightened Leadership and it has a Robust Resilience Lifecycle, meaning that while not perfect, it has most of the eight elements present. The likelihood of better and more consistent and predictable financial and nonfinancial results is greater and the likelihood of stakeholder satisfaction is also greater, adding to the sustainability of the organization (Figure 8.17).

What does this mean in the real world? It means that this type of organization is humming along on all cylinders with the mission, vision, strategy and culture all pretty much aligned and moving on a positive and constructive path that generally benefits key stakeholders. This includes well-organized businesses, well-functioning democratic governments and highly effective nonprofits like the ones singled out in Table 8.6.

The Responsible Organization sports the kind of leadership we all want for our organizations – Responsible or Enlightened Leaders who take their role seriously, support in both word and deed the ESGT programs important to the organization and its stakeholders.

Table 8.6 Examples of "Responsible" organizations

Business	• Merck • Eli Lilly • Apple • Starbucks
Nonprofit or Non-Governmental	• American Cancer Society • AARP
Governmental	• Germany (under Merkel) • Scandinavian Countries • The World Food Programme

8.2.b.v.A MERCK & CO. INC.

Merck is a company with consistently good-to-great leadership for decades and a conscientiousness about corporate responsibility and responsibility to primary stakeholders above and beyond shareholders. The early 2000s handling of the Vioxx crisis by then-CEO Ray Gilmartin illustrates this responsibility to stakeholders well. Under Gilmartin, Merck opted to withdraw the medication from the market voluntarily and in the process lost a major source of revenues (as opposed to their competitor Pfizer, which had a similar product and did not withdraw it from the market). This is an example of a CEO placing the welfare of a key stakeholder – the patient – before the primary stakeholder – the shareholder – because of the duty of care that he and the company felt under the circumstances.

Current CEO Ken Frazier has done something similar on a different issue in the face of untenable pronouncements by President Trump after the violence and racism that occurred in Charlottesville, Virginia, in the summer of 2017. Ken Frazier led the wave of resignations by corporate CEOs from President Trump's Presidential Corporate Councils because of the President's apparent support of both sides of the issue – the peaceful and the violent, the democratic and the neo-nazi.

On diversity, ethics and compliance programs, Merck has also been long recognized as a leading if not the leading pharmaceutical corporate program with Jacqueline E. Brevard, Esq., as the lead architect of their innovative program. She was asked in 1995 by CEO Ray Gilmartin to start one of the first ethics and compliance programs in the pharmaceutical industry, which became a gold standard for the industry.

A look at the RepRisk Index Trend for Merck shows a company with relatively low reputational risk exposure which is more or less evenly distributed between the ESG categories (Figure 8.18).

8.2.b.v.B STARBUCKS

Starbucks deserves its place in this category because its founder and former CEO and chairman Howard Schultz set the tone for leadership and culture at this company from the start and was able to maintain it over his entire leadership of the company.

When the company faced one of its most trying challenges with an incident at a Philadelphia Starbucks (as discussed in Chapter 7), the new CEO was quick and ready to do all the right things immediately and was humble in recognizing the mistakes made. If he is able to maintain the strong foundational culture and continue to display responsible or even enlightened leadership qualities, while continuing to be as innovative and consistent as they have been on ESGT issues, Starbucks could make it into the next step on the organizational evolutionary ladder – the Transformative Organization.

Figure 8.19 shows the RepRisk Index Trend of Starbucks, which has had its ups and downs – including from earlier in 2018 when the Philadelphia incident occurred.

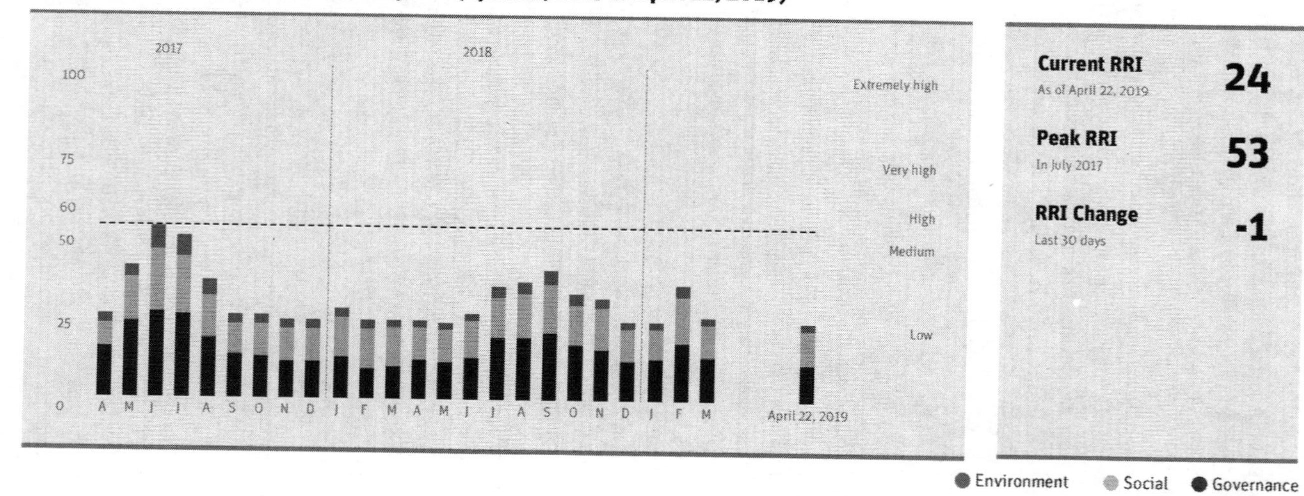

Figure 8.18 RepRisk Index Trend for Merck & Co. Inc. 2017–2019.

Source: RepRisk.

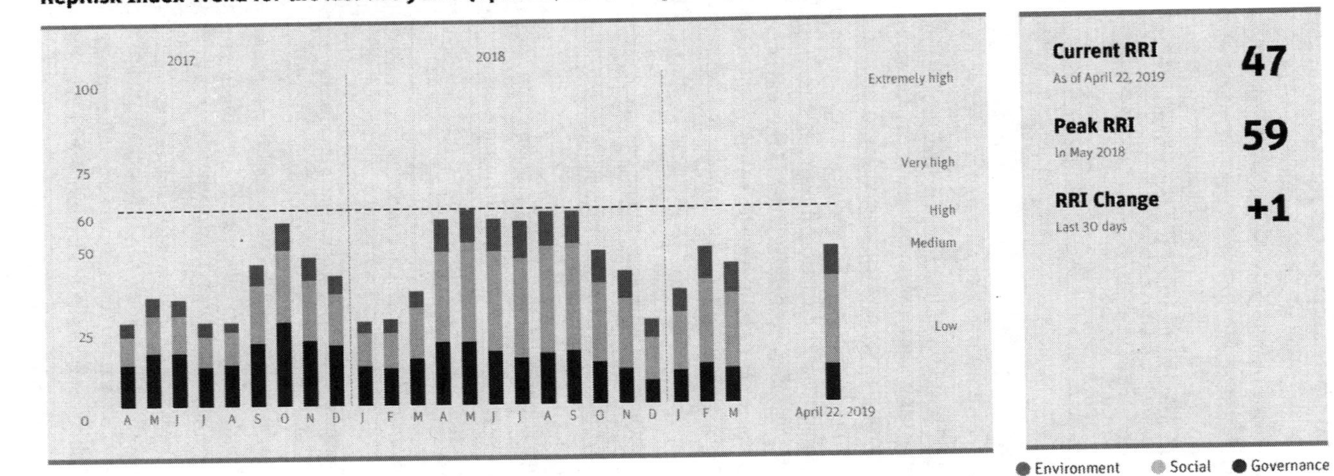

Figure 8.19 RepRisk Index Trend for Starbucks 2017–2019.

Source: RepRisk.

8.2.b.v.C ELI LILLY

Historically, Eli Lilly encountered two waves of serious issues that tested their resilience on ESGT, first in the late 1990s and then again in the early 2000s. While their first attempt didn't quite resolve the health, safety and quality issues that arose in the first place, the second time around their new leadership did restructure their entire operation to include ESGT resilience-building measures where the coordination between strategy, health and safety, risk ethics, compliance and continuous improvement were underscored most dramatically.

A look at Figure 8.20, which shows their RepRisk Index Trend, reveals how this pharma company has been able to maintain a good balance over the past few years.

8.2.b.v.D APPLE

As covered in greater detail in Chapter 6 on Technology, CEO Tim Cook and Apple have acquitted themselves in a more responsible manner than most of the big technology companies, being a first mover on issues of data privacy and now championing that cause in terms of developing proper regulatory oversight. While one can say that this provides them with a competitive advantage, one can also say that they are intelligently transforming risk into value for themselves and for one of their most important stakeholders – the customer.

While not low, Apple's RepRisk Index Trend as seen in Figure 8.21 demonstrates steadiness, and if compared to other tech giants is somewhat more favorable, as we saw in Figure 6.6 in Chapter 6.

8.2.b.vi The Transformative Organization

And, finally, we make it to the top of the evolutionary ladder – to that rarified place where both leadership and Resilience Lifecycle management are not only done really well but are synchronized, coordinated and integrated in such a way that the products and services (and primary mission, vision and strategy) of the organization enable it to create new and added value to the bottom line – both financially and reputationally.

This integration of ESGT into products and services is what truly distinguishes the Transformative Organization from all others, including the very laudable Responsible Organization type. What they don't do that the Transformative Organization and its Enlightened leaders do is go the extra mile to not only understand and fully deploy an ESGT program but also integrate these issues into their products and services for the benefit of an array of key stakeholders.

The Transformative Organization (see Figure 8.22) is the kind of organization that has that rare confluence of a number of positive things that favor not only their top stakeholder but all of their key stakeholders. The Transformative Organization has the best possible leadership – which at the end of the day makes

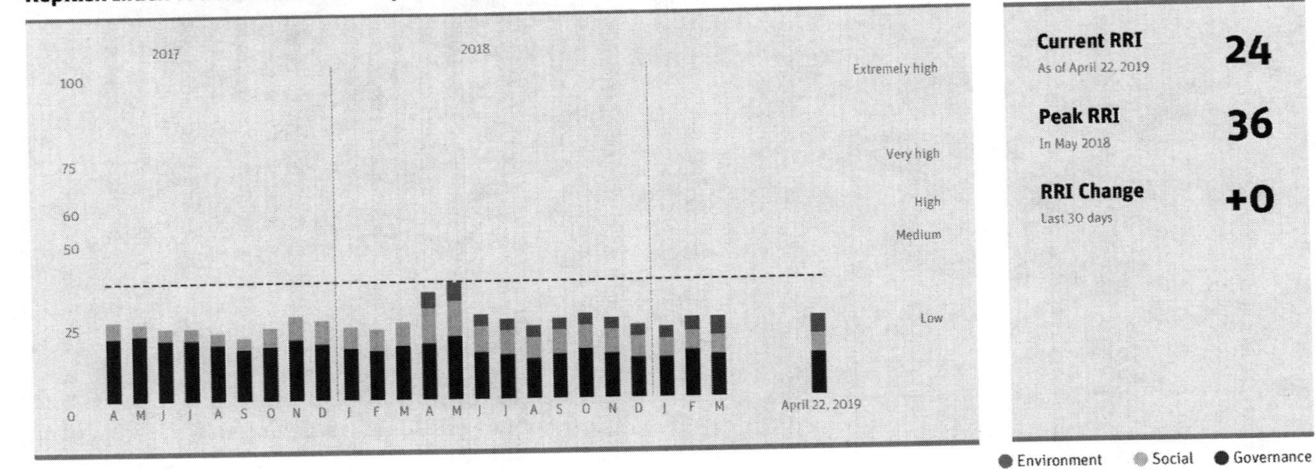

Figure 8.20 RepRisk Index Trend for Eli Lilly 2017–2019.

Source: RepRisk.

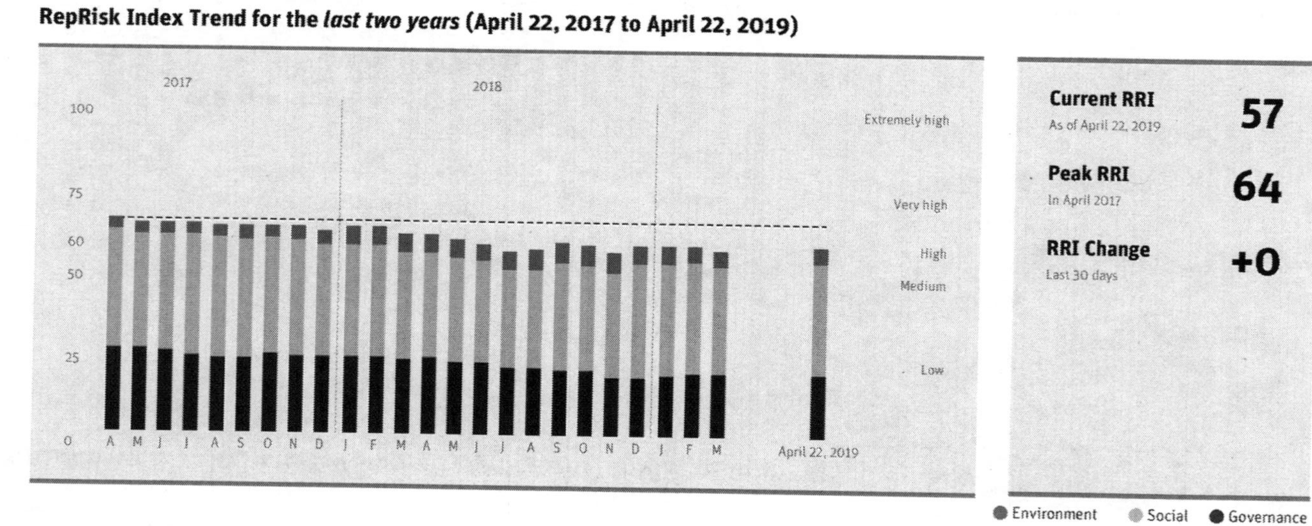

Figure 8.21 RepRisk Index Trend for Apple 2017–2019.

Source: RepRisk.

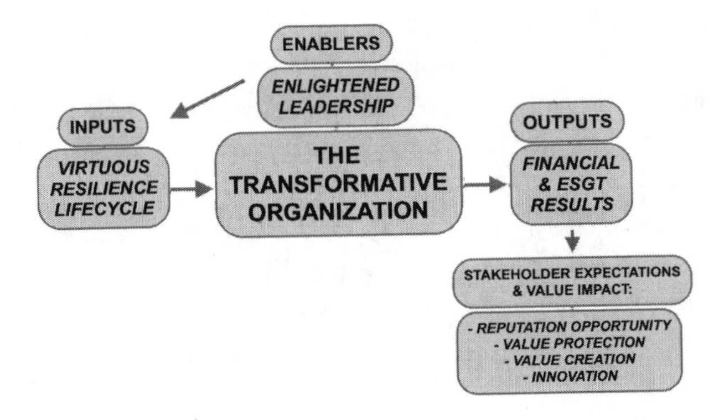

Figure 8.22 The Transformative Organization.

Source: Author and GEC Risk Advisory.

it so special – the Enlightened kind. What makes Enlightened leaders so special is that they don't just believe in the ESGT package of issues, risks and opportunities, they actually bake ESGT into their strategy, their products and their services.

The Transformative Organization and its Enlightened Leadership are not simply content to be responsible – they are actually trying to change the world:

- These are companies, for example, that create new products and services that are aligned with the SDGs.
- These are companies that create new products for their customers which they have also tested and developed internally for themselves.
- These are government agencies that don't rest with serving their citizens well but look for product, service and process improvements that turbocharge and change the way their services impact their stakeholders.
- These are educational institutions that completely break the mold of educational delivery to accommodate the great variety and diversity and location of their students, wherever they are, whatever they need.
- These are nonprofits that while serving their stakeholders and constituents to the maximum also create new products and services that benefit the primary beneficiaries.

Table 8.7 provides examples of businesses, nonprofits and governments that fit this laudable category. Let's now examine some of them.

Table 8.7 Examples of "transformative" organizations or entities

For-Profit	• Salesforce (under founder Marc Benioff) • Unilever (under Paul Polman) • Microsoft (under Satya Nadella) • L'Oréal (under Paul Agon)

(*Continued*)

Table 8.7 Continued

Nonprofit	• Epic Theatre Ensemble (under R. Russell) • Ford Foundation (under D. Walker)
Governmental	• US Special Counsel Robert S. Mueller III and team • Ethiopia (under A. Ahmed) • New Zealand (under J. Adern)

Looking at the elements of the "Virtuous Resilience Lifecycle" and applying them to a Transformative Organization – the following observations would ring true:

- **Governance:** It is robust and fully formed – with a diverse, proactive, informed lean-in board of directors, exercising many best practices such as CEO and chairman separation, independent directors dominate, turn-over happens regularly, etc. This kind of governance also proactively supervises CEO performance management and coordinates all other management activity well.
- **Culture:** Leadership is fully engaged with, reflective of and dedicated to a culture of openness, learning, speak-up/listen-up, inclusiveness and continuous self-improvement. Almost more than anything else, the CEO is emotionally intelligent (in addition to being great at the core mission of the organization).
- **Stakeholders:** In its formulation and implementation of strategy, leadership is mindful and inclusive of not only the primary stakeholder but of all key stakeholders.
- **Risk:** A fully formed, adaptive and customized enterprise risk and strategic risk management is in place.
- **Strategy:** ESGT issues are not only fully integrated into strategy formulation but are an integral part of the creation and development of new products and services.
- **Performance:** Performance management is not only keyed to financial and economic metrics but values, culture and ethical metrics – at the top and throughout the organization.
- **Crisis:** Crisis planning is integral to this organization – there is a team, there is a plan and there is scenario training and constant improvement taking place.
- **Improvement:** Innovation is a hallmark of the Transformative Organization – it not only has embedded continuous improvement processes within the organization but also deploys ESGT issues, risks and opportunities into the creation and deployment of new products and services.

A special note on the governments and governmental agencies listed in Table 8.7; they are categorized as "Transformative" for various reasons:

- In the case of Prime Minister Jacinda Ahern of New Zealand – showing the world what it means to be the top female official governing a country and becoming a new mom and then handling in the most emotionally intelligent and healing manner that a leader could the worst possible national crisis short of war – killings by a self-proclaimed white nationalist of dozens of Muslim worshippers in their sanctuaries.
- In the case of new Ethiopia's Prime Minister Abiy Ahmed, the most-watched new leader in Africa, moving quickly on ceasing two decades of hostilities with Eritrea and implementing a fast-track to full democratization and economic development.
- In the case of US Special Counsel Robert S. Mueller III and his team – assembling the team and conducting one of the most difficult, complex and explosive investigations ever of the most powerful office in the world in complete secrecy, and professionalism, with timely and devastating findings and results.

8.2.b.vi.A SALESFORCE

Salesforce lands in the Transformative Organization category because its leader, Marc Benioff, has not only succeeded as the founder and CEO of a very successful business for a sustainable period of time but also because the company exhibits the signs of a Robust or even Virtuous Resilience Lifecycle in terms of the stewardship of the interests of its stakeholders. Salesforce has received almost every "best company" category, including best places to work. They embrace a broad view of who their stakeholders are. And put their money where their mouth is.

Indeed, in recent times, Benioff has made himself a name by spearheading the discussion on the impact of wealthy Silicon Valley business interests on the needs and capabilities of lower-income working people and the growing inequality and scourge of homelessness taking place especially in and around San Francisco and Palo Alto, where the conspicuous consumption and expensive lifestyle of the technology elite not only stands in stark contrast to these growing social problems but exacerbates them (e.g., real estate is becoming unaffordable to anyone but the wealthy). Salesforce is also at the forefront of creating an "ethical and humane use" office to deal with all the new challenges the world is facing, so many of which we have dealt with under the moniker of ESGT.

Salesforce's RepRisk Index Trend is displayed in Figure 8.23 and is astoundingly low, certainly for a tech company but in general for any kind of company. They are clearly doing something right.

8.2.b.vi.B L'ORÉAL

Why is L'Oréal a Transformative Organization? As a company, L'Oréal set out over a decade ago to become a leader in transforming beauty products from products that catered only to white-skinned customers (mainly women) to catering to all shades of skin color around the world. In other words, diversity became

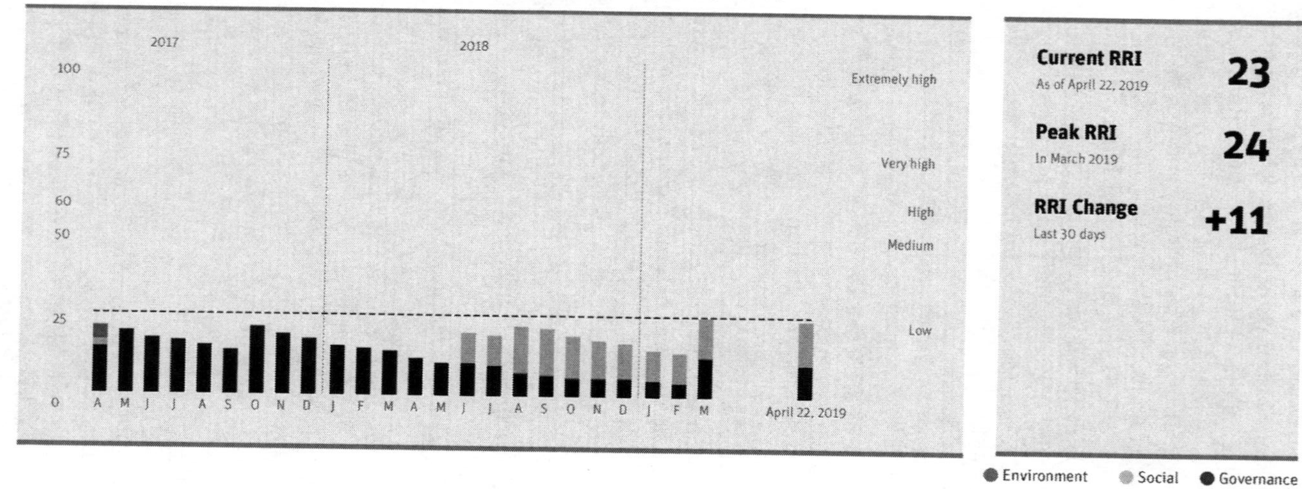

Figure 8.23 RepRisk Index Trend for Salesforce.com 2017–2019.

Source: RepRisk.

a core principle that applied not only to employees and internal company culture but also to products and services. Born out of a sense of diversity, such a move also provided L'Oréal with a first-mover competitive advantage and with a win/win strategy that combined doing the right thing with a very smart business strategy of growing revenues from formerly underserved parts of the world.

Moreover, also over a decade ago, L'Oréal launched one of the most forward-thinking and still most-advanced, innovative and sustainable ethics programs I have seen in any company globally. Once again, it started with leadership – with the current CEO, Agon, choosing an internal executive, Emmanuel Lulin, to create and spearhead an ethics and compliance program for the company globally.

Together they have developed a gold standard in global corporate ethics and compliance programs consisting of highly customized, high touch, tone from the top, including the chief ethics and compliance officer reporting directly to the CEO, reporting regularly to the board of directors, and traveling closely with the CEO all over the world to various preselected locations to do joint town halls, meetings and Q&A sessions not only with local management but also with staff. They have consistently won a number of important awards for this program, including being listed as an Ethisphere Most Ethical Company for over a decade and receiving awards from the Ethics Research Center and Ethics and Compliance Association.

Figure 8.24 shows the RepRisk Index Trend for L'Oréal, which once again (for a very large, visible, complex, global consumer goods company that could easily have a multitude of issues and risks) demonstrates a controlled and relatively low ESG reputation risk exposure over time.

8.2.b.vi.C MICROSOFT

We've already reviewed the culture change and progress that Satya Nadella has achieved as CEO of Microsoft in Chapters 2 and 6. The reason I include Microsoft here as a Transformative Organization is that above and beyond the transformation that the CEO has achieved in the internal culture, broadly and deeply enhancing the experience of employees there, the company has also taken a leadership role in representing the interests of key stakeholders in the privacy debate and the debate about developing ethical solutions to technological dilemmas, including AI. Their president, Brad Smith, has been visibly championing the stakeholder and external discussion on this topic. Moreover, Microsoft has also upgraded its products and services, especially those that touch on cyber-security, positioning themselves on the front lines of providing public–private cyber-security and awareness services in this space.

As one of the largest tech companies in the world (on April 26, 2019, the company briefly achieved US$ trillion capitalization – the third company to do so, so far), Microsoft displays a lower (though still relatively high) RepRisk Index than some of its peers, as Figure 8.25 demonstrates and as previously seen in Figure 6.6 in Chapter 6.

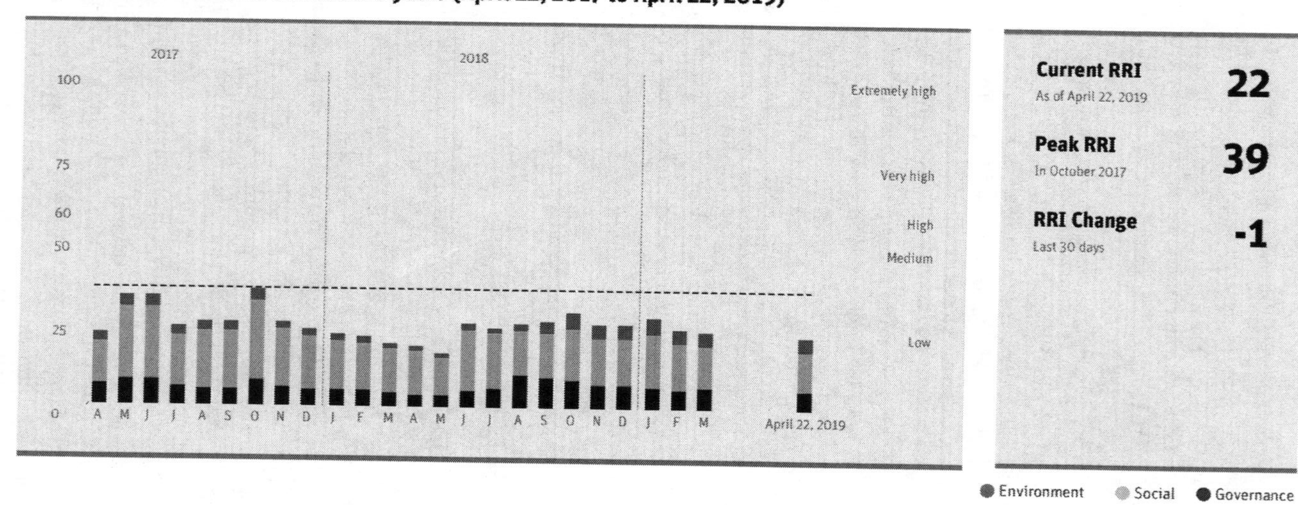

Figure 8.24 RepRisk Index Trend for L'Oréal 2017–2019.

Source: RepRisk.

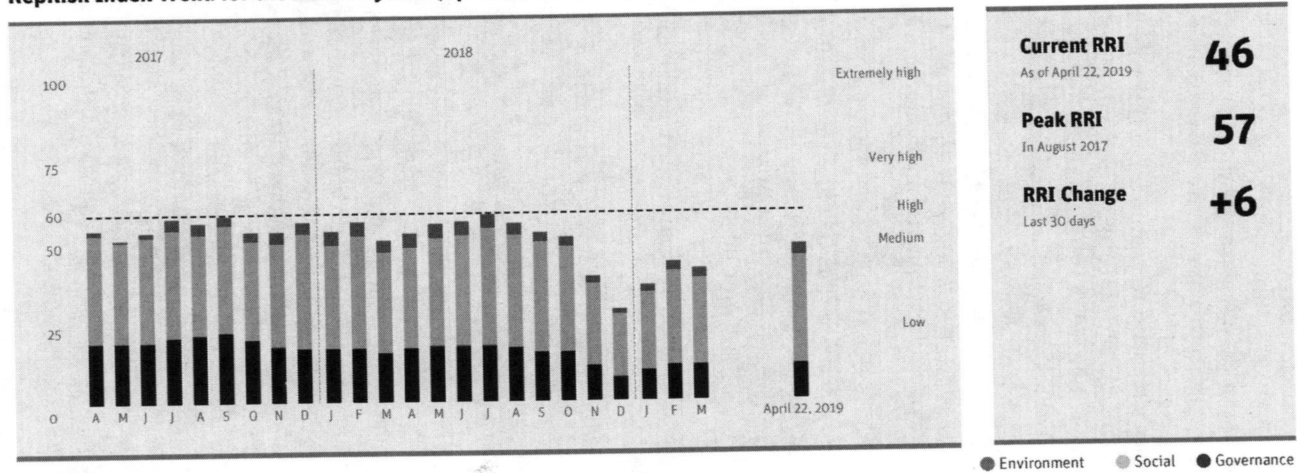

Figure 8.25 RepRisk Index Trend for Microsoft 2017–2019.

Source: RepRisk.

8.2.b.vi.D EPIC THEATRE ENSEMBLE

Epic Theatre Ensemble is a small nonprofit local theatre ensemble in New York City born on 911 (September 11, 2001) amid the chaos of that day in New York City to serve a passionate mission:

> Epic's mission is to create bold work with and for diverse communities that promotes vital discourse and social change.
>
> We accomplish this mission by:
>
> - Inspiring students to be creative and engaged citizens.
> - Presenting compelling topics that transform the way people think.
> - Collaborating with artists, students and thought leaders to produce plays about key issues.

An example of its accomplishments as a small but mighty institution serving its primary stakeholder – at-risk high school students in underserved neighborhoods of New York City – Epic has made strides from year to year including:

> Epic NEXT students, and most REMIX seniors, receive individualized guidance on college search/choice, application development, and financial aid, including visits from Admissions officers, ACT prep, and 12 overnight college visits. This initiative has been very successful: 100% of Epic NEXT grads go to college, and of the $1.1M they would pay in tuition, room, and board in 2017–18, $965,000 is in need- and merit-scholarships Epic helped them earn through auditions, interviews, and interventions.[12]

While most of my readers will never know Epic Theatre, I have had the privilege of sitting on their board of directors for a number of years, and it is the small but vital voices of transformative organizations like Epic that actually have an impact on key stakeholders.

8.2.c A cautionary note: beware of transforming value into unacceptable risk

A note of caution is important as we complete our discussion of these positive examples of Transformative and Responsible Organizations. Nothing lasts forever, and neither is the guarantee that an organization that has achieved one of these two top forms of organization will remain there. Indeed, the opposite is likely, because one of the toughest acts to pull off is the continuation of great leadership and great culture once the leaders who provided such leadership and culture are gone.

No one has yet found the elixir of how to institutionalize sustainable great culture over a long period of time – though Starbucks is giving it a try. Look what happened to the great example of HP – Hewlett Packard and the "HP Way". Once the founders left, the culture continued for a while, and then the scandals and problems began to occur.

Another recent example seems to be happening at GE. Long the darling of investors, professors, ethics and compliance practitioners and many others (including stockholders), GE started a decline under Jeff Immelt that only became clearer in the last few years of his long tenure. Indeed, the ethics and compliance profession has long held GE up as the gold standard for ethics and compliance programs. Whatever else you might say about him as a leader, it was Jack Welch who originated the idea of the defense industry creating a voluntary ethics and compliance program back in the mid-1980s – the Defense Industry Initiative (DII) – in response to a major wave of defense industry fraud scandals.

I believe that part of the answer to this question lies not only in the institutionalization of great practices and programs (à la Virtuous Resilience Lifecycle) but also in the stakeholders of an organization remaining vigilant, proactive, awake and alive vis-à-vis the organization – something along the following lines:

- Just as business stakeholders get up in arms when things go wrong, stakeholders must be vigilant and proactive in times of prosperity as well. Don't just protest an airline on Twitter when you see a nasty incident on board; support airlines that have a great track record by using them, recommending them and tweeting about them when they do the right thing.
- Just as citizens rise up in times of political danger, citizens should be vigilant and proactive at times of relative calm and the apparent well-functioning of institutions. Citizens should always think about their civic duty – not only in the bad times but in the good – and not become complacent as have many of us living in the relative safety and peace of our democracies over the past couple of decades.
- Just as donors are happy to donate to their favorite charities as long as those charities don't seem to get bad press, those donors should maintain vigilance about the proper governance of their charity in the good times as well – asking questions, prodding, getting involved.
- Boards of directors and oversight bodies in any type of organization should always be questioning management – in the good, the bad and the mediocre times – peeling back layers of the organizational onion to go beyond the obvious financial and ESGT results to understand the culture, the values and the stakeholders, and not simply take in the glossy, superficial information they may be offered by management.

8.2.d Applying the organizational typology to governments

One could extend this concern to political regimes – to functioning democracies until they no longer function consistently with their founding charter (their constitutional principles). How does a country institutionalize long-term sustainable democratic practices? One of the longest-standing and most consistent democracies of all time with a deeply admired constitution – the US – is currently experiencing a serious constitutional crisis or even a series of constitutional crises under Trump. And not in isolation, as a number of other long-standing democracies in Western Europe are experiencing similar crises of democracy.

But there is hope and there is opportunity even in a time of relative gloom. Despite his recent setback on ethical issues, look at the case of Canadian democracy under Prime Minister Justin Trudeau. At the beginning of his administration he was able to create a cabinet that truly reflected the diversity of the nation – with 50% women in ministerial positions and a heretofore unprecedented ethnic balance (something I believe no other regime has consciously done before). Despite severe pressures from its southern border from the Trump administration, the Canadian government also maintained a balanced and proactive approach on other critical national and international policies, including significantly, broadening its immigration policy in the face of the dramatic, and some would say willful, mean-spirited and racist restrictions taking place under the Trump administration in the US.

Of course, some of the earlier cautionary comments apply here too – all this can also disappear when and if there is less than transparent leadership behavior, misbehavior or the mishandling of important ESGT topics. So while the Trudeau regime can be commended for some of its earlier groundbreaking policies on gender and immigration, its prime minister is now under severe pressure – in an election year no less – for alleged ethical and possibly legal violations.

So while Figures 8.26 and 8.27 attempt to illustrate what several current political regimes might look like on our Typology of Resilience and Sustainability x/y axis and evolutionary spectrum, even I would look at these graphics with a skeptical eye as things can change so quickly as the current Canadian case illustrates – what was transformative yesterday might have become compromised or even irresponsible today.

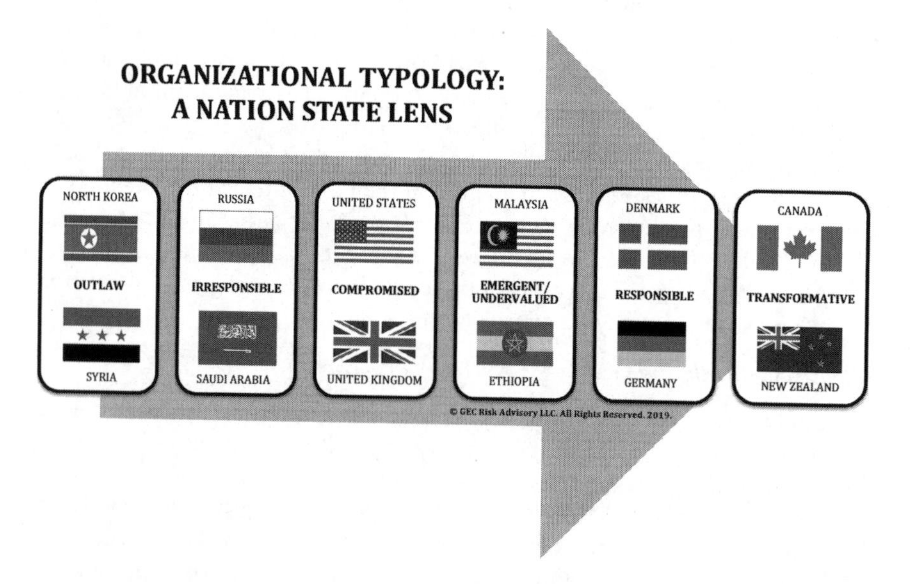

Figure 8.26 Organizational typology: the nation state lens.

Source: Author and GEC Risk Advisory.

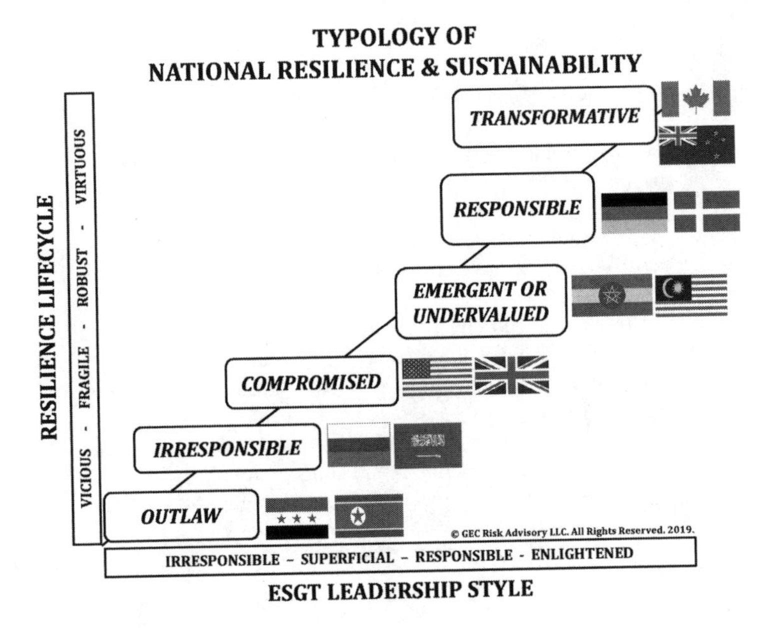

Figure 8.27 Typology of national resilience and sustainability.

Source: Author and GEC Risk Advisory.

8.3 The value of "values": the new multi-stakeholder bottom line

8.3.a What is "value"? What are "values"?

The essence of our business case for ESGT, and everything that goes with it, is that entities and their leaders that don't look at their ESGT issues, risks and opportunities through the lens of the Virtuous or Robust Resilience Lifecycle are leaving value – literally and figuratively – on the table. They are missing a major, material and potentially strategic part of value preservation and creation by not including a systematic and disciplined approach to ESGT issue, risk and opportunity management within their organization. Indeed, they are acting in a manner that could be considered derelict or perhaps even a violation of the duty of care and loyalty that executives and directors owe their organizations and ultimately their stakeholders. Such actions may lead to neither preserving nor creating value but to value destruction.

Table 8.8 Definitions of "value" and "values"[13]

Economic Definition	A measure of the benefit that may be gained from goods or services.
Marketing Definition	The difference between a customer's evaluation of benefits and costs.

(*Continued*)

Table 8.8 Continued

Investor Definition	Value investing – an investment paradigm.
Ethics Definition	Treating actions themselves as abstract objects, putting value to them.
Social Definition	A set of values, institutions, laws, and symbols common to a particular social group.
"Values"	In ethics, value denotes the degree of importance of some thing or action, with the aim of determining what actions are best to do or what way is best to live (normative ethics) or to describe the significance of different actions. A concept that is studied in many different disciplines, including anthropology, behavioral economics, business ethics, corporate governance, moral philosophy, political science, social psychology, sociology and theology.
"Personal Values"	Personal values provide an internal reference for what is good, beneficial, important, useful, beautiful, desirable and constructive.
"Cultural Values"	Values of a society can often be identified by examining the level of honor and respect received by various groups and ideas.

Source: Wikipedia.

So what do we mean by "value"? We can of course summon up some good definitions, and we do so in Table 8.8. But what do we really mean by "value" when it comes to building resilience and sustainability in organizations and entities? I would argue that the definition of value needs to be designed and customized by each organization – whether business, nonprofit or governmental – by a thorough understanding of the particular value that they seek to create, not only from the perspective of their primary stakeholder (shareholders, beneficiaries and citizens, respectively) but of other key stakeholders.

Table 8.9 provides examples of what "value" might mean not only to the primary stakeholder of each entity but also to other key stakeholders. The point here is that no entity should wear horse-blinders or myopically focus exclusively on the interests of only their primary stakeholder.

Table 8.9 Different stakeholders: different meanings of "value"

Organization/ Sector	*Stakeholder*	*What "Value" means to this stakeholder*
For-Profit	**Shareholder**	Return on equity, stock appreciation, return on investment, reputation.
	Employee	Pay, job opportunity, culture, company's reputation.
	Customer	Product or service quality, economic value, safety.

(*Continued*)

Table 8.9 Continued

Organization/ Sector	*Stakeholder*	*What "Value" means to this stakeholder*
Nonprofit	**Beneficiary**	Receiving the benefit offered (health, education, food, other).
	Donor	Money reaching and benefiting beneficiaries; tax deduction; reputation of nonprofit.
	Partners	Visibility of partnering with a reputable charity.
Governmental	**Citizens**	Rule and protection of law; constitutional rights; right to vote; independent media.
	Residents, transients and immigrants	Rule and protection of law; freedom from arbitrary, extra-legal or illegal government action.
	Neighboring countries	Peace; respect for borders; economic cooperation, trade and peaceful movement of peoples.

Source: Author.

If an entity hasn't done its homework to understand who beyond the principal stakeholder is in the panoply of other key stakeholders applicable to it as an entity, the entity and its leadership will not have a serious or complete understanding of value – whether value creation, preservation or destruction. Thus, the concept of "value" must be informed and driven by this stakeholder understanding and analysis:

- If you are a narrow-minded business, you will think of "value" narrowly as the return on investment or the stock price appreciation for your shareholders and owners.
- If you are a more broadly minded business, you will expand your notion of "value" to include the value created from benefiting and meeting the expectations of other stakeholders beyond your primary stakeholder – like employee loyalty, customer satisfaction or supplier diversity.
- If you are a government agency – say, the Department of Justice in the US (DOJ) – your primary value might be to prosecute and convict criminal wrongdoers within your federal mandate.
- But if the DOJ is being run by an enlightened attorney general, perhaps the "value" of the DOJ goes beyond law enforcement to include crime prevention and the carrying out of investigations in a professional manner that obeys all protections afforded under the Constitution and creates citizen trust in the system.
- If you are the American Cancer Society (ACS), your primary stakeholders are people with cancer and their families who will benefit from cancer-fighting research and health initiatives sponsored by the ACS.
- But if you are forward-looking and looking to benefit society as a whole, part of the ACS strategy will be to build internal resilience that will benefit employees and third parties that are providing services to ACS in support of and furtherance of research.

The idea is that every organization needs to look at what their value proposition is and then augment it with a consideration of who the other primary, key and critical stakeholders are. While there may be a singular notion of value that can be defined for each organization – whether governmental, for-profit or nonprofit – there are also a series of additional values that may or may not be quantitatively measurable that are associated with the organization and its license to operate that are important or critical to the well-being of stakeholders and others.

An additional lens or perspective on this issue is illustrated in Table 8.10, where we parse out the possible value associated with each of the ESGT issues we have reviewed in this book from the standpoint of the three types of organization we have been examining as well – for-profit, nonprofit and governmental. In essence – value is in the eye of the stakeholder.

Table 8.10 What does "value" mean in different contexts? Value is in the eye of the stakeholder

	Government Stakeholders	*For-Profit Stakeholders*	*Nonprofit Stakeholders*
Environmental Issues, Risks and Opportunities	Protection of air, land and water from pollution through laws and law enforcement.	Production of goods and services in compliance with environmental laws and norms.	Delivery of services to beneficiaries in compliance with environmental laws and norms.
Social Issues, Risks and Opportunities	Maintenance of the peace, freedom from war and law enforcement subject to rule of law.	Running of the business with an effective code of conduct that respects human rights.	Delivery of services to beneficiaries in compliance with labor and employment laws and norms.
Governance Issues, Risks and Opportunities	Constitutional protections such as freedom of speech, of association, voting rights, and separation of powers with checks and balances.	Development of new business without engaging in corruption bribery and fraud.	Governance of the nonprofit in accordance with high governance standards of similar to best run for profits.
Technology Issues, Risks and Opportunities	Protection of privacy & other rights of individuals and society against technology abuse and theft and safety against cyber-attacks on infrastructure.	Creation of new products and services without violating the privacy data of consumers, employees and users.	Ensuring that all donors, employees and others' data privacy is not violated in furtherance of the beneficiary focused mission.

Source: Author.

The point of Table 8.10 is that "value" can show up in many shapes and forms, and just because we might have a knee-jerk reaction to what value might mean to a given organization doesn't mean that we shouldn't dig a little deeper to understand other "value" and "values" that might be embedded in that organization's activities. A full understanding of the organization's key stakeholders and their expectations will help greatly in understanding such "value" or "values". Indeed, the ESGT dimensions of any of these organizations provide a wealth of additional concepts and even metrics that organizations could unleash and use to measure organizational performance and the individual performance of their leaders and individual employees in accordance with their job descriptions.

Let us turn now to several excellent resources from the investment, scholarly and business community that I believe provide further fodder and hope for a more integrated holistic concept of "value" and "values" to be deployed at all kinds of organizations in every sector.

8.3.b The value of values: the business case for ESGT, organizational resilience and sustainability

In the spring of 2018, Larry Fink, founder, CEO and chairman of BlackRock, the largest asset manager in the world, wrote a letter to CEOs that succinctly summarized and encapsulated much of what we have said in greater detail in this book. BlackRock's perspective of course is that of a critical stakeholder in the business world – the globe's largest asset manager.

Part of Larry Fink's letter stated the following:

> We also see many governments failing to prepare for the future, on issues ranging from retirement and infrastructure to automation and worker retraining. As a result, society increasingly is turning to the private sector and asking that companies respond to broader societal challenges.
>
> Indeed, the public expectations of your company have never been greater. Society is demanding that companies, both public and private, serve a social purpose. To prosper over time, every company must not only deliver financial performance, but also show how it makes a positive contribution to society. Companies must benefit all of their stakeholders, including shareholders, employees, customers, and the communities in which they operate.
>
> Without a sense of purpose, no company, either public or private, can achieve its full potential. It will ultimately lose the license to operate from key stakeholders. It will succumb to short-term pressures to distribute earnings, and, in the process, sacrifice investments in employee development, innovation, and capital expenditures that are necessary for long-term growth.[14]

There is a growing body of work from leading think tanks, investors, businesses and scholars expanding the concept of value and the metrics that might be associated with such an expanded view beyond strictly financial metrics. While what Larry Fink states in his CEO letter of 2018 and further materials that BlackRock has been issuing since then is at the vanguard of what investors and others are saying about ESG and diversity generally, it is also a reflection of reality that has now emerged: that ESG (and in my opinion ESGT) is and should be an integral and critical part of every organization's ethos.

What follows is by no means exhaustive but a survey of a diversity of examples of how ESGT issues, risks and opportunities can provide value – sometimes verifiable and quantifiable – to organizations and their stakeholders. The following array of "value" examples and research in essence underscores a major message of this book: that *values create value.*

We start with some of the bigger-picture studies that look at how the increase in sustainability, ethics and corporate responsibility programs is associated with value creation and then we hone in on a few more focused studies that look at aspects of such resilience-building ESGT programs that add value to their organizations.

8.3.b.i Some big-picture ESG "value" studies

8.3.b.i.A ECCLES, IOANNOU AND SERAFEIM SUSTAINABILITY STUDY

A study that caught my eye a few years ago and exhibits staying power is one that looked over an 18-year period at 180 publicly listed companies and compared companies with corporate responsibility and sustainability programs against companies without such programs. The economic metrics of their performance over time are dramatic, to say the least. Table 8.11 demonstrates the clear value delta that the high-sustainability companies had versus the low-sustainability companies over a very significant (18-year) period of time.

Table 8.11 Long-term positive financial impact of sustainability

	An Investment in 1993 of:	*Results in 2010 for High-Sustainability Companies*	*Results in 2010 For low/no Sustainability Companies*
Investment	$1	$22.6	$15.4
Return on Equity	$1	$31.7	$25.7
Return on Assets	$1	$7.1	$4.4

Source: Eccles, Ioannou and Serafeim.

The same coauthors – Eccles, Ioannou and Serafeim – later added to this initial study, and the abstract of their study says it all – pay particular attention to the bold words regarding boards of directors:

> We investigate the effect of corporate sustainability on organizational processes and performance. Using a matched sample of 180 US companies, we find that corporations that voluntarily adopted sustainability policies by 1993 – termed as High Sustainability companies – exhibit by 2009, distinct organizational processes compared to a matched sample of firms that adopted almost none of these policies – termed as Low Sustainability companies. **We find that the boards of directors of these companies are more likely to be formally responsible for sustainability and top executive compensation incentives are more likely to be a function of sustainability metrics.** Moreover, High Sustainability companies are more likely to have established processes for stakeholder engagement, to be more long-term oriented, and to exhibit higher measurement and disclosure of nonfinancial information. **Finally, we provide evidence that High Sustainability companies significantly outperform their counterparts over the long-term, both in terms of stock market as well as accounting performance.**[15]

8.3.b.i.B UNIVERSITY OF OXFORD AND ARABESQUE PARTNERS "FROM STOCKHOLDER TO STAKEHOLDER"

In another broadly based study by the University of Oxford with the support of Arabesque Partners, a number of fascinating findings to further the idea of increased value from sustainability were made, as Table 8.12 summarizes.

Table 8.12 University of Oxford and Arabesque Partners: From the stockholder to the stakeholder study findings[16]

1. **Sustainability is one of the most significant trends in financial markets** for decades.
2. This report represents the most comprehensive knowledge base on sustainability to date … **based on more than 200 academic studies**, reports, etc.
3. 90% of studies on cost of capital show that **sound sustainability standards lower the cost of capital** of companies.
4. 88% of research shows that solid **ESG practices result in better operational performance** of firms.
5. 80% of studies show that **stock price performance of companies is positively influenced by good sustainability** practices.
6. Based on economic impact, it is in the **best interest of investors and corporate managers to incorporate sustainability** considerations into their decision-making processes.
7. Active ownership **allows investors to influence corporate behavior** and benefit from improvements in sustainable business practices.
8. **The future of sustainable investing is likely to be active ownership by multiple stakeholder groups.**

Source: University of Oxford and Arabesque Partners.

8.3.b.i.C ACCENTURE'S "THE BOTTOM LINE ON TRUST"

In another big-picture study demonstrating the sustainability and profitability of ESGT issues generally in the long run, Accenture (also with the help of Arabesque) recently produced a study demonstrating the connection between value protection and a critical nonfinancial or ESGT issue – trust. The study is based on their concept of "competitive agility", defined as follows:

> The Competitive Agility Index is a measure of a company's competitiveness developed by Accenture Strategy to measure the value of an interdependent strategy targeting growth, profitability, and sustainability and trust. Our analysis shows that companies can no longer rely solely on traditional or historical gauges like market cap and total shareholder return to paint the full picture of their future competitiveness.

Their main findings are summarized in Table 8.13. Note that they made a dramatic finding about the disproportionate impact of a deficit of trust: "While trust accounts for a fraction of a company's total score, it disproportionately impacts revenue and EBITDA".

Table 8.13 Accenture strategy "the bottom line on trust": main findings[17]

- 54% of companies on Accenture's Competitive Agility Index experienced a material drop in trust and conservatively lost a collective US$180 Billion.
- A US$30 Billion retail company experiencing a material drop in trust stands to lose US$4 Billion in future revenue.
- When we compared companies that experienced a trust drop to those that did not, we saw a trend. Those that had a drop in trust saw their Competitive Agility Index scores decrease more than those that did not. While trust accounts for a fraction of a company's total score, it disproportionately impacts revenue and EBITDA.
- While the average percentage varies by industry, when a drop in trust occurs, companies in all industries will experience a material decline in both revenue and EBITDA.
- It's clear that trust has become material. If we define "material" as anything that could change the perceived value of a company, trust is now a bona fide poster child for materiality. Trust declined in 10 of 15 industry sectors in 2017, signaling that companies must better position themselves for resiliency from trust incidents. Companies with a higher Competitive Agility Index score, however, are more resilient when their trust score drops—seeing less of an overall impact on competitiveness.

Source: Accenture.

Table 8.14 presents an overview of the potential impact on EBITDA growth from a two point drop in Accenture's Competitive Agility Index on a wide variety of sectors in which "trust" (or a deficit thereof) as mentioned earlier plays a disproportionate role.

Table 8.14 Potential impact on EBITDA growth from a 2-point drop in the Competitive Agility Index score (ACCENTURE STUDY)[18]

Industry sector	*EBITDA growth drop (%)*
Global average (includes all sectors below and others)	*–9.8*
Industrial services	–16.2
Electronics and high tech	–14.8
Travel & transportation	–13.4
Consumer goods & services	–11.8
Energy	–10.2
Communications	–10.0
Media	–9.4
Software, platforms, services	–8.8
Retail	–8.0
Utilities	–7.8
Manufacturing	–3.9
Insurance	–3.0

Source: Accenture.

8.3.b.i.D BOSTON CONSULTING GROUP "TOTAL SOCIETAL IMPACT"

An extensive study undertaken by BCG (Boston Consulting Group) around a concept of Total Societal Impact (TSI) looks at measuring the nonfinancial performance captured by ESG metrics on the valuation of companies. Their main conclusions include the following:

- Nonfinancial performance (as captured by the ESG metrics) was statistically significant in predicting the valuation multiples of companies in all the industries we analyzed.
- In each industry, investors rewarded the top performers in specific ESG topics with valuation multiples that were 3% to 19% higher, all else being equal, than those of the median performers in those topics.
- Top performers in certain ESG topics had margins that were up to 12.4 percentage points higher, all else being equal, than those of the median performers in those topics.[19]

8.3.b.i.E BSR GLOBESCAN "STATE OF SUSTAINABILITY 2018"

In terms of major themes and trends, leading organizations Business for Social Responsibility (BSR) and GlobeScan produced their 2018 sustainability and business report, and the findings continue to support the trend lines we have examined throughout this book. They are summarized in Table 8.15.

Table 8.15 BSR and GlobeScan "The State of Sustainable Business 2018": key findings[20]

Companies Are Defining a new Sustainability Agenda	• Corporate integrity and diversity and inclusion, while long-standing corporate issues, are top priorities for sustainability efforts in 2018 –perhaps a reflection of recent political, technological, and social transformations that have accelerated socially responsible activism. • Climate change and human rights remain in the top four priority issues, while less than half of companies are prioritizing inclusive growth or public policy frameworks. • Disruptive technologies, such as artificial intelligence, concern over data privacy and ownership, and disruptions to climate and energy systems are shaping future business strategies. • Priority issues are still more driven by risk management than value creation.
Sustainability Needs to Be Integrated into Strategy	• Three-quarters of practitioners observe that effectively navigating global megatrends means ensuring that sustainability is a mainstream business issue, necessitating both organizational integration and new approaches to strategy and governance.
SDGs Are Driving Strategy	• There has been a significant increase in companies using the SDGs to inform their goals.
Companies Have Limited Focus on Value Chain Impacts	• Companies take an inconsistent approach to addressing key issues across their value chains, with efforts to go beyond their own operations still limited.
There Is a Need for More Cross-Functional Collaboration	• Sustainability teams still struggle to get traction with strategic planning and core business functions. There is surprisingly limited engagement with investor relations, marketing, or human resources—despite the recognized significance of investors, customers, and employees as key drivers of sustainability.
There Is Room to Improve Communications	• Fewer than half of BSR members find their own sustainability communications to customers or consumers to be effective.

Source: Business for Social Responsibility and GlobeScan.

8.3.b.i.F BAML ESG INVESTMENT STRATEGY REPORT

Finally, a late 2018 comprehensive study of the performance and future impact of ESG investment strategies on companies and markets by Bank of America Merrill Lynch (BAML) provides the following powerful conclusion:

> ESG is a better signal of earnings risk than any other metric we have found.

The study states that ESG investment strategies not only outperform on shareholder value but also provide a highly accurate prediction of success for a series of other resilience and sustainable metrics like the following:

- Less likely to go bankrupt over the next five years.
- Less likely to have large price declines.
- Less likely to have earnings declines or increased EPS volatility.
- More likely, if small cap, to become high-quality stocks in future.
- More likely to see extreme inflows over the next few decades.
- Delivering three-year returns significantly better than their peers.[21]

8.3.b.ii More specific ESGT "value" data points

Larry Fink of Blackrock has also spoken specifically on the topic of diversity in the same 2018 CEO letter quoted earlier:

> We ... will continue to emphasize the importance of a diverse board. Boards with a diverse mix of genders, ethnicities, career experiences, and ways of thinking have, as a result, a more diverse and aware mindset. They are less likely to succumb to groupthink or miss new threats to a company's business model. And they are better able to identify opportunities that promote long-term growth.[22]

Let's turn to some of these examples to illustrate how value may be created through specific types of programs or practices that may or should be part of a Robust or Virtuous Resilience Lifecycle linking such concepts as diversity, culture, ethics, resilience, corporate responsibility and reputation to value creation, preservation or protection. Hardly radical thought leaders and mainstays of capitalism like McKinsey, for instance, are increasingly showing the deep connection between diversity, on the one hand, and greater corporate resilience and profitability, on the other.

8.3.b.ii.A MCKINSEY AND CATALYST ON THE VALUE OF DIVERSITY

A number of leading think tanks and institutions have delved into the question of whether greater diversity at the highest levels of an organization – the executive team and/or the board – may be correlated with greater long-term value creation. Catalyst and Credit Suisse, respectively, produced studies over the past few years that looked at this issue. And so has McKinsey, which in its ongoing major research project called "Delivering through Diversity" shows a continuing strong link between diversity (defined broadly) and performance. McKinsey's research demonstrates a correlation between gender and ethnic/cultural diversity and value creation over time. According to McKinsey:

> We found that companies with the most ethnically/culturally diverse boards worldwide are 43% more likely to experience higher profits. We also found a positive correlation between ethnic/cultural diversity and value creation at both the executive team and board levels.[23]

Catalyst has collected and sponsored a vast array of research and has published exhaustively on its website on the correlation of the matters summarized in Table 8.16.

Table 8.16 Catalyst compendium of research demonstrating correlation between diversity and value creation[24]

Innovation and Group Performance

- Diverse organizations are more successful in retaining talent.
- Inclusive workplaces maximize talent and productivity.
- Diverse teams are critical for innovation.
- Diverse management teams are innovative and earn a premium for their innovation.
- Diversity reduces groupthink and enhances decision-making.
- Inclusion is key to team performance.

Reputation and Responsibility

- Gender diverse companies have excellent reputations.
- Mixed gender boards have fewer instances of fraud.
- Boardroom diversity strengthens corporate social responsibility performance.

Financial Performance

- Diversity is associated with improved financial performance, including in the following ways:
  - Accounting returns.
  - Cash flow return on investment.
  - Earnings per share.
  - Earnings before interests and taxes (EBIT) margins.
  - Gross and net margins.
  - Investment performance.
  - Market performance.
  - Market value.
  - Return on Assets (ROA).
  - Return on Equity (ROE).
  - Return on Sales (ROS).
  - Revenue.
  - Sales growth.
  - Share price performance.
  - Tobin's Q".

Source: Catalyst.

8.3.b.ii.B NATIONAL ASSOCIATION OF CORPORATE DIRECTORS (NACD) "CULTURE AS AN ASSET"

NACD Blue Ribbon Commission "Report on Culture as a Strategic Asset" is the first major governance study on the value and importance of an effective corporate culture as an asset (and potential liability) to a company.[25] While this study does not provide quantitative research on this topic, it does reflect the increasing importance of intangibles, ESG and culture issues to boards and directors and is a very welcome addition to the dialogue.

8.3.b.ii.C ETHICAL SYSTEMS "ETHICS PAYS"

Ethical Systems is an NYU-based behavioral ethics think tank that collects and promotes studies from a wide array of academics nationally and globally, all focused on behavioral ethics and cultural issues. They have an "Ethics Pays" webpage containing a vast array of leading academic and practical research on the upside to companies of ethical behavior, sustainability and CSR programs.[26]

8.3.b.ii.D THE VALUE OF WHISTLEBLOWER REPORTING

Academics Stephen Stubben and Kyle Welch deployed a vast treasure trove of whistleblower hotline and helpline data supplied by Navex and found in very general terms that active and robust whistleblower reporting was a sign of a healthy company.

Specifically, some of their key findings of more whistleblower reporting correlated with:

- Fewer material lawsuits.
- Lower litigation costs.
- Fewer external whistleblower reports.
- Greater profitability and productivity – "Companies with higher levels of reporting activity had ROA as much as 2.8 percent higher than comparable companies with lower levels of hotline activity".
- Less potential for earnings management.[27]

The following quote from their research encapsulates almost everything we have been trying to demonstrate about the value of ESGT generally in this book – and I agree with them that robust and healthy whistleblower reporting within an organization is truly a reflection of the culture of the organization that denotes a leadership that is willing to listen and allow employees and others to speak up without fear of retaliation, a dramatic element of a healthy culture:

> We found that companies that more actively use their internal reporting systems can identify and address problems internally before litigation becomes likely. Significantly, our analysis shows that a one-standard-deviation increase in the use of an internal reporting system is associated with 3.9% fewer pending material lawsuits in the subsequent year and 8.9% lower aggregate legal settlement amounts. Over a three-year period, a one-standard-deviation increase was associated with 6.9% fewer lawsuits and 20.4% lower settlement amounts. Avoiding lawsuits is important for reasons beyond just the direct financial costs of legal defense and settlements. Although settlement costs can often be in the hundreds of millions

of dollars, the hit to brand reputation and stock price can easily exceed all other out-of-pocket expenses. We found that in addition to reduced legal exposure, firms that more actively use their internal reporting systems are typically more profitable and have been in business longer.

8.3.b.ii.E ETHISPHERE "ETHICS PREMIUM RESEARCH"

Ethisphere has conducted what they call "Ethics Premium Research" showing that publicly traded companies with robust ethics and compliance programs show higher stock value compared to companies without such programs.[28] Ethisphere has tracked the stock value of publicly traded companies on its annual Ethisphere Most Ethical Companies list and found what they call a three-year "ethics premium" with the stock value of these companies on average 4.88% higher than that of their competitors.

8.3.b.ii.F S&P GLOBAL THE VALUE OF "ESG AND PREPAREDNESS"

Reflecting the growth of the importance of ESG in the investor and business community overall, S&P Global has rolled out a new product called the ESG Evaluation Assessment in which their analysts not only look at the ESG profile of a company to provide it a rating but combine the concept of ESG with the concept of preparedness – very similar to what we have done in this book – combining the concepts of ESGT and organizational resilience through the Organizational Resilience Lifecycle.

Indeed, in their new ESG materials, S&P Global cite the data in Figure 8.28 to demonstrate the connection between ESG and financial performance from aggregate evidence from more than 2,000 empirical studies showing an overwhelmingly positive correlation of ESG and financial performance:

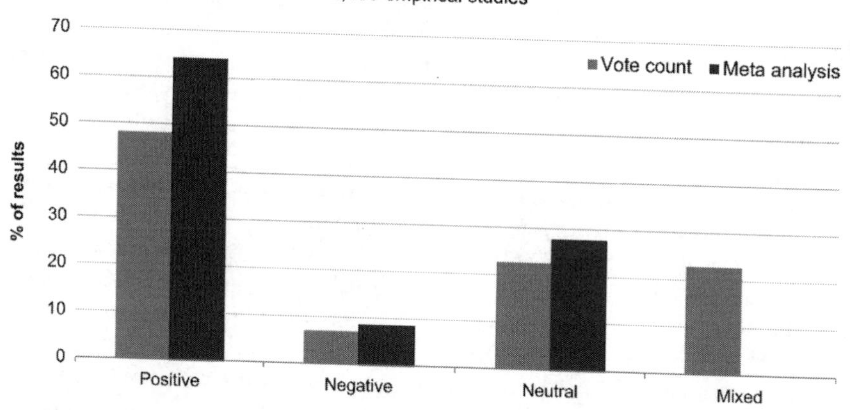

Figure 8.28 ESG and corporate financial performance study findings.

Source: S&P Global.

8.3.b.ii.G COMPARITECH: THE VALUE OF CYBER PREPAREDNESS

An interesting cyber-breach-focused study shows some very interesting value-based patterns, as summarized in Table 8.17.

Table 8.17 Relative stock performance of 28 NYSE companies that had suffered and reported cyber-data breaches against average NASDAQ share price over three-year period[29]

	One year after breach	*Two years after breach*	*Three years after breach*
Average Share Price Growth Nasdaq	12.23%	29.13%	44.29%
Average Share Price Growth of NYSE Breached Companies	8.53%	17.78%	28.71%
Net Under-Performance with Nasdaq Average	*−3.7%*	*−11.35%*	*−15.58%*

Source: Comparitech.

8.3.b.ii.H STEEL CITY RE: REPUTATION EFFECTS ON EQUITY RETURNS AND MEASURING "REPUTATION VALUE"

Finally, Steel City Re has done interesting work in the reputation risk, value and financial performance space. What follows are two charts in which, on the one hand, they measured the trailing 6-month and trailing 12-month stock performance of companies considered reputation leaders (based on a selection of best performers they created for FT Agenda in late 2018), and, on the other hand, a similar 6 and 12 month trailing period for companies selected by GEC Risk Advisory for their reputational hits and impairment. In addition, the S&P500 6-Month and 12-month performance are included in Figure 8.29 to show the comparable performance of that overall index against the high reputation and low reputation companies profiled.[30]

Table 8.18 lists the high and low reputation value companies profiled by Steel City Re for Agenda and GEC Risk Advisory for the evaluations set forth on Figures 8.29 and 8.30.

Table 8.18 High and low reputation companies selected for reputation value and reputation effects on equity returns evaluations (set forth in Figures 8.29 and 8.30)

Sector	*High reputation value*	*Low reputation value*
Commercial Services	S&P Global	Equifax
Finance	Public Storage	Wells Fargo
Producer Manufacturing	Roper Technologies	General Electric Company
Technology Services	Global Payments	Facebook
Utilities	Nextera Energy	PG&E

Source: Steel City Re, GEC Risk Advisory.

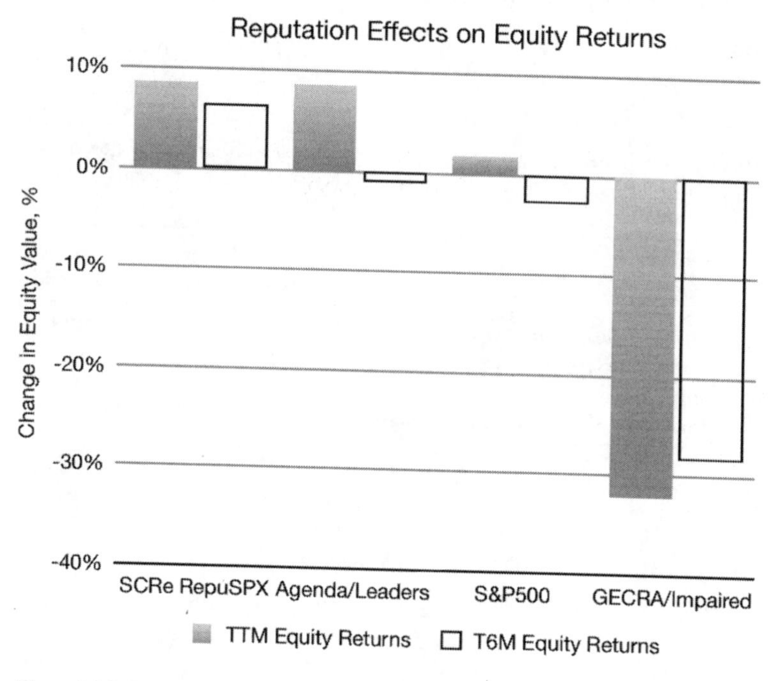

Figure 8.29 Reputation effects on equity returns.[31]

Source: Steel City Re.

Dr. Nir Kossovsky, CEO of Steel City Re, explains the information in Figures 8.29 and 8.30 as follows:

> Reputation is a soft power. Nevertheless, it can be measured, managed, and insured. How and when stakeholders appreciate and value it creates opportunities. Some firms have it but its value is not appreciated immediately by equity investors. These are the firms Steel City Re, an insurer of reputation risk, exposes with its synthetic measures of reputation value and places annually into its RepuSPX Reputation Arbitrage equity portfolio. After nearly 16 years, that portfolio is outperforming the parent S&P500 by around 400%. Figure 8.29 shows the current portfolio assembled in January 2018 was returning at the time of this writing 6.41% over the trailing 6 months and 8.66% over the trailing 12 months compared to the S&P500's returns of −2.69% and 1.66%, respectively.
>
> Other firms squander it. Five firms with high-profile reputation scandals selected by GEC Risk Advisory—Equifax, Wells Fargo, General Electric,

Facebook and PG&E—had returns of −28.26% and −32.34% over the trailing 6 and 12 months, respectively.

Usually the market discovers fair value and rewards firms for the reputational value excellence. Five sector-matched firms from a portfolio of reputation value leaders assembled from the S&P500 in June 2018 for an article by Tony Chapelle writing for Agenda, the *Financial Times* service—Agenda/Leaders—showed returns of −1% and 8.6% over the trailing 6 and 12 months, respectively.

Steel City Re attributes the greater volatility of this latter group, in part, to their challenges in reputation risk management. The changes in reputation value of this group, Agenda/Leaders, are compared to the changes in reputation value for the firms that squandered it, GECRA/Impaired, are shown in Figure 8.30 – Reputation Value Changes. Notwithstanding equity market volatility at the time of this writing, the reputation value of the Agenda/Leaders showed changes of only −1.7GU% and −0.3GU% over the trailing 6- and 12-month periods. The reputation value of the GECRA/Impaired group showed changes of −9.3GU% and −12.6GU% respectively.[32]

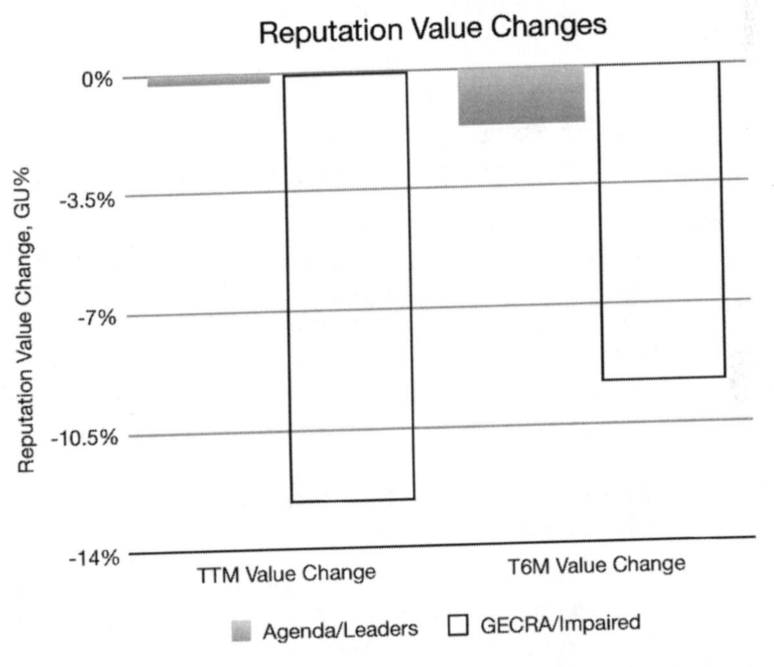

Figure 8.30 Reputation value changes in low v. high reputation value companies.[33]

Source: Steel City Re.

8.3.b.ii.I NEW RESEARCH: RESILIENCE PAYS OVER TIME

A fascinating new study that was recently published in the *Harvard Business Review* makes the case that companies that are prepared for future downturns – i.e., know that a recession will be coming at some point and act accordingly in being prepared (on a variety of levels) – are not only more resilient when the time comes but their compound annual growth rate (CAGR) is substantially higher than that of companies that don't prepare, to the tune of 17% CAGR for prepared companies versus 4% CAGR for unprepared companies.[34]

8.3.b.iii The value of values

With this whirlwind review of big-picture and more focused research and studies on the value of values (as I like to phrase it), we pretty much come to the end of our Gloom to Boom journey.

The demonstration of the value of values will continue to evolve. Even just the slice of work that is being done that I have presented briefly in this chapter should encourage everyone who believes in the value of ESGT issues, risks and opportunities and good leadership and culture at the helm of making these changes. We are getting there in terms of hard-core quantitative and qualitative proof that ESGT is good for the stakeholder bottom line, no matter what organization we are talking about.

And as Figure 8.31 from the 2019 Edelman Trust Barometer shows, "[C]ompanies can improve society – and also do well" according to the 73% of the global population they surveyed. Speaking of transforming risk into opportunity!

Finally, we now turn to the last section of this chapter (and of this book!) to underscore one of the more important endeavors we should all be participating in – as citizens, workers, leaders, students, board members, government decision-makers, politicians or social activists – the future of technology. The next section looks specifically at what organizational leaders should do to transform tech fear into tech trust, one of the greatest areas of risk and of opportunity globally today.

8.4 Future-proofing value creation: how leaders transform tech fear into tech trust

The Fourth Industrial Revolution has ushered in a brave new world few of us can begin to fathom – including those who fancy themselves cutting-edge techno-geeks, super-experts and digital business geniuses. Change is happening exponentially faster than any human can process and in a qualitative manner that no machine can understand.

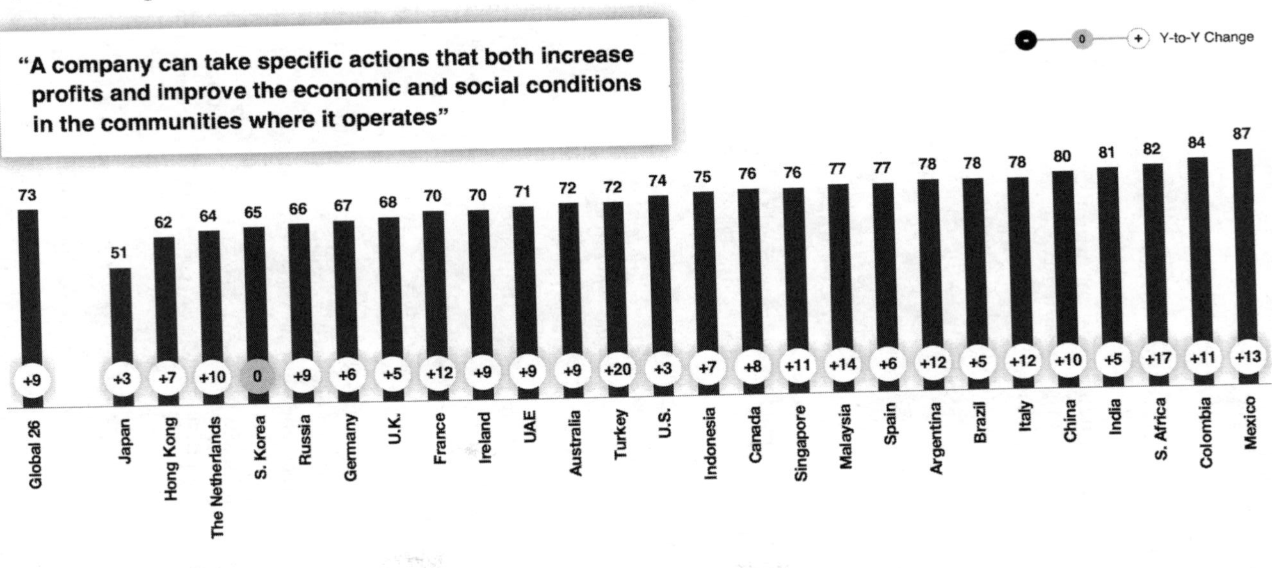

Figure 8.31 Edelman Trust Barometer 2019.

Source: 2019 Edelman Trust Barometer.

It will take the focus and collaboration of all kinds of people – from traditional social scientists and futurists to highly specialized, bleeding edge technologists and everyone in between – across silos, disciplines, sectors and barriers, with the assistance of every kind of new machine we create to navigate the coming technological storm safely and successfully. The world will need deep horizontal and vertical collaboration at every level – the individual, communal, local, social, corporate, national and global ... maybe even galactic if Elon Musk gets his way ....

That said, in this concluding portion of the book, I want to share a critical aspect of the coming daunting challenge to transform technology risk into technology safety, resilience and opportunity. What follows is a consideration of what the average organization (executives and board members) can do today to help their company, or nonprofit, or other type of entity successfully navigate these churning, changing waters in a way that builds longer-term resilience and sustainability to the benefit of shareholders and other key stakeholders.

Besides the usual business development, numbers crunching and technological nitty-gritty that go into any organization's business planning and long-term strategy, there are four actionable technology governance approaches that boards and executives – especially those who are not presently deeply engaged in new technologies and digital change – should adopt immediately at the very least for the sustainability and survival of their organizations and their commitments to their stakeholders:

1 Adopt *Robust, Triangular Governance*
2 Engage in *Agile, Strategic Risk Management*
3 Ensure *Ethical* and *Responsible Innovation*
4 Understand *Stakeholder Trust and Reputation Risk*

If management and boards pay attention to these four critical components of technology governance, their organizations will have a better chance of understanding the current maelstrom and surviving it and maybe even thriving in this age of disruption that we have only just begun to experience and still barely understand.

As we review these four components, we will reference a particular new technology – artificial intelligence (AI) – as our "go to" example to concretize what we mean. AI has already become pervasive (in ways that we don't even know) and is predicted by experts to be as disruptive and far-reaching as electricity once was. Indeed, experts say that AI and related technologies will have a much larger and transformative impact than the age of computing ushered in a few decades ago.[35]

8.4.i Adopt lean-in triangular technology governance

What is triangular technology governance? Very simply – it's making sure that the board (or other appropriate oversight body), executive management and operational and functional leadership of an organization are working in unison, collaboratively and in an integrated manner on the formulation, oversight and implementation of an appropriate technology strategy for the company (see Figure 8.32).

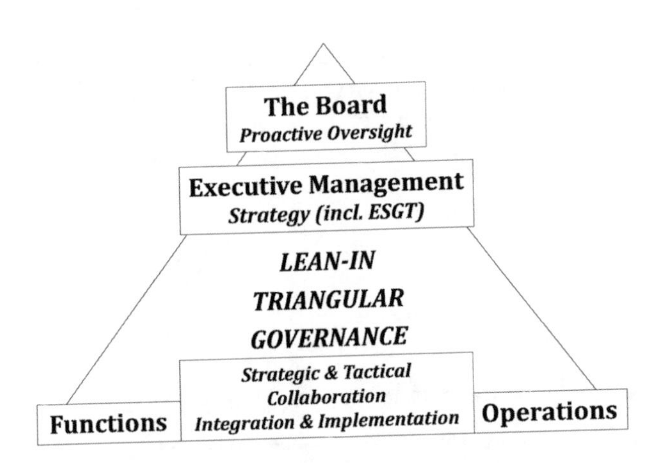

Figure 8.32 Lean-in triangular technology governance.

Source: Author and GEC Risk Advisory.

Simply put, boards and management must get technology governance right, and it starts with the board making sure management is engaging effectively in analyzing and integrating relevant technological trends and analysis into business planning and strategy formulation. While it is of course the primary responsibility for management to understand the contours and effects of the marketplace, competitive forces, research and development, product and services central to their business plan and strategy, it is incumbent that the board also become knowledgeable and proactive on how new technological developments will affect the integrity and long-term competitiveness and survival of the organization regardless of its type – whether business, NGOs, nonprofits, academia, government agencies, international bodies, etc.

The core questions every such organization (its management from a strategy-making standpoint and its board or other oversight body from a strategy-oversight perspective) needs to ask include:

- Should new technologies (e.g., AI) play a role in the development of our products and services?
- If so, are those technologies part of our current strategic planning?
- Do we have the appropriate data and the internal talent to understand and deploy the data?
- Are these technologies part of our traditional competitors' business plan and strategy?

- Are there disruptors in the marketplace from other sectors (or a brand-new one) that are about to eat our lunch?
- What are the risks and opportunities of deploying a new technology strategy in our business or organization?

The illustration in Figure 8.33 is one I have used to depict what boards need to do for cyber-security oversight as well as oversight of AI, which applies just as well to any form of technology transformation oversight – the point is that boards need to go from sit-back to lean-in technology governance and that means that the board itself needs to be equipped to do so (have the right directors, proper education and skill set, etc.).[36]

AI GOVERNANCE:
FROM SIT-BACK TO LEAN IN

1. KNOW THE BASICS ABOUT AI IN YOUR SECTOR & BUSINESS

2. OVERSEE ORGANIZATION'S AI ANALYSIS & PREPAREDNESS

3. LEAN INTO YOUR AI STRATEGY

Figure 8.33 AI governance: from sit-back to lean-in.
Source: Author and GEC Risk Advisory.

8.4.ii Engage in agile, strategic risk management: think "black-swan" or maybe "purple techno-swan"?

Besides fully understanding what the possibilities might be to introduce AI into a business or other type of venture, it is critically important for management and the board to think about the risks and opportunities – to proactively plan for an AI or other technology Black-Swan event that might obliterate the current business model – and that goes for nonprofits and other types of organizations as well.

Think about what happened to Kodak when they invented digital photography and decided to shelve it as a commercial product – the rest is history, as they say. The same can happen many times over in this crazy age of invention, reinvention and rapid product obliteration. It's do-or-die time and everyone needs to think about not only what the competition is doing but even more importantly where the new and differentiated competition is coming from.

Despite how it has stumbled, look at the disruption Uber has achieved through technology at the severe expense of the taxi industry globally. Uber in turn knows that their existing business model may well be obliterated by self-driving cars, and that's why they're planning actively on morphing eventually into a fleet of self-driving cars and other mobility services.

Some of the basic questions boards and management need to ask about risk and opportunity include some basic things like:

- Does our organization currently effectively identify, analyze and mitigate its risks, including, very importantly, environmental, social, governance and technology (ESGT) risks?
- Is the risk management approach a holistic and strategic approach (such as through enterprise risk management) that not only identifies risks but also looks at opportunities – i.e., understands how to transform risk into a value creation strategy?
- Does the company have the right talent within management and within the organization itself to effectively integrate technology risk and opportunity into its strategy and everyday tactics?
- Has the organization engaged in an effective technology risk identification exercise that has been integrated into business planning and product or service development?
- Does our board have the proper, diverse mix of director-level talents and skills needed for today and tomorrow's ESGT strategic risk and opportunity oversight?

8.4.iii Ensure ethical and responsible innovation

One of the critical activities management and boards must engage in is understanding the overall ESGT issues that affect their company. And, sadly, few but the most enlightened leaders consider these issues to be a core part of their remit. That time is coming quickly to a close with the critical strategic issues we are seeing in the ESGT space globally – climate change, #MeToo, AI ethics, labor displacement, human trafficking, corruption, fraud and others.

Thus, it is incumbent on oversight bodies and management to understand how technology affects their organization's key ESGT issues – new technologies deeply intersect with key ethical, social and geopolitical issues, and what follows is a simple and broad overview of how some of these ESG issues intersect specifically with AI (though one could extrapolate this table to a few other technologies as well) that was extracted from Chapter 1 of *The Artificial Intelligence Imperative*:

- Designing "safe" AI into robotic and IoT products.
- Understanding the impact of on sustainability issues such as effects on air, land and water resource issues.
- Designing AI products that facilitate conservation and sustainability.
- Creating intersections between AI products and services and electrification.
- Data and information privacy in the age of AI.
- Impact of AI on labor and employment.

- Ensuring AI products and services are not created in biased or discriminatory ways.
- Need for a lifelong learning education policy shift.
- AI requires new governance parameters from inception in products and services.
- The paradox of ethical AI in military products and services.

Table 8.19 contains the abstract of an experiment conducted by MIT modeled vaguely around the traditional "Trolley" ethical experiment that asked, among other questions: If you had the power to divert a runaway train at a fork on the train tracks, whom should the runaway train kill – one person on one track or several persons on another track? MIT crowdsourced the experiment globally to 40 million participants in 200 plus countries and were able to aggregate cultural differences on whom would receive the short end of the stick, which appeared to dovetail strongly with three cultural clusters around the world. In some locations, the elderly were singled out for oblivion, in others the young, but overall the experiment demonstrated that while we have some local cultural differences, we also have a grave need for the development of global social ethical principles for the development of AI and other technologies.

Table 8.19 The MIT moral machine experiment: whom should the autonomous vehicle kill – the baby or the elderly?[37]

With the rapid development of artificial intelligence have come concerns about how machines will make moral decisions, and the major challenge of quantifying societal expectations about the ethical principles that should guide machine behavior. To address this challenge, we deployed the moral machine, an online experimental platform designed to explore the moral dilemmas faced by autonomous vehicles. This platform gathered 40 million decisions in ten languages from millions of people in 233 countries and territories. Here we describe the results of this experiment. First, we summarize global moral preferences. Second, we document individual variations in preferences, based on respondents' demographics. Third, we report cross-cultural ethical variation, and uncover three major clusters of countries. Fourth, we show that these differences correlate with modern institutions and deep cultural traits. We discuss how these preferences can contribute to developing global, socially acceptable principles for machine ethics.

To meet these complex and sometimes alarming new sets of ethical questions and challenges, leaders and organizations must expand their horizons around ethical and responsible innovation and ask more than the standard questions they might have been asking to date. Now they should consider additional angles such as:

- What is the state of your internal ethics and compliance program?
- What is the state of your corporate responsibility or sustainability program?

- Do you have a defined set of values that fuel your mission, vision and strategy?
- Do you have a stated culture?
  - If you do, is it window dressing and an exercise in public relations or does it have teeth (is it measurable and actionable)?
- If technology is part of your equation now, have you integrated CSR and ethical implications into your development and/or use of technology and the design of your products and services?
- Do your defined ethics, values and/or culture have anything to do with how you create, design, develop and implement your products and services? If not, why not?
- How do you know whether the data and technological portions of your products are being programmed and created free of discrimination and bias?
- How do you know whether your technological content (whether home-grown or purchased from third parties) has been ethically created and programmed?

8.4.iv Understand stakeholder trust and reputation risk

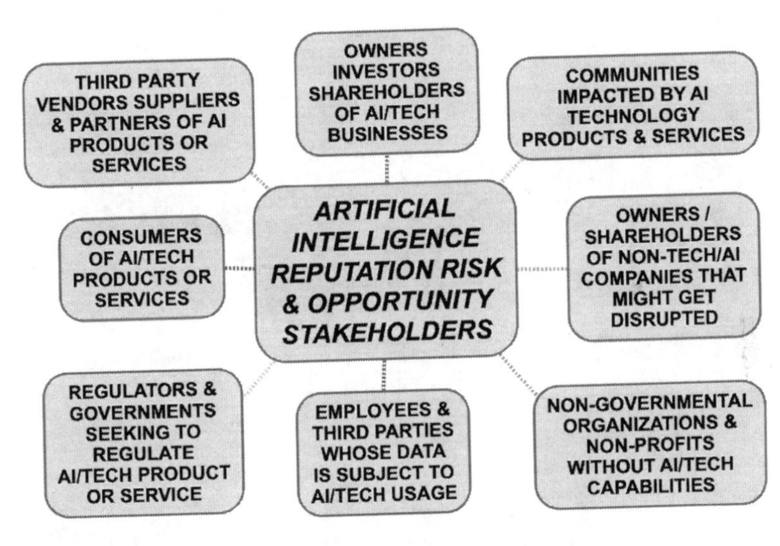

Figure 8.34 AI reputation risk and opportunity stakeholders.

Source: Author and GEC Risk Advisory.

A final category boards and management should be actively considering as part of their technology governance is to pay close attention to key stakeholders, going beyond shareholders to understand who the primary stakeholders like customers, employees, regulators and partners are. And beyond knowing who they are – do you know how they feel about the critical ESGT issues that affect your business? Do they trust you? Do you have a good reputation with your stakeholders? If you are not transparent with key stakeholders, will they punish you when the truth comes out in this age of social media? Figure 8.34 provides a sampling of who some of the AI reputation risk and opportunity stakeholders might be.

We live in a world where stakeholder trust is of the essence and has definite reputational and financial consequences. Following are two stock performance charts and one reputation performance chart for Facebook which has found itself in a reputation maelstrom – of its own making - of late. The two stock charts (Figures 8.35 and 8.36), one for one year and the other for five years show Facebook's stock shock after the Cambridge Analytica fiasco first came to light which adversely affected the data of its key stakeholders – users.

Table 8.20 shows selective reputation rankings of technology companies from 2018 and 2019 by Axios and Harris Poll with Facebook's plummeting reputation with stakeholders over the past year made that much more stark in contrast to other technology giants, some of whom (Microsoft comes to mind) have been doing relatively well on this front.

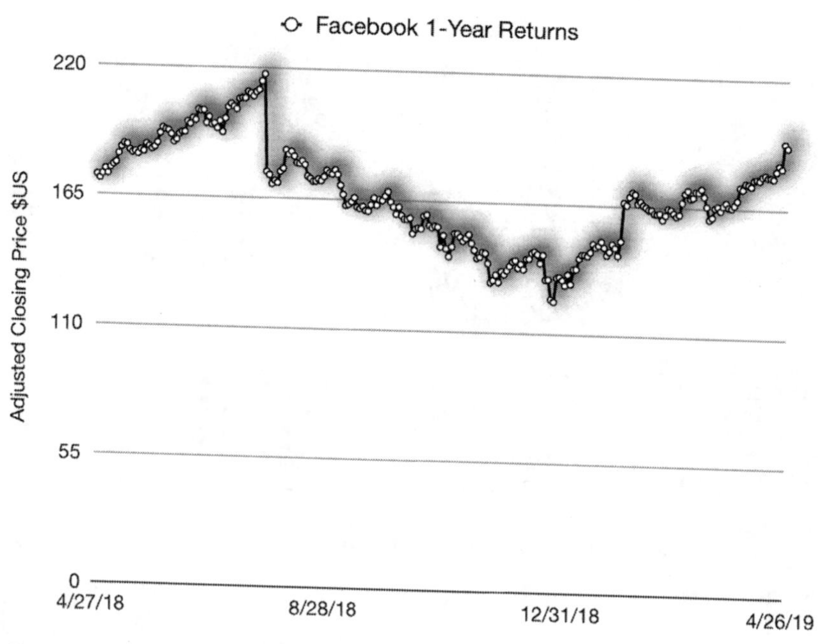

Figure 8.35 Facebook stock performance 1 year.

Source: Steel City Re.

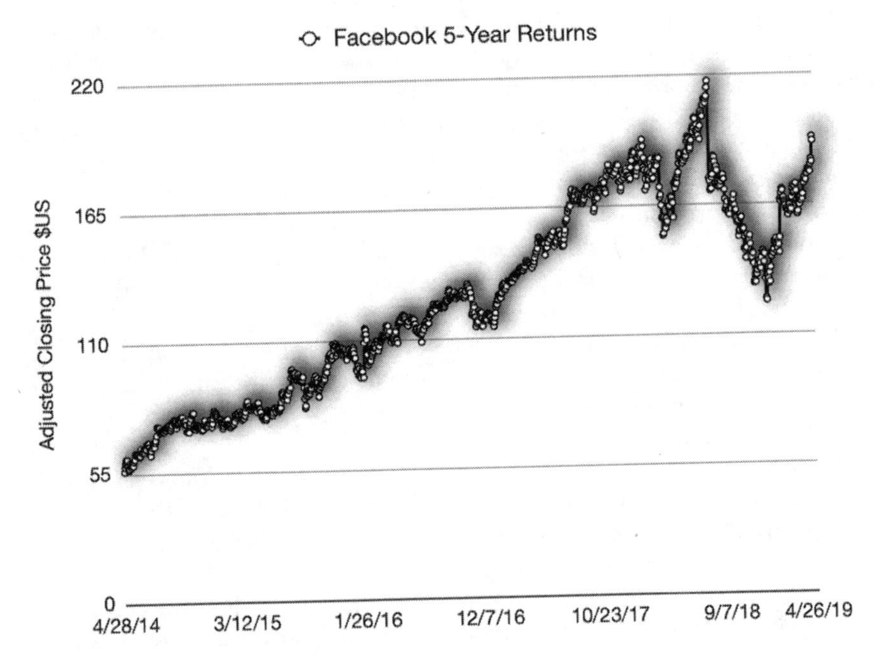

Figure 8.36 Facebook stock performance 5 years.

Source: Steel City Re.

Table 8.20 Axios Harris Poll 100: 2019 corporate reputation rankings[38]

Technology company and rank in 2019	*2019 Score*	*Change in rank from 2018*
#2 Amazon	82.3	Down 1 from #1
#9 Microsoft	79.7	Up 2 from #11
#24 Netflix	77.3	Down 3 from #21
#32 Apple	76.4	Down 3 from #29
#41 Google	75.4	Down 13 from #28
#42 Tesla	75.4	Down 39 from #3
#78 Uber	67.3	Down 2 from #76
#89 Twitter	61.9	No data
#94 Facebook	58.1	Down 43 from #51

Source: Axios, Harris Poll.

Instead of addressing their trust issues proactively and constructively, over a lengthy period of time, Facebook and its leaders dug themselves into even bigger holes with key stakeholders globally – especially regulators and governments in the US, UK and EU – even after the Cambridge Analytica revelations came to light. If Facebook does not make long-term governance, culture and leadership changes, they may become a classic business case study of what went wrong. One key measure that would restore trust (which will probably never happen) would be for Zuckerberg to resign as CEO and for a new outside professional technology CEO to come in (as did the Uber's new CEO post-Kalanick). Another trust restoring measure (highly unlikely as well) would include major governance reforms including the separation of CEO and chairman roles. As a stakeholder in Facebook – both a shareholder of their stock on and off and a (minor) user of their platform, I am not, however, holding my breath that they will do the right thing certainly not in the absence of another major scandal.

But in this story lies opportunity too – opportunity for Facebook to reverse course and opportunity for a competitor to come in and start a new, more ethical and trustworthy version of Facebook.

Let's see what happens.

We are in a more complex and deeply interconnected, intersected and unpredictable time than humankind has experienced before. It is incumbent on all of us as individuals and members of society to educate ourselves. It is even more urgent for leaders – like executives and boards of businesses – to educate themselves immediately on the disruption that is here and now and that will upend business, environment, society and governance as we know them.

Hope is not lost, but much work needs to be done in the boardroom and the C-Suite. Opportunity is knocking like crazy on the door, but so is disruption. What boards and management need to do now to avoid being disrupted out of their current products, services and even existence is to put on their collective thinking caps to understand the current and future landscape and the big "what ifs" that are now staring them in the face and pivot to a wining long-term strategy.

8.5 Coda: the Oracle of Omaha and the Japanese concept of "Ikigai"

We have now completed our journey from Gloom to Boom, we have navigated successfully through the both treacherous and promising waters between Scylla and Charybdis. I hope my readers have found this journey to be at times enlightening, other times disturbing and even frightening but most of the time useful, constructive, energizing and hopeful.

If there is one wish I would have from having written this, my most elaborate and lengthiest work so far, it is that I helped to open hearts and minds especially among those tougher leaders who are so often skeptical, don't have enough time, don't think it's useful or valuable to dwell on the collection of issues, risks and opportunities we have focused on in this tome.

To these leaders I wish a fond "Auf Wiedersehen" and "Hasta luego" because I am sure I will encounter you from time to time in my travels along future Scyllas and Charybdises of ESGT. I will continue to try to convince you with every breath and brain cell that I have that doing good is valuable in of itself – because doing so creates inherent value and because values bring their own very unique "value".

That said, I would like to offer a couple of final words of wisdom not from myself but from much wiser sources – one is from Warren Buffett, the legendary businessman, and one is from Japanese culture.

Here is a very recent quote (2018) from the Oracle of Omaha:[39]

> Basically, when you get to my age, you'll really measure your success in life by how many of the people you want to have love you actually do love you. That's the ultimate test of how you have lived your life. The trouble with love is that you can't buy it. You can buy sex. You can buy testimonial dinners. But the only way to get love is to be lovable. It's very irritating if you have a lot of money. You'd like to think you could write a check: I'll buy a million dollars' worth of love. But it doesn't work that way. The more you give love away, the more you get.
>
> (Warren Buffett 2018)

Finally, I want to leave my readers with the Japanese concept of "Ikigai" which wiser people than I have written and talked about. Figure 8.37 provides a beautiful visualization of its meaning, which I encourage my readers to spend some time absorbing because it is really worth it and it applies not only to each and every one of us individually but also to our organizations – think about that as you review this final Figure in this book.

This is Ikigai:[40]

> Your *ikigai* is at the intersection of what you are good at and what you love doing. . . . Just as humans have lusted after objects and money since the dawn of time, other humans have felt dissatisfaction at the relentless pursuit of money and fame and have instead focused on something bigger than their own material wealth. This has over the years been described using many different words and practices, but always hearkening back to the central core of meaningfulness in life.
>
> (Hector Garcia, co-author of *Ikigai: The Japanese Secret to a Long and Happy Life*)

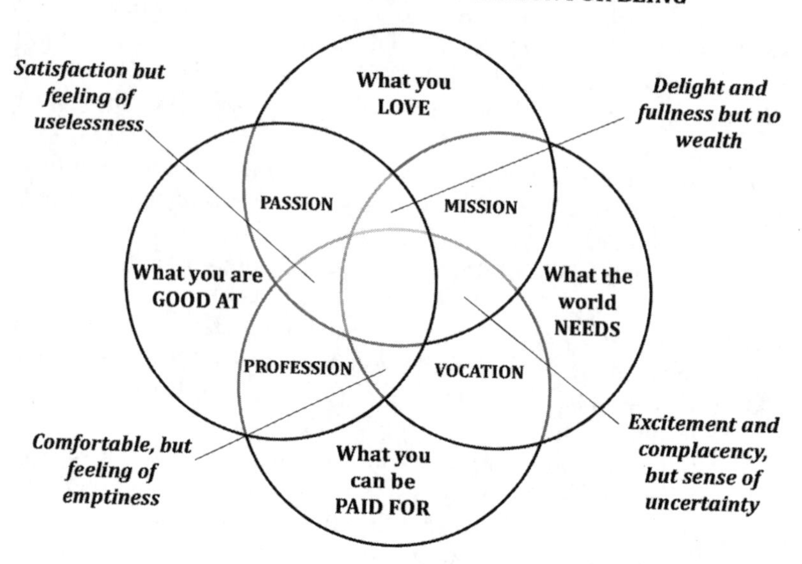

Figure 8.37 Ikigai – A Japanese concept meaning "A reason for being".

Source: Toronto Star and Thomas Oppong.

By writing this book, I feel like I have reached my own personal version of Ikigai and for that I am forever grateful. Thank you for reading!

Notes

1 Rebecca M. Henderson. "What Would It Take for Businesses to Focus Less on Shareholder Value?". *Harvard Business Review*. August 21, 2018. https://hbr.org/2018/08/what-would-it-take-to-get-businesses-to-focus-less-on-shareholder-value. Accessed on August 21, 2018.

2 Dalai Lama. Twitter. September 14, 2018. https://twitter.com/DalaiLama/status/1040533247577546752. Accessed on September 6, 2019.

3 Dalai Lama. Twitter. November 30, 2018. https://twitter.com/DalaiLama/status/1068452282109386753. Accessed on September 6, 2019.

4 Maya Kosoff. "Why the Google Walkout Could Be a Turning Point for Tech". November 2, 2018. *Vanity Fair*. https://www.vanityfair.com/news/2018/11/why-the-google-walkout-could-be-a-turning-point-for-tech. Accessed on November 3, 2018.

5 Abigail Disney. "It's Time to Call Out My Family's Company and Anyone Else Rich off Their Workers' Backs". *The Washington Post*. April 24, 2019. https://www.washingtonpost.com/opinions/its-time-to-call-out-my-familys-company--and-anyone-else-rich-off-their-workers-backs/2019/04/23/5d4e6838-65ef-11e9-82ba-fcfeff232e8f_story.html?utm_term=.fb4cb160da15. Accessed on April 28, 2019

6 Danny Palmer. "Notorious Cyber Crime Gang Behind Global Bank Hacking Spree Returns with New Attacks". August 30, 2018. ZDNet.com. https://www.zdnet.com/

article/notorious-cyber-crime-gang-behind-global-bank-hacking-spree-returns-with-new-attacks/. Accessed on November 4, 2018.

7 Julia Horowitz. "How Goldman Sachs Got Tied Up in Malaysia's $4.5 Billion Embezzlement Scandal". CNN. November 15, 2018. https://www.cnn.com/2018/11/15/business/goldman-sachs-1mdb-scandal/index.html. Accessed on November 18, 2018.

8 Anthony Faiola. "The Corruption Scandal Started in Brazil. Now It's Wreaking Havoc in Peru". *The Washington Post*. January 23, 2018. https://www.washingtonpost.com/world/the_americas/the-corruption-scandal-started-in-brazil-now-its-wreaking-havoc-in-peru/2018/01/23/0f9bc4ca-fad2-11e7-9b5d-bbf0da31214d_story.html?utm_term=.97cf9578689b. Accessed on November 10, 2018.

9 Linda Pressly. "The Largest Foreign Bribery Case in History". April 22, 2018. BBC.com. https://www.bbc.com/news/business-43825294. Accessed on November 18, 2018.

10 Gideon Long. "Colombian Deaths Form Latest Twist in Odebrecht Scandal". November 16, 2018. *Financial Times*. https://www.ft.com/content/2aba897c-e7a7-11e8-8a85-04b8afea6ea3. Accessed on November 18, 2018.

11 "Making Sense of the Suicide of Alan Garcia, a Former President of Peru". *The Economist*. April 27, 2019. https://www.economist.com/the-americas/2019/04/27/making-sense-of-the-suicide-of-alan-garcia-a-former-president-of-peru. Accessed on April 28, 2019.

12 Epic Theatre Ensemble. http://www.epictheatreensemble.org/. Accessed on November 10, 2018.

13 Sources: Wikipedia "Value". https://en.wikipedia.org/wiki/Value. Accessed on October 31, 2018; "Values". https://en.wikipedia.org/wiki/Value_(ethics). Accessed on October 31, 2018.

14 Larry D. Fink. "Letter to CEOs". Blackrock. 2018. https://www.blackrock.com/corporate/investor-relations/larry-fink-ceo-letter. Accessed on November 10, 2018.

15 Robert G. Eccles, Ioannis Ioannou, George Serafeim. "The Impact of a Corporate Culture of Sustainability on Corporate Behavior and Performance". 2012. Working Paper; "The Impact of Corporate Culture of Sustainability on Corporate Processes and Performance". 2012. https://www.hbs.edu/faculty/Publication%20Files/SSRN-id1964011_6791edac-7daa-4603-a220-4a0c6c7a3f7a.pdf. Accessed on October 31, 2018.

16 Arabesque. Partners & The University of Oxford. "From the Stockholder to the Stakeholder: How Sustainability Can Drive Financial Outperformance". 2015. https://arabesque.com/research/From_the_stockholder_to_the_stakeholder_web.pdf. Accessed on October 31, 2018.

17 Accenture Strategy. "The Bottom Line on Trust". November 1, 2018. https://www.accenture.com/t20181029T113120Z__w__/us-en/_acnmedia/Thought-Leadership-Assets/PDF/Accenture-Competitive-Agility-Index.pdf. Accessed on November 3, 2018.

18 Adapted from: Accenture Strategy. "The Bottom Line on Trust". November 1, 2018. https://www.accenture.com/t20181029T113120Z__w__/us-en/_acnmedia/Thought-Leadership-Assets/PDF/Accenture-Competitive-Agility-Index.pdf. Accessed on November 3, 2018.

19 BCG. Total Societal Impact. October 2017. https://www.bcg.com/publications/2017/total-societal-impact-new-lens-strategy.aspx. Accessed on September 18, 2018.

20 BSR & GlobeScan. "The State of Sustainable Business 2018". September 18, 2018. https://www.bsr.org/en/our-insights/report-view/state-of-sustainable-business-2018-bsr-globescan. Accessed on November 4, 2018.

21 Oliver Schutzmann. "ESG Investing: It's Here, Are You Ready". *IR Magazine*. October 12, 2018. https://www.irmagazine.com/esg/esg-investing-its-here-are-you-ready. Accessed on November 4, 2018.

22 Blackrock. "Larry Fink CEO Letter". https://www.blackrock.com/corporate/investor-relations/larry-fink-ceo-letter. Accessed on October 30, 2018.

23 McKinsey. "Delivering through Diversity". January 2018. https://www.mckinsey.com/~/media/McKinsey/Business%20Functions/Organization/Our%20Insights/

Delivering%20through%20diversity/Delivering-through-diversity_full-report.ashx. Accessed on October 30, 2018.
24 Catalyst. "Why Diversity and Inclusion Matter". https://www.catalyst.org/knowledge/why-diversity-and-inclusion-matter. Accessed on November 11, 2018.
25 National Association of Corporate Directors Blue Ribbon Commission Report. "Culture as a Corporate Asset". October 2017. https://www.nacdonline.org/insights/publications.cfm?itemNumber=48252. Accessed on October 30, 2018.
26 Ethical Systems. "Ethics Pays". https://www.ethicalsystems.org/content/ethics-pays. Accessed on October 30, 2018.
27 Stephen Stubben and Kyle Welch. "Research: Whistleblowers Are a Sign of Healthy Companies". November 14, 2018. https://hbr.org/2018/11/research-whistleblowers-are-a-sign-of-healthy-companies. Accessed on November 15, 2018.
28 Ethisphere. "Companies that Shine Focus on Ethics. May 9, 2017. https://insights.ethisphere.com/lessons-from-the-global-ethics-summit-how-companies-that-shine-never-lose-focus-on-culture-of-ethics/. Accessed on October 30, 2018
29 Catalin Cimpanu. "Data Breaches Affect Stock Performance in the Long Run, Study Finds". ZDNet.com. September 14, 2018. https://www.zdnet.com/article/data-breaches-affect-stock-performance-in-the-long-run-study-finds/. Accessed on September 21, 2018; citing study available here: https://www.comparitech.com/blog/information-security/data-breach-share-price-2018/. Accessed on September 22, 2018.
30 Tony Chapelle. "Listed: Companies with Best Reputations". *Agenda.* July 23, 2018.
31 Steel City Re. Extended Legend and Notes: TTM=Trailing 12 months; T6M=Trailing 6 months.

- SCRe RepuSPX: Steel City Re S&P500© Reputation Arbitrage Portfolio comprising member of the S&P500 equity index whose stock prices relative to their reputational value indicate arbitrage opportunities. The measures of reputational value and the constituent components of the portfolio are determined by Steel City Re, which provides parametric reputation risk insurance and risk management solutions.
- Agenda/Leaders: Sector-sample from the portfolio of reputational value described in Chapelle, T: *Listed: Companies with Best Reputations.* Agenda Week (A Financial Times Services), 23 July 2018.
- S&P500: The Standard & Poor's 500, an equity index based on the market capitalizations of 500 large companies having common stock listed on the NYSE or NASDAQ. The constituent components and their weightings are determined by S&P Dow Jones Indices.
- GECRA/Impaired: Portfolio of companies with highly publicized reputational scandals. The constituents are determined by GEC Risk Advisory.

32 Interview and materials from discussion between the author and Dr. Nir Kossovsky.
33 Steel City Re. Extended Legend and Notes: TTM=Trailing 12 months; T6M=Trailing 6 months. GU%=Gerken Unit percentile, a measure of reputational value developed by Steel City Re, which provides parametric reputation risk insurance and risk management solutions.

- Agenda/Leaders: Sector-sample from the portfolio of reputational value described in Chapelle, T: *Listed: Companies with Best Reputations.* Agenda Week (A Financial Times Services), 23 July 2018.
- GECRA/Impaired: Portfolio of companies with highly publicized reputational scandals. The constituents are determined by GEC Risk Advisory.

34 Mark Kovak and Jamie Cleghorn. "What Sales Teams Should Do to Prepare for the Next Recession". November 23, 2018. *Harvard Business Review.* https://hbr.org/2018/11/what-sales-teams-should-do-to-prepare-for-the-next-recession. Accessed on November 23, 2018.

35 My coauthor, A. Lauterbach, and I delve into all of the reasons, antecedents, background, competitive landscape and business planning in our new book recently published by Praeger, *The Artificial Intelligence Imperative: A Practical Roadmap for Business*. Praeger 2018.

36 Andrea Bonime-Blanc. "A Strategic Cyber-Roadmap for the Board". The Conference Board. November 2016. https://gecrisk.com/wp-content/uploads/2016/11/ABonimeBlanc-Strategic-Cyber-Roadmap-for-the-Board-November-17-2016.pdf. Accessed on august 28, 2018; and in Anastassia Lauterbach & Andrea Bonime-Blanc. *The Artificial Intelligence Imperative*. Praeger 2018.

37 "The Moral Machine Experiment". Nature. https://www.nature.com/articles/s41586-018-0637-6. Accessed on November 5, 2018; Charlie Osborne. "MIT Reveals Who Self-Driving Cars Should Kill: The Cat, the Elderly or the Baby". October 26, 2018. https://www.zdnet.com/article/mit-reveals-who-self-driving-cars-should-kill-the-cat-the-elderly-or-the-baby/?utm_medium=email&utm_source=topic+optin&utm_campaign=awareness&utm_content=20181105+ai+nl&mkt_tok=eyJpIjoi-TnpObFpEUmpNVEE0WTJZeSIsInQiOiJ0TUdVWmhIK29JZGd0TG5NYitrbEtYeTlcL0ExSGtHeEFvWFBwYkpudnpcL0kzSEFCVWFWU3ZnQTRPbTlmeGZNRlVXWGh1aDdJbHcwK1RqcWRYZlROMlBJckNXTzh5XC9hblFpK0lFaUI3dlwvM2x3SWkya0V4ZEVrR0lYN21FTlBnRnEifQ%3D%3D

38 Axios Harris Poll 100. https://theharrispoll.com/axios-harrispoll-100/. Accessed on April 28, 2019.

39 Marcel Schwantes. "Warren Buffett Says Your Greatest Measure of Success At the End of Your Life Cones Down to one Word". *Inc Magazine*. September 17, 2018. https://www.inc.com/marcel-schwantes/warren-buffett-says-it-doesnt-matter-how-rich-you-are-without-this-1-thing-your-life-is-a-disaster.html. Accessed on October 21, 2018.

40 Thomas Oppong. "Ikigai: The Japanese Secret to a Long and Happy Life Might just Help You Live a More Fulfilling Life". January 10, 2018. Medium/Thrive Global. https://medium.com/thrive-global/ikigai-the-japanese-secret-to-a-long-and-happy-life-might-just-help-you-live-a-more-fulfilling-9871d01992b7. Accessed on November 18, 2018.

Index

Italic page numbers indicate figures, while page numbers in **bold** indicate tables.